Mathematical
Surveys
and
Monographs

Volume 200

Nonlinear Elliptic Equations and Nonassociative Algebras

Nikolai Nadirashvili
Vladimir Tkachev
Serge Vlăduţ

American Mathematical Society
Providence, Rhode Island

EDITORIAL COMMITTEE

Ralph L. Cohen, Chair
Robert Guralnick
Michael A. Singer

Benjamin Sudakov
Michael I. Weinstein

2010 *Mathematics Subject Classification.* Primary 17Cxx, 17Dxx, 35J60;
Secondary 16H05, 17A35, 49Q05, 53C38.

For additional information and updates on this book, visit
www.ams.org/bookpages/surv-200

Library of Congress Cataloging-in-Publication Data

Nadirashvili, Nikolai, 1955– author.
  Nonlinear elliptic equations and nonassociative algebras / Nikolai Nadirashvili, Vladimir Tkachev, Serge Vlăduţ.
    pages cm. — (Mathematical surveys and monographs ; volume 200)
  Includes bibliographical references and index.
  1. Jordan algebras. 2. Nonassociative rings. 3. Differential equations, Elliptic. I. Tkachev, Vladimir, 1963– author. II. Vlăduţ, S. G. (Serge G.), 1954– author. III. Title.

QA252.5.N33 2014
512′.48—dc23
                                                                              2014028806

Copying and reprinting. Individual readers of this publication, and nonprofit libraries acting for them, are permitted to make fair use of the material, such as to copy select pages for use in teaching or research. Permission is granted to quote brief passages from this publication in reviews, provided the customary acknowledgment of the source is given.

Republication, systematic copying, or multiple reproduction of any material in this publication is permitted only under license from the American Mathematical Society. Permissions to reuse portions of AMS publication content are handled by Copyright Clearance Center's RightsLink® service. For more information, please visit: http://www.ams.org/rightslink.

Send requests for translation rights and licensed reprints to reprint-permission@ams.org.

Excluded from these provisions is material for which the author holds copyright. In such cases, requests for permission to reuse or reprint material should be addressed directly to the author(s). Copyright ownership is indicated on the copyright page, or on the lower right-hand corner of the first page of each article within proceedings volumes.

© 2014 by the American Mathematical Society. All rights reserved.
The American Mathematical Society retains all rights
except those granted to the United States Government.
Printed in the United States of America.

∞ The paper used in this book is acid-free and falls within the guidelines
established to ensure permanence and durability.
Visit the AMS home page at http://www.ams.org/

10 9 8 7 6 5 4 3 2 1   19 18 17 16 15 14

Contents

Preface	v
Chapter 1. Nonlinear Elliptic Equations	1
1.1. Elliptic equations	1
1.2. Viscosity solutions	8
1.3. Linear elliptic operators of nondivergence form	11
1.4. Nonlinear equations with smooth solutions	16
1.5. Degenerate elliptic equations	22
1.6. Homogeneous solutions	27
1.7. Liouville type theorems. Removable singularities	38
Chapter 2. Division Algebras, Exceptional Lie Groups, and Calibrations	45
2.1. The Cayley-Dickson construction. Octonions	46
2.2. Clifford algebras and spinor groups	48
2.3. The group G_2. Trialities	51
2.4. Cross products, associative and coassociative calibrations	53
2.5. Cross products, normed algebras, and Cayley calibrations	58
Chapter 3. Jordan Algebras and the Cartan Isoparametric Cubics	63
3.1. Isoparametric cubics and the eiconal equation	63
3.2. Jordan algebras	67
3.3. The eiconal equation and cubic Jordan algebras	77
Chapter 4. Solutions from Trialities	85
4.1. The simplest nonclassical construction	85
4.2. The octonionic construction	96
4.3. Hessian equations in twelve dimensions	104
4.4. Isaacs equations	114
Chapter 5. Solutions from Isoparametric Forms	117
5.1. The spectrum	117
5.2. Nonclassical and singular solutions	125
5.3. Singular solutions in five dimensions	132
5.4. Other isoparametric forms	143
5.5. Two more applications	150
Chapter 6. Cubic Minimal Cones	155
6.1. Brief overview of the main results	155
6.2. Algebraic minimal hypercones in $\mathbb{R}^n$ and radial eigencubics	159
6.3. Metrized and Freudenthal-Springer algebras	161
6.4. Radial eigencubic algebras	163

6.5.	Polar and Clifford REC algebras	169
6.6.	The harmonicity of radial eigencubics	176
6.7.	The Peirce decomposition of a REC algebra	180
6.8.	Triality systems and a hidden Clifford algebra structure	185
6.9.	Jordan algebra $V_c^\bullet$	189
6.10.	Reducible REC algebras	194
6.11.	Exceptional REC algebras	198

Chapter 7. Singular Solutions in Calibrated Geometries — 205
   7.1. Calibrated geometries — 205
   7.2. Singular coassociative 4-folds — 209
   7.3. Singular solutions of special Lagrangian equations — 211
   7.4. More singular solutions and a failure of the maximum principle — 218

Bibliography — 223

Notation — 235

Index — 239

Preface

The present volume contains some applications of noncommutative and nonassociative algebras to constructing unusual (nonclassical and singular) solutions to fully nonlinear elliptic partial differential equations of second order. Here the solutions are to be understood in a weak (viscosity) sense. Using such algebras to construct exotic or specific analytic and geometric structures is not new. One can mention here, for instance, the constructions of exotic spheres by Milnor [**163**], of singular solutions to the minimal surface system by Lawson and Osserman [**144**], the ADHM and constuction of instantons by Atiyah, Drinfel'd, Hitchin, and Manin [**17**], and the construction of singular coassociative manifolds by Harvey and Lawson [**103**], all four using quaternions, as well as the recent constructions of unusual solutions of the Ginzburg-Landau system by Farina and Ge-Xie [**86**], [**93**], using isoparametric polynomials and thus, implicitly, Jordan or Clifford algebras.

However, our applications of quaternions, octonions, and Jordan algebras to elliptic partial differential equations of second order are new; they allow us to solve a longstanding problem of the existence of truly weak viscosity solutions, which are not smooth (= classical) ones. Moreover, in some sense, they give (albeit along with some other arguments) an almost complete description of homogeneous solutions to fully nonlinear elliptic equations. In fact, a major part of the book is devoted to the simplest class of fully nonlinear uniformly elliptic equations, namely those of the form

$$(0.1) \qquad F(D^2 u) = 0,$$

F being a nonlinear sufficiently smooth functional on symmetric matrices and $D^2 u$ being the Hessian of a putative solution u. Those are "constant coefficient" fully nonlinear elliptic equations. Moreover, often we impose a rather drastic condition that F depends only on the eigenvalues of the Hessian (so-called "Hessian equations"). In that case F is a function of only n values of symmetric functions of $D^2 u$ rather than of $n(n+1)/2$ partial derivatives, n being the dimension of the ambient space. Our methods show that even in that very restricted setting in five and more dimensions (some of) those equations admit homogeneous δ-order solutions with any $\delta \in]1, 2[$, that is, of all orders compatible with known regularity results by Caffarelli and Trudinger [**37**], [**255**] for viscosity solutions of fully nonlinear uniformly elliptic equations, proving the optimality of these regularity results. To the contrary, the situation in four and fewer dimensions is completely different. First of all, in two dimensions the classical result by L. Nirenberg [**186**] guarantees the regularity of all viscosity solutions, homogeneous or not. In Section 1.6 we prove that in four (and thus three) dimensions there are no homogeneous order 2 solutions to fully nonlinear uniformly elliptic equations, at least in the analytic setting, which suggests strongly that there are no nonclassical homogeneous solutions at all in four and three dimensions. If so, we get a complete list of dimensions where nonclassical homogeneous solutions to fully nonlinear uniformly elliptic equations do exist. One can compare this with the situation of, say, ten years ago, when the very existence of nonclassical viscosity solutions was not known.

We should repeat once more that this result of fundamental importance for the theory of partial differential equations is obtained by applications of relatively elementary algebraic (and differential geometric) means, thus stressing once more that studying relations between apparently disconnected mathematical areas can often be very fruitful.

Furthermore, there are some cases where (singular) solutions of some classes of nonlinear elliptic equations and some nonassociative algebras are interrelated even more strongly, leading in certain circumstances to the equivalence of those objects. A study of these relations and their applications to classifying both classes of objects is the second theme of the book, intimately related to the previous one.

Our exposition is as follows. Since we hope that our work can be of use to a rather diversified mathematical audience, we devote the first three chapters to the basics of nonlinear elliptic equations and of noncommutative and nonassociative algebraic structures used in our constructions.

In Chapter 1 we recall basic facts about nonlinear elliptic equations and their viscosity solutions. The material in the first five sections is quite traditional in many papers devoted to viscosity solutions. However, in Section 1.2 we also formulate two recent results on partial regularity of solutions, in Section 1.3 we expose a recent result concerning the difference of viscosity solutions, and in Section 1.4 we give a recent result on the regularity of solutions to axially symmetric Dirichlet problems for Hessian equations. Section 1.6 is devoted to recent results and conjectures for homogeneous solutions to fully nonlinear uniformly elliptic equations. Section 1.7 gives some Liouville type results and various results on removable singularities for solutions of fully nonlinear elliptic equations, including a recent result describing viscosity solutions of a uniformly elliptic Hessian equation in a punctured ball.

Chapter 2 is devoted to the construction and elementary properties of the real division algebras $\mathbb{H}, \mathbb{O}$, Clifford algebras, spinor groups, and some exceptional Lee groups, especially G_2. We also discuss cross products in the algebra $\mathbb{O}$ and the resulting calibrations (in their algebraic form).

In Chapter 3 we give an overview of Jordan algebras in their relation to special cubics and some partial differential equations (of first order). Most of its material is classical, but some new facts concerning relations between cubic Jordan algebras and the so-called eiconal differential equation, $|\nabla f(x)|^2 = c|x|^4$, are proven.

In Chapters 4 and 5 we give our main constructions of nonclassical and singular solutions to fully nonlinear, uniformly elliptic equations, often of Hessian or of Isaacs type. In fact, all our nonclassical solutions are of the form $P(x)/|x|^\alpha$ with a homogeneous polynomial $P(x)$, $x \in \mathbb{R}^n$, of degree 3, 4, or 6 and a suitable α. Chapter 4 contains the constructions based on trialities, which use real division algebras: quaternions and octonions; there $n = 12$ or 24, $\deg P = 3$, $\alpha \in [1,2[$. Chapter 5 gives constructions based on isoparametric polynomials $P(x)$, $x \in \mathbb{R}^n$, $n \geq 5$, of degrees 3, 4, or 6 coming from Jordan and Clifford algebras. The constructions of Chapter 4 are more elementary in that they use less of algebraic theory but need more calculations than those of Chapter 5. The arguments in these chapters are based on several closely related criteria for solutions of fully nonlinear uniformly elliptic equations in terms of appropriate combinations of the spectrum for their Hessians. Those conditions are extremely restrictive and one needs rather elaborated arguments and/or calculations to verify them, which is obtained partially by using MAPLE calculations. One notes, however, that the calculations in Chapters 4 and 5 (and in Chapter 7) which use MAPLE extensively are completely rigorous since there MAPLE is used to verify algebraic identities, albeit rather cumbersome ones.

Chapter 6 is devoted to a classification of cubic minimal cones, that is, the simplest nontrivial solutions to the minimal surface system which is (almost) complete under a natural additional condition (the case of radial eigencubics). The main method there is to construct a certain nonassociative algebra from a given minimal cubic cone in such a way that the differential-analytical structure of the cones becomes transparent from the algebraic side, and vice versa. The main tool for this is the so-called Freudenthal multiplication. It associates to any fixed cubic form u on a vector space V carrying a symmetric nondegenerate bilinear form Q the multiplication $(x, y) \to xy$ by setting $\partial_x \partial_y u|_z = Q(xy; z)$. The algebra $V(u)$ defined in this way is called the Freudenthal-Springer algebra of the cubic form u. In the basic case of a radial eigencubic the corresponding Freudenthal-Springer algebra leads to a so-called radial eigencubic algebra, or just a *REC algebra*. Thus, the classification of radial eigencubics becomes equivalent to that of REC algebras. There exist two principal classes of REC algebras, namely those coming from Clifford and Jordan structures, respectively. Applying standard methods of nonassociative algebra such as Pierce decomposition and a thorough study of certain defining relations in REC algebras, one eventually gets their complete classification. Note, however, that the algebraic techniques of this chapter are elaborated more than in the other chapters and assume more advanced knowledge of the nonassociative algebraic systems.

In Chapter 7 we treat elliptic equations arising in calibrated geometry [**103**], namely, the special Lagrangian, associative, coassociative, and Cayley equations; they are not uniformly, but only strictly, elliptic, and we recall briefly their constructions in Section 7.1. One notes, however, that the construction of singular coassociative 4-folds given by Harvey and Lawson in [**103**] and recalled in Section 7.2 resembles strongly the constructions in Chapter 4. It would be very interesting to understand a possible common ground of constructions in Chapters 4 and 7 (and, presumably, in Chapters 5 and 6) and eventually find some other situations where it works. Sections 7.3 and 7.4 are devoted to constructions of some singular solutions to the special Lagrangian equations (SLE) in the nonconvex case, in three dimensions. Note that in the convex (or concave) case those solutions are smooth in any dimension by [**48**] and that in two dimensions these equations cannot be nonconvex. These constructions also lead to examples of a failure of the maximum principle for the Hessian of solutions to a uniformly elliptic equation in three and more dimensions as well as to examples of solutions to the minimal surface system with a notably low regularity.

Acknowledgements. We would like to thank Luis Caffarelli, Charles Smart, and Yu Yuan for their interest in and very fruitful discussions of the subject of this book. We also thank the anonymous reviewers for their pertinent remarks allowing us to improve the exposition. We would also like to thank A. Rüland for her thorough reading of the manuscript, leading to many improvements of the text.

CHAPTER 1

Nonlinear Elliptic Equations

In this chapter we give a brief introduction to the theory of second-order elliptic equations, principally fully nonlinear ones. After defining them in Section 1.1 and giving some basic examples, we formulate the principal problems concerning the solutions of such equations, namely, their existence, uniqueness, and regularity. The main properties of the viscosity (weak) solutions giving a partial answer to these problems are discussed in Section 1.2. In addition to the now classical foundational results, we formulate there two recent results on partial regularity of solutions, namely Theorems 1.2.5 and 1.2.6. In Section 1.3 we consider linear elliptic operators of nondivergence form; in addition to classical results, we expose a recent result concerning the difference of viscosity solutions, Theorem 1.3.6. Section 1.4 is devoted to nonlinear equations with smooth solutions, i.e., those with a convex functional F; we describe the Evans-Krylov theory, which guarantees that under the convexity assumption the viscosity solutions are classical, i.e., really verify the equation. Note that one of our principal aims in this book is to show that without the convexity assumption this regularity *does not* hold. We also give a recent result, Theorem 1.4.4, which guarantees the same property for axially symmetric Dirichlet problems for Hessian equations. Section 1.5 is devoted to degenerate elliptic equations; first we expose a geometric approach to degenerate elliptic equations proposed recently by Harvey and Lawson [**104**] and then discuss various regularity results for them. In Section 1.6 we formulate some results and conjectures for homogeneous solutions to fully nonlinear uniformly elliptic equations. Finally, in Section 1.7 we give some Liouville type theorems and also various results on removable singularities for solutions of fully nonlinear elliptic equations, including a recent result, Theorem 1.7.6, describing viscosity solutions of a uniformly elliptic Hessian equation in a punctured ball.

The literature devoted to the topic is overwhelming; the basic references concerning Sections 1.1–1.4 are [**94**], [**42**], [**71**], [**133**], [**134**]; see also [**47**], [**46**], [**48**], [**83**], [**137**], [**37**], [**253**], [**254**], [**255**], [**36**].

1.1. Elliptic equations

1.1.1. Definition and examples. Throughout this book we consider second-order partial differential equations of the form

(1.1.1) $$F(D^2u, Du, u, x) = 0.$$

Here D^2u denotes the Hessian of the function $u : \Omega \longrightarrow \mathbb{R}$, Du being its gradient, $x \in \Omega \subset \mathbb{R}^n$ for a domain (= open connected set) Ω. The functional is of the form

$$F = F(X, r, p, x) : A \times \mathbb{R}^n \times \mathbb{R} \times \Omega \longrightarrow \mathbb{R}$$

where A is a domain in $\operatorname{Sym}_n(\mathbb{R})$, the space of symmetric real $n \times n$-matrices over $\mathbb{R}$. The degenerate ellipticity condition is given by

(1.1.2) $$F(X, r, p, x) \leq F(Y, r, p, x) \text{ if } X \leq Y,$$

i.e., the matrix $Y - X$ is nonnegatively defined; we also suppose that F is continuous. Often we will demand more on the functional F.

If F is a C^1-function in X, (1.1.2) yields the inequality

$$F_X \geq 0,$$

i.e., the matrix of the first derivatives of F with respect to the X variables is nonnegative. If in addition $A = \operatorname{Sym}_n(\mathbb{R})$, the last two inequalities are equivalent.

Let us give several well-known examples of elliptic equations which describe important natural processes, geometrical problems, and the like.

EXAMPLE 1.1.1 (Laplace's equation).

$$\Delta u - c(x)u = f(x).$$

The corresponding F is given by $F(X; p; r; x) = \operatorname{tr}(X) - c(x)r + f(x)$. As particular cases we get the classical Laplace equation $\Delta u = 0$ defining harmonic functions and the Poisson equation $\Delta u = f(x)$.

EXAMPLE 1.1.2 (Degenerate elliptic linear equations). Degenerate elliptic linear equations are of the form

$$\sum a_{ij}(x) \frac{\partial^2 u}{\partial x_i \partial x_j} - \sum b_i(x) \frac{\partial u}{\partial x_i} - c(x)u(x) = f(x),$$

where the matrix $A(x) = a_{ij}(x)$ is symmetric; the corresponding F is

$$F(X; p; r; x) = \operatorname{tr}(A(x)X) - \sum b_i(x)p_i - c(x)r - f(x).$$

In this case, F is degenerate elliptic if and only if $A(x) \geq 0$. If a constant $C > 0$ exists such that $CI \geq A(x) \geq C^{-1}I$ for all $x \in \Omega$ where I is the identity matrix, F is said to be *uniformly elliptic*. If $C(x)I \geq A(x) \geq C(x)^{-1}I$ for $C(x) > 0$ and any $x \in \Omega$, F is called *strictly elliptic*.

EXAMPLE 1.1.3 (Quasilinear elliptic equations in divergence form). An equation

(1.1.3) $$\sum \frac{\partial}{\partial x_i}(a_i(Du, x)) - b(Du, u, x) = 0$$

is elliptic if the vector field $a(p, x)$ is monotone in p regarded as a mapping from $\mathbb{R}^n$ to itself. If the coefficients are differentiable, one can rewrite (1.1.3) as (1.1.1) with

$$F(X, p, r, x) = \operatorname{tr}((D_p a(p, x))X) - b(p, r, x) + \sum \frac{\partial a_i(p, x)}{\partial x_i}.$$

EXAMPLE 1.1.4 (p-Laplace equation). Let $p \geq 1$. The equation

(1.1.4) $$\Delta_p u = \operatorname{div}(|\nabla u|^{p-2} \nabla u) = 0$$

is an important example of a quasilinear elliptic equation in divergence form; one easily calculates that

$$\Delta_p u = |\nabla u|^{p-4} \left\{ |\nabla u|^2 \Delta u + (p-2) \sum_{i,j} \frac{\partial u}{\partial x_i} \frac{\partial u}{\partial x_j} \frac{\partial^2 u}{\partial x_i \partial x_j} \right\}.$$

For $p = 1$ one gets
$$-\Delta_1 u = -\operatorname{div}\left(\frac{\nabla u}{|\nabla u|}\right) = H$$
where H is the mean curvature operator. For $p = 2$ one returns to the Laplacian: $\Delta_2 u = \Delta u$.

When $p = n$ is the dimension of the ambient space, the operator Δ_n becomes conformally invariant.

For $p = \infty$ one gets the limit ∞-*Laplacian*,
$$\Delta_\infty u = \sum_{i,j} \frac{\partial u}{\partial x_i} \frac{\partial u}{\partial x_j} \frac{\partial^2 u}{\partial x_i \partial x_j}.$$

EXAMPLE 1.1.5 (Quasilinear elliptic equations in nondivergence form). The equation
$$\sum a_{ij}(p,x)\frac{\partial^2 u}{\partial x_i \partial x_j} - b(Du, u, x) = 0,$$
where $A(p,x) = a_{ij}(p,x) \in \operatorname{Sym}_n(\mathbb{R})$, corresponds to
$$F(X, r, p, x) = \operatorname{tr}(A(p,x)X) - b(p, r, x).$$

EXAMPLE 1.1.6 (Pucci's equations). These are uniformly elliptic equations important in many applications, especially in the theory of viscosity solutions for fully nonlinear elliptic equations. They are of the form
$$\mathcal{M}^-(D^2 u) = f(x), \quad \mathcal{M}^+(D^2 u) = f(x)$$
where $\mathcal{M}^-$ and $\mathcal{M}^+$ are *Pucci's extremal operators* defined as follows.

Let $A \in \operatorname{Sym}_n(\mathbb{R})$ and let $\lambda \in]0, \Lambda]$. Define
$$\mathcal{M}^-(A, \lambda, \Lambda) = \mathcal{M}^-(A) = \lambda \sum_{\lambda_i > 0} \lambda_i + \Lambda \sum_{\lambda_i < 0} \lambda_i,$$
$$\mathcal{M}^+(A, \lambda, \Lambda) = \mathcal{M}^+(A) = \lambda \sum_{\lambda_i < 0} \lambda_i + \Lambda \sum_{\lambda_i > 0} \lambda_i,$$
where λ_i, $i = 1, 2, \ldots, n$, are the eigenvalues of A.

An equivalent definition is given by
$$\mathcal{M}^-(A, \lambda, \Lambda) = \inf_{M \in M_{\lambda, \Lambda}} \operatorname{tr}(MA), \quad \mathcal{M}^+(A, \lambda, \Lambda) = \sup_{M \in M_{\lambda, \Lambda}} \operatorname{tr}(MA),$$
$M_{\lambda, \Lambda}$ being the set of all symmetric matrices with all eigenvalues in $[\lambda, \Lambda]$.

One verifies without difficulty that $\mathcal{M}^-$, $\mathcal{M}^+$ are uniformly elliptic with the ellipticity constant $C := \max\{n\lambda, n\Lambda\}$ and that $\mathcal{M}^-$ is concave and $\mathcal{M}^+$ is convex.

EXAMPLE 1.1.7 (Hamilton-Jacobi-Bellman and Isaacs equations). These are the fundamental partial differential equations for stochastic control and stochastic differential games. The natural setting involves a collection of elliptic operators of second-order depending either on one parameter α (in the Hamilton-Jacobi-Bellman case) or two parameters α, β (in the case of Isaacs equations). These parameters lie in some index sets.

Let us define for $a_{ij}^\alpha(x), a_{ij}^{\alpha,\beta}(x) \in \operatorname{Sym}_n(\mathbb{R})$

(1.1.5) $\quad \mathcal{L}^\alpha u := \sum a_{ij}^\alpha(x)\frac{\partial^2 u}{\partial x_i \partial x_j} - \sum b_i^\alpha(x)\frac{\partial u}{\partial x_i} - c^\alpha(x)u(x) + f^\alpha(x),$

(1.1.6) $\quad \mathcal{L}^{\alpha,\beta} u := \sum a_{ij}^{\alpha,\beta}(x)\frac{\partial^2 u}{\partial x_i \partial x_j} - \sum b_i^{\alpha,\beta}(x)\frac{\partial u}{\partial x_i} - c^{\alpha,\beta}(x)u(x) + f^{\alpha,\beta}(x),$

all the coefficients being uniformly bounded and the linear operators (1.1.5) and (1.1.6) being elliptic. Hamilton-Jacobi-Bellman equations are of the form

(1.1.7) $$\sup_{\alpha} \mathcal{L}^{\alpha} u = 0,$$

and Isaacs equations are

(1.1.8) $$\sup_{\alpha} \inf_{\beta} \mathcal{L}^{\alpha,\beta} u = 0.$$

Notice that for (1.1.7) the corresponding F is concave in the variables (X, p, r) while for (1.1.8) it is not generally the case.

More concrete examples of geometric origin include

EXAMPLE 1.1.8 (General Monge-Ampère equations). The general Monge-Ampère equation is

$$\det D^2 u = \psi(Du, u, x)$$

where ψ is a given function on $\mathbb{R}^n \times \mathbb{R} \times \mathbb{R}^n$. This equation is elliptic on the set of convex functions. The most important case is the prescribed Gauss curvature equation

$$\det D^2 u = K(u, x)(1 + |Du|^2)^{\frac{n+2}{2}},$$

the function K being given, and the equation means that K is the Gauss curvature of the graph of u (with respect to an upwards directed normal).

EXAMPLE 1.1.9 (Transport Monge-Ampère equations). Initially the Monge-Ampère equation came into mathematics as a solution of an applied problem of the optimal mass transportation. Let G, G' be domains in $\mathbb{R}^n$ and let $c: G \times G' \to \mathbb{R}$ be a cost function expressing the cost of transportation from a point of G to G'. Let f and g be measures on G and G' such that $\int_G f = \int_{G'} g$. For a measure-preserving transportation function $T: G \to G'$ the cost is given by

$$C(T) = \int_G c(x, T(y)).$$

Problem (Monge, 1784). *Find a transportation function which minimizes the cost.*

Solution. A cost-minimizing function T satisfies $T = \nabla u$ for some function u on G. Moreover, the function u satisfies the so-called *transport Monge-Ampère equation*:

$$\det(D^2 u - c_{xx}(x, T(x))) = \frac{f(x)}{g(T(x))}.$$

As was shown by Brenier [**32**] for the quadratic cost function, the last equation is reduced to

$$\det D^2 u = \frac{f(x)}{g(\nabla u(x))}.$$

EXAMPLE 1.1.10 (Complex Monge-Ampère and Donaldson's equations). The *complex Monge-Ampère equation* has the form

$$\det\left(\frac{\partial^2 u}{\partial z_i \partial \bar{z}_j}\right) = f,$$

where u is a function of complex variables $z_1, \ldots, z_n$ and $f > 0$. When the solution u is a C^2-function, u is a pluri-subharmonic function, i.e., the Hermitian form $\partial_{i\bar{j}} u dz_i d\bar{z}_j$ is positive. The complex Monge-Ampère equation

$$(1.1.9) \qquad \det\left(g_{i\bar{j}}\frac{\partial^2 u}{\partial z_i \partial \bar{z}_j}\right) \det(g_{i\bar{j}})^{-1} = e^{tu+f}$$

defined on a compact Kähler manifold M with metric $g_{i\bar{j}}$, $t > 0$, plays an important role in the study of the geometry of M; see [**18**], [**276**]. Regularity of solutions of (1.1.9) implies for manifolds with negative first Chern class the existence of a Kähler-Einstein metric.

Donaldson's equation. Let X be a surface. Donaldson, [**75**], considered for a given real parameter ε the following fully nonlinear equation on $X \times [0, 1]$:

$$(1.1.10) \qquad u_{tt}(1 - \Delta_x u) - |\nabla_x u_t|^2 = \varepsilon,$$

$x \in X$, $t \in [0, 1]$. The equation (1.1.10) is the equation of geodesics on infinite dimensional space of "Kählerian potentials". It can also be formally considered as a Nahm's equation of motion of a particle on an infinite-dimensional Lie group of area-preserving diffeomorphisms of a surface X.

When X has dimension 1, (1.1.10) is the real Monge-Ampère operator. When X is of dimension 2, (1.1.10) can be reduced to a complex Monge-Ampère operator on $X \times \mathbb{S}^1 \times (0, 1)$.

EXAMPLE 1.1.11 (Hessian equations). The Monge-Ampère equation is a special case of a *Hessian equation*,

$$(1.1.11) \qquad F(D^2 u) := f(\lambda(D^2 u)) = \psi(Du, u, x)$$

where f is a given symmetric function of n variables and

$$\lambda(D^2 u) = \lambda = (\lambda_1, \ldots, \lambda_n)$$

denotes the eigenvalues of $D^2 u$. If the function f is continuously differentiable, then the ellipticity of (1.1.11) is equivalent to the condition $f_{\lambda_i} > 0$ for all $i = 1, 2, \ldots, n$. Typical examples of functions f are the elementary symmetric functions

$$\sigma_k(\lambda) = \sum_{1 \leq i_1 < \cdots < i_k} \lambda_{i_1} \cdots \lambda_{i_k}$$

and their quotients

$$\sigma_{k,l}(\lambda) := \frac{\sigma_k(\lambda)}{\sigma_l(\lambda)}, \quad 1 \leq l < k \leq n,$$

restricted to the positivity set of the denominator. In the case $k = 1$ we return to (the slightly generalized) Example 1.1.1. The ellipticity of these operators is not obvious and depends on the properties of some functions σ_k. However, one easily checks that all these operators are elliptic on locally uniformly convex functions (all $\lambda_i > 0$).

EXAMPLE 1.1.12 (Equations linear in symmetric functions σ_k). These are Hessian equations of the form

$$\sum_{k=0}^{n} a_k \sigma_k(\lambda) = 0, \qquad a_k \in \mathbb{R}, \ k = 0, \ldots, n,$$

which are strictly elliptic for appropriate constants a_k; they appear in some problems of differential geometry. However, they are never uniformly elliptic. From the

results of [**208**] one can deduce that these equations can be rewritten in a variational manner. Special Lagrangian equations (see Section 7.1),
$$\operatorname{Im}\{e^{-i\theta}\det(I+iD^2u)\}=0$$
belong to this class. Note that the equation of the form
$$\sigma_k(\lambda)=1$$
sometimes is called the σ_k-*equation*.

EXAMPLE 1.1.13 (Curvature equations). The general form of a curvature equation (or so-called *Weingarten equation*) in Euclidean space is
$$F(u):=f(\varkappa(u))=\psi(Du,u,x)$$
where now $\varkappa(u)=\varkappa=(\varkappa_1,\ldots,\varkappa_n)$ denotes the principal curvatures of the graph of u and again f is a given symmetric function of n variables. Since $(\varkappa_1,\ldots,\varkappa_n)$ are the eigenvalues of the Hessian D^2u with respect to the metric $I+Du\otimes Du$, this equation is an equation of the form (1.1.1). The *Gauss curvature equation* corresponds to the case $f(\varkappa)=\sigma_n(\varkappa)=\prod_i \varkappa_i$. Other important examples are the *mean curvature*, $\sigma_1(\varkappa)$, yielding a quasilinear elliptic equation, the *scalar curvature* $\sigma_2(\varkappa)$, and the *harmonic curvature* $\sigma_{n,1}(\varkappa)$.

EXAMPLE 1.1.14 (Conformal Hessian equations). Let $n\geq 3, u>0$, and let
$$F[u]:=f(\lambda(A^u))=\psi(u,x),$$
f again being a symmetric function and $\lambda(A^u)=(\lambda_1,\ldots,\lambda_n)$ being the eigenvalues of the conformal Hessian
$$A^u:=uD^2u-\frac{1}{2}|Du|^2 I.$$
In this case F is invariant under conformal mappings $T:\mathbb{R}^n\longrightarrow\mathbb{R}^n$, i.e., transformations which preserve angles between curves. In contrast to the case $n=2$, for $n\geq 3$ any conformal transformation of $\mathbb{R}^n$ is decomposed into a family of finitely many Möbius transformations, that is, mappings of the form
$$Tx=y+\frac{kA(x-z)}{|x-z|^a},$$
with $x,z\in\mathbb{R}^n, k\in\mathbb{R}, a\in\{0,2\}$, and an orthogonal matrix A. In other words, each T is a composition of a translation, a homothety, a rotation, and (maybe) an inversion.

1.1.2. Uniqueness, existence, and regularity problems. Let us then discuss solutions of nonlinear elliptic equations. There are many problem types for them, but we will study only the most simple (and most fundamental) formulation, namely, the following *Dirichlet problem*:

(1.1.12) $$\begin{cases} F(D^2u,Du,u,x)=0 & \text{in }\Omega,\\ u=\varphi & \text{on }\partial\Omega, \end{cases}$$

where $\Omega\subset\mathbb{R}^n$ is a bounded domain with a smooth boundary $\partial\Omega$ and φ is a continuous function on $\partial\Omega$.

A function u is called a *classical* solution of (1.1.12) if $u\in C^2(\Omega)$ and u satisfies (1.1.12). Actually, any classical solution of (1.1.12) is a smooth ($C^{\alpha+2}$) solution, provided that F is a C^α function of its arguments with $\alpha>1, \alpha\notin\mathbb{N}$.

Assuming that $\partial F/\partial u \leq 0$, it is not difficult to prove that (1.1.12) has no more than one classical solution and thus classical solutions verify the fundamental condition of uniqueness. The basic problem is the existence of such classical solutions, and there is no hope of getting such an existence for a sufficiently general class of nonlinear elliptic equations at least for $n \geq 3$ (for $n = 2$ the solutions are classical for uniformly elliptic equations by Nirenberg [**186**]).

The only way out is to define a class of generalized (weak) solutions for the problem (1.1.12) in such a manner that the unicity and existence of solutions may be verified in this class. That is possible for nonlinear elliptic equations (NLEE), which constitutes a major breakthrough in the theory of partial differential equations resulting in the theory of viscosity solutions described in the next section. These solutions are by definition merely continuous functions, and this leads to another major question, the regularity problem, namely, what can be said about differentiability and continuity properties of those generalized solutions. This last problem is very far from a satisfactory answer except for some specific classes of NLEE.

However, there do exist some classes of NLEE with advanced regularity properties, sometimes giving classical solutions. A major part of these results is obtained using the continuity method, a priori bounds, and maximum principles. Now we comment briefly on those fundamental methods.

1.1.3. Continuity method, a priori bounds, and maximum principle(s). The general setting of the *continuity method* is as follows. To prove the existence of a regular solution to a certain elliptic equation, one chooses a continuous family F_t of equations parametrized, say, by a unit interval $t \in [0,1]$ in such a way that for F_0 regular solutions do exist, F_1 being the initial equation. For example, one can often choose F_0 to be the Laplacian equation $\Delta u = 0$ for which one has very precise and complete information. One then considers the maximal subset $S \subset [0,1]$ such that for any $t \in S$ the equation F_t has an appropriate solution; therefore $0 \in S$. If one can prove that S is open and closed, then $S = [0,1]$ by connectedness of the interval and thus the problem is solved. Therefore, to solve the problem one needs to prove

1) S is open in $[0,1]$,
2) S is closed in $[0,1]$.

The first point is usually much simpler than the second; often it is a consequence of some general results on implicit functions in appropriate functional spaces. To prove 2) one often uses theorems of Arzelà–Ascoli type; to apply these theorems one needs some upper bounds of appropriate norms of smooth solutions to the initial equation. To be useful, those bounds should not be dependent on the solutions themselves, but on other data such as the ellipticity constant, boundary data, the domain's geometry, etc. Such bounds are called a priori and their proof is usually the only possibility of proving the existence of sufficiently smooth solutions. One finds some examples of a priori bounds in the next section.

Another essential tool of the theory is given by *maximum principles*, the most simple being the maximum principle for harmonic functions, i.e., for Laplace's equation.

THEOREM 1.1.1. *Let u be a solution to the Dirichlet problem* (1.1.12) *for Laplace's equation* $\Delta u = 0$. *Then u attains its supremum on the boundary of Ω.*

In the nonlinear context one often uses the Alexandrov-Bakelman-Pucci (ABP) maximum principle; see Theorem 1.3.1 below.

1.2. Viscosity solutions

In defining the weak solution to fully nonlinear equations, we have to decide what property of the classical solutions we want to keep instead of considering smooth functions which satisfy the equation. For equations written in variational form one immediately gets an integral identity for the solutions and it is natural to keep such an identity as a characteristic property of the weak solutions. For general fully nonlinear equations that does not work. Fortunately, there is another universal property of the solutions which one can try to use. We will suppose here and below that the functional F depends only on the Hessian D^2u of the unknown function and thus equation (1.1.1) becomes (0.1) of the Preface; the same is supposed for the Dirichlet problem (1.1.12).

Let F be an elliptic operator and let $u \in C^2(\Omega)$ be a *subsolution* $(F(D^2u) \leq 0)$ in a domain Ω. Let $v \in C^2(\Omega)$ be a *supersolution* $(F(D^2u) \geq 0)$ in Ω. Then $u - v$ attains its supremum on the boundary of Ω. That is the maximum principle for classical solutions of fully nonlinear equations. The idea of viscosity solutions is to extend the notions of sub/supersolutions to a large set of nonsmooth functions preserving the maximum principle.

Notice that weak solutions do not necessarily increase the set of classical solutions. In some situations they are automatically classical. For instance, consider the Laplace operator. The weak solutions in a variational sense are defined as functions $u \in H^1(\Omega)$ which satisfy the following integral identity:

$$\int_\Omega \nabla u \nabla \psi dx = 0,$$

for any smooth function ψ vanishing on $\partial\Omega$. By classical results of Weyl, weak solutions in the sense of this integral identity are smooth functions in any open domain and satisfy the Laplace equation; see, e.g., [**84**]. Therefore, one should verify that the idea of weak solutions works in the general case, increasing sufficiently the set of possible solutions.

One of the classical methods for the solution of the Dirichlet problem for the Laplace equation was suggested by Perron in his 1923 paper [**200**]. He defined a solution to the Dirichlet problem in a bounded domain Ω for the Laplace operator taking the infimum of superharmonic functions which are on the boundary $\partial\Omega$ greater than or equal to the boundary data. Perron proved that there is a unique such infimum which gives a solution to the Dirichlet problem. The *viscosity solutions* to fully nonlinear equations can be defined in a similar way:

DEFINITION 1.2.1. Let G be a bounded domain. A continuous function u in G is a viscosity subsolution of

$$F(D^2u) = f$$

in G if the following condition holds: For any $y \in G$, $\phi \in C^2(G)$ such that $u - \phi$ has a local maximum at y one has

$$F(D^2\phi(y)) \geq f(y).$$

Correspondingly a continuous function u is a viscosity supersolution if $-u$ is a subsolution. If u is both a sub- and supersolution, then u is called a viscosity solution of the equation.

The notion of viscosity solution was introduced by Crandall, Lions, and Evans initially for Hamilton-Jacobi equations; see [**72**], [**70**].

Consider now a Dirichlet problem (1.1.12). Applying the Perron method to the nonlinear elliptic operator (1.1.1), Ishii proved the existence of a viscosity solution of the Dirichlet problem (1.1.12), [**116**], and Jensen [**122**] proved the uniqueness of the viscosity solution of the Dirichlet problem (1.1.12). Finally in [**42**], [**71**] the following general result was obtained :

THEOREM 1.2.1. *For a uniformly elliptic operator F and a continuous function g on ∂G, Dirichlet problem (1.1.12) has a unique viscosity solution u. For a strictly elliptic operator F the Dirichlet problem (1.1.12) has a unique viscosity solution if the domain G is convex.*

The case of strictly elliptic operators was clarified by Harvey and Lawson in [**104**]. We discuss their approach in Section 1.5 below.

Safonov gave an alternative way to define viscosity solutions to (1.1.12) as follows. Denote by U^+ (resp. U^-) the set of C^2 supersolutions (resp. C^2 subsolutions) of the problem (1.1.12):

$$U^+ = \{u^+ : u^+ \in C^2(G), F(D^2 u^+) \leq 0 \text{ and } u^+|_{\partial \Omega} \geq \varphi\},$$
$$U^- = \{u^- : u^- \in C^2(G), F(D^2 u^-) \geq 0 \text{ and } u^-|_{\partial \Omega} \leq \varphi\}.$$

The following important result is essentially due to M. Safonov; see [**173**]:

THEOREM 1.2.2. *Let u be a viscosity solution of (1.1.12). Then there are uniformly converging sequences $u_n^+ \in U^+$, $u_n^- \in U^-$ such that*

$$u = \lim_{n \to \infty} u_n^+ = \lim_{n \to \infty} u_n^-.$$

PROOF. Let u be a continuous function in G and let $H \Subset G$ be an open set. Define, for $\varepsilon > 0$, the upper ε-envelope of u:

$$u^\varepsilon(x_0) = \sup_{x \in H} \{u(x) + \varepsilon - |x - x_0|^2/\varepsilon\}.$$

According Jensen's result [**122**] (cf. [**42**, Sec. 5.1]), if u is a viscosity subsolution of $F(D^2 u) = 0$, then the upper ε-envelopes u^ε are also viscosity subsolutions of $F(D^2 u) = 0$. Moreover, $u^\varepsilon \to u$ as $\varepsilon \to 0$, uniformly on compact subsets, and the u^ε are $C^{1,1}$. In particular, they have second differential almost everywhere and $F(D^2 u^\varepsilon) \geq 0$ a.e.

Let η^δ be a smooth nonnegative function with the support in B_δ and the total integral 1. Consider the standard mollifiers $u^{\varepsilon,\delta} = u^\varepsilon * \eta^\delta$, which are smooth and satisfy

$$D^2 u^{\varepsilon,\delta} \to D^2 u^\varepsilon$$

almost everywhere, as $\delta \to 0$. Since the functions $u^{\varepsilon,\delta} + C|x|^2$ are convex, with $C = C(\varepsilon)$, we have

$$0 \leq f^{\varepsilon,\delta} := (F(D^2 u^{\varepsilon,\delta}))_- \leq C = C(\varepsilon),$$

and $f^{\varepsilon,\delta} \to 0$ a.e. as $\delta \to 0$.

Let $v^{\varepsilon,\delta}$ be a classical solution of the Dirichlet problem of the concave minimal Pucci operator and a Lipschitz right side $f^{\varepsilon,\delta}$:
$$\mathcal{M}^-(D^2 v^{\varepsilon,\delta}) = f^{\varepsilon,\delta} \text{ in } G, \quad v^{\varepsilon,\delta} = 0 \text{ on } \partial G.$$

By the APB estimates, see Theorem 1.3.1 below, one has $0 \geq v^{\varepsilon,\delta} \to 0$ as $\delta \to 0$, uniformly on G. Then one can choose small positive ε, δ, and c_ε such that the function $w^{\varepsilon,\delta} := u^{\varepsilon,\delta} + v^{\varepsilon,\delta} - c_\varepsilon < u$, and it can be made arbitrarily close to u. Finally
$$F(D^2 w^{\varepsilon,\delta}) \geq F(D^2 u^{\varepsilon,\delta}) + \mathcal{M}^-(D^2 v^{\varepsilon,\delta})$$
$$= F(D^2 u^{\varepsilon,\delta}) + (F(D^2 u^{\varepsilon,\delta}))_- \geq 0.$$

Thus we proved that a viscosity solution can be uniformly approximated from below by classical subsolutions. The "upper" approximation is quite similar. □

As we will see below, the viscosity solution is not necessarily a C^2 function in an open domain. The best known regularity for the viscosity solutions of uniformly elliptic equations is due to Caffarelli [**37**] and Trudinger [**255**]; see also [**42**].

THEOREM 1.2.3. *Let u be a viscosity solution of $F(D^2 u) = 0$ in B_2, F being uniformly elliptic. Then $u \in C^{1,a}(B_1)$ and*
$$\|u\|_{C^{1,a}(B_1)} \leq C(\|u\|_{L_\infty(B_1)} + |F(0)|),$$
where $a \in\,]0,1[$, $C > 0$ depend on the ellipticity constant of the operator F.

Recently in [**12**] a partial regularity of the viscosity solutions was proved, i.e., if u is a viscosity solution to (0.1) in a domain G, then u is a smooth function on an open dense subset of G and satisfies there the equation. More precisely,

THEOREM 1.2.4. *Let u be a viscosity solution of $F(D^2 u) = 0$ in G, F being uniformly elliptic. Then there exist $\varepsilon > 0$ depending on the ellipticity constant of F and a closed set $E \subset G$ such that the Hausdorff dimension of E is at most $n - \varepsilon$ and u is a $C^{2,\alpha}$ function on $G \setminus E$.*

Notice that just the partial regularity of solutions to $F(D^2 u) = 0$ without estimates of the Hausdorff dimension of the set E can be deduced directly from Theorem 1.2.5 and Corollary 1.3.3 of Section 1.3 below.

The proof of Theorem 1.2.4 is based on a deep result of Savin [**218**], which asserts that any viscosity solution of (0.1) that is sufficiently close to a quadratic polynomial must be $C^{2,\alpha}$:

THEOREM 1.2.5. *Let u be a viscosity solution of a strictly elliptic equation $F(D^2 u) = 0$ in a domain Ω. Assume that $F(0) = 0$. Let $G \Subset \Omega$. If the L_∞-norm of u in Ω is sufficiently small, then u is a $C^{2,\alpha}$ function on G.*

Using this result, the authors of [**224**] proved the regularity near the boundary of viscosity solutions to the Dirichlet problem (1.1.12).

THEOREM 1.2.6. *Let u be a solution of the Dirichlet problem (1.1.12). Assume that $\partial\Omega$ is of class $C^{2+\varepsilon}$, $\varphi \in C^{2+\varepsilon}$, $F \in C^{1+\varepsilon}$, for some $\varepsilon > 0$. Then u is a classical solution in a neighborhood of the boundary $\partial\Omega$; more precisely, $u \in C^{2+\varepsilon}(\Omega_\delta)$ for a positive δ depending on F, Ω, and the $C^{2+\varepsilon}$-norm of φ where Ω_δ is an intersection of Ω with the δ-neighborhood of $\partial\Omega$.*

1.3. Linear elliptic operators of nondivergence form

Let u be a solution of the fully nonlinear elliptic equation
$$F(D^2 u) = 0$$
in a domain $G \subset \mathbb{R}^n$. Let v be a partial derivative of u, $v = \partial u/\partial x_i$. Then v is a solution of a linear elliptic equation in nondivergence form:

$$(1.3.1) \qquad Lv = \sum a_{ij}(x)\frac{\partial^2 v}{\partial x_i \partial x_j} = 0.$$

In general we have no a priori information about the coefficients $a_{ij}(x)$ except of their pointwise ellipticity for a strictly elliptic operator F,

$$0 < \sum a_{ij}(x)\xi_i \xi_j < C|\xi|^2,$$

or, in the case of uniformly elliptic operator F, the $a_{ij}(x)$ satisfy the uniform ellipticity condition,

$$(1.3.2) \qquad C^{-1}|\xi|^2 < \sum a_{ij}(x)\xi_i \xi_j < C|\xi|^2.$$

There are not so many universal results for the equation (1.3.1) which are valid for equations with measurable coefficients. But each of these universal results appears to be very useful for applications to fully nonlinear problems. Apart from the maximum principle for (1.3.1), there are two very deep estimates for the uniformly elliptic equations of the form

$$(1.3.3) \qquad Lv = \sum a_{ij}(x)\frac{\partial^2 v}{\partial x_i \partial x_j} = f.$$

The first result is known as the *Alexandrov-Bakelman-Pucci (ABP) maximum principle*:

THEOREM 1.3.1. *Let v be a solution of the uniformly elliptic equation (1.3.3) defined in a unit ball $B \subset \mathbb{R}^n$. Assume that $f \in L_n(B)$. Then*

$$\sup_B |v| \leq C_0 \|v\|_{L_n(B)} + \sup_{\partial B} |v|,$$

where constant C_0 depends only on the ellipticity constant C.

One can regard the maximum principle as an a priori C^0 estimate for solutions of (1.3.1). For uniformly elliptic equations it is possible to improve this estimate up to C^ε-norm for $\varepsilon \in]0, 1[$, on subdomains of Ω. The following result is due to Krylov and Safonov [**137**]:

THEOREM 1.3.2. *Let v be a solution of a uniformly elliptic equation (1.3.3) defined in a unit ball $B \subset \mathbb{R}^n$. Assume that $f \in L_n(B)$. Then*

$$\|v\|_{C^\varepsilon(B_{1/2})} \leq C_0 \|v\|_{L_n(B)} + \sup_{\partial B} |v|,$$

where positive constants $\varepsilon \in]0, 1[$, C_0 depend only on the ellipticity constant C.

As was shown by Safonov [**215**] in general, one cannot expect better regularity than C^ε in Theorem 1.3.2.

Surprisingly enough, in two dimensions much stronger results hold. First of all, the maximum principle for the equation (1.3.1) holds for the gradient of solutions:

THEOREM 1.3.3. *Let v be a solution of a homogeneous equation (1.3.1) defined in a bounded 2-dimensional domain. Then any partial derivative $\partial v/\partial x_i$ attains its supremum on the boundary of the domain.*

Again for uniformly elliptic equations in two dimensions we have very strong universal estimates of solutions on subdomains:

THEOREM 1.3.4. *Let v be a solution of a uniformly elliptic equation (1.3.1) defined in a unit ball $B \subset \mathbb{R}^2$. Assume that $f \in L_p(B)$, $p > 2$. Then*

$$\|D^2 v\|_{L_{2+\varepsilon}(B_{1/2})} \leq C_0 \|v\|_{L_p(B)} + \sup_{\partial B} |v|,$$

where positive constants ε, C_0 depend only on p and the ellipticity constant C.

As an immediate corollary we have

COROLLARY 1.3.1. *Let v be a solution of a uniformly elliptic equation (1.3.1) defined in a unit ball $B \subset \mathbb{R}^2$. Assume that $f \in L_p(B)$, $p > 2$. Then*

$$\|v\|_{C^{1+\varepsilon}(B_{1/2})} \leq C_0 \|v\|_{L_p(B)} + \sup_{\partial B} |v|,$$

where the positive constants ε, C_0 depend only on p and the ellipticity constant C.

Let us return to arbitrary dimension. Let $\Omega \subset \mathbb{R}^n$ be a smooth bounded domain and let L be a linear uniformly elliptic operator (1.3.1) defined in Ω with the ellipticity constant C.

We consider a Dirichlet problem in Ω:

$$(1.3.4) \quad \begin{cases} Lv = 0 & \text{in } \Omega, \\ u = \varphi & \text{on } \partial\Omega. \end{cases}$$

By Theorem 1.3.1 the solution of the Dirichlet problem (1.3.4) in $W^{2,n}(\Omega)$ is unique. On the other hand, from the example of Safonov [**215**] one can easily see that the Dirichlet problem (1.3.4) is not always solvable in $W^{2,n}(\Omega)$. That creates certain difficulties in the definition of weak solutions to Dirichlet problem (1.3.4). However, one can define viscosity (weak) solutions to (1.3.4) as follows.

First of all, due to Jensen's result [**122**] the two following definitions of the viscosity solutions to the Dirichlet problem (1.3.4) are equivalent.

DEFINITION 1.3.1. *A function v is a viscosity solution of (1.3.4) if $v = \lim v_k$, where*

$$Lv_k = \sum a_{ij}^k(x) \frac{\partial^2 v_k}{\partial x_i \partial x_j} = 0,$$

the a_{ij}^k are continuous, and $a_{ij}^k \to a_{ij}$ in $L_1(\Omega)$.

DEFINITION 1.3.2. *A function v is a viscosity solution of (1.3.4) if*

$$(\mathrm{i}) \qquad \lim_{\varepsilon \to +0} \sup \varepsilon^{-n} \int_{|x-y|<\varepsilon} \left[\sum a_{ij}(y) \left(\frac{\partial^2 \phi}{\partial x_i \partial x_j} + \eta \delta_{ij} \right) \right]^+ dy > 0$$

1.3. LINEAR ELLIPTIC OPERATORS OF NONDIVERGENCE FORM

for all $\eta > 0$, whenever $x \in \Omega$ and $\phi \in C^2(\Omega)$ are such that

$$0 = (v - \phi)(x) \geq (v - \phi)(y)$$

for all $y \in \Omega$, and if

(ii) $\qquad \lim_{\varepsilon \to +0} \sup \varepsilon^{-n} \int_{|x-y|<\varepsilon} \left[\sum a_{ij}(y) \left(\frac{\partial^2 \phi}{\partial x_i \partial x_j} - \eta \delta_{ij} \right) \right]^- dy > 0$

for all $\eta > 0$, whenever $x \in \Omega$ and $\phi \in C^2(\Omega)$ are such that

$$0 = (v - \phi)(x) \leq (v - \phi)(y)$$

for all $y \in \Omega$, where $[t]^+$ denotes $\max\{0, t\}$ and $[t]^-$ denotes $\max\{0, -t\}$.

We notice that the uniqueness problem of the viscosity solution to the Dirichlet problem (1.3.4) has a natural setting in the framework of diffusion processes. Strook and Varadhan [**237**] stated it as a *martingale problem* for the elliptic operator L: Is it true that for any $x \in \mathbb{R}^n$ there exists a *unique* probability measure P_x on $C([0, \infty[, \mathbb{R}^n)$ such that

$$P(\xi(0) = x) = 1$$

while

$$\psi(\xi(t)) - \psi(\xi(0)) - \int_0^t L\psi(\xi(s))ds$$

is a martingale for all $\psi \in C_0^2(\mathbb{R}^n)$?

If the diffusion P is defined, we immediately have a solution of the Dirichlet problem (1.3.4) by the formula $u(x) = E_x(\phi(\tau))$, where τ is the first time when the path starting at x leaves Ω.

In addition to the a priori Hölder regularity for solutions of the homogeneous equation (1.3.1) the *Harnack inequality* holds.

THEOREM 1.3.5. *Let v be a positive solution of uniformly elliptic equation (1.3.1) defined in a unit ball $B \subset \mathbb{R}^n$. Then*

$$\sup_{B_{1/2}} v < C_0 \inf_{B_{1/2}} v,$$

where the positive constant C_0 depends only the ellipticity constant C.

The main tools in the proof of Theorem 1.3.5 are Theorem 1.3.1 and the two following lemmas:

LEMMA 1.3.1. *If v is a nonnegative supersolution of (1.3.1) in B_1, then*

$$\left| \left\{ v > t \inf_{B_{1/2}} v \right\} \cap B_{1/4} \right| \leq C_0 t^{-\varepsilon},$$

where C_0, ε depend on the ellipticity constant of F.

LEMMA 1.3.2. *If v is a supersolution of (1.3.1) in B_1 and $v < 1$, then*

$$\sup_{B_{1/2}} v \leq C_0 |\{v > 0\} \cap B_{3/4}|,$$

where C_0 depends on the ellipticity constant of F.

Though Definition 1.3.1 implies the existence of the viscosity solution to the Dirichlet problem (1.3.4), the principal question on the uniqueness of the viscosity solution remains open. For a general uniformly elliptic operator (1.3.1), in [**172**] it is shown that viscosity solutions to the Dirichlet problem (1.3.4) are not unique (see also some extension of the result in [**216**]). However under certain restrictions on the coefficients of operator L some uniqueness results are known; see, e.g., the paper [**135**].

Viscosity solutions defined for the linear operator L have an important connection [**173**] with viscosity solutions for the elliptic fully nonlinear equations (0.1):

THEOREM 1.3.6. *Let u_1, u_2 be two viscosity solutions to a fully nonlinear uniformly elliptic equation (0.1) defined in a domain $\Omega \subset \mathbb{R}^n$. Set $v = u_1 - u_2$. Then v is a viscosity solution to a linear uniformly elliptic equation (1.3.1) with measurable coefficients $a_{ij}(x)$ satisfying the inequality (1.3.2).*

We need the following results from [**9**].

LEMMA 1.3.3. *Let $B \subset \mathbb{R}^n$ be a unit ball. Let $u \in W^{2,n}(B)$ be such that*

$$Lu \geq -1 \quad in\ B,$$

where L is the uniformly elliptic operator (1.3.1) and $u_{|\partial B} \leq 0$. Let U be the convex envelope of the graph of u^+ and let $\nu : U \to \mathbb{S}^n$ be the Gauss normal map. Let ds be the element of the surface area of U and let Jds be the Jacobian of the map ν. Then

$$J \leq C_1,$$

where a positive constant C_1 depends on the ellipticity constant of the operator L. Moreover the support of the function J is a subset of the coincidence set of the graphs of the functions u^+ and J.

As an immediate corollary we get

LEMMA 1.3.4. *Let $B \subset \mathbb{R}^n$ be a unit ball. Let $u_n \in W^{2,n}(B)$, $n = 1, 2, \ldots$, be such that*

$$L_n u_n \geq -1 \quad in\ B,$$

where L_n is a sequence of uniformly elliptic operators of the form (1.3.1) admitting a joint ellipticity constant. Assume that the sequence converges in $C^0(B)$, $u_n \to u$, and $u_{|\partial B} \leq 0$. Let U be the convex envelope of the graph of u^+ and let $\nu : U \to \mathbb{S}^n$ be the Gauss normal map. Let ds be the element of the surface area of U and let Jds be the Jacobian of the map ν. Then

$$J \leq C_1,$$

where the positive constant C_1 depends on the ellipticity constant of the operators L_n.

The following two results are due to Trudinger [**255**], [**257**]:

LEMMA 1.3.5. *Let u be a viscosity solution of the fully nonlinear uniformly elliptic equation of the form (0.1). Then for almost every point $y \in \Omega$ there exists a second-order polynomial $p_y(x)$ such that $u(x) - p_y(x) = o(|x-y|^2)$ and $F(D^2 p_y) = 0$. Furthermore, for some $\delta > 0$ we have $u \in C^{1,\delta}$.*

PROOF OF THEOREM 1.3.6. Since by Lemma 1.3.5 the functions u_1, u_2 have almost everywhere *approximative second differentials* which satisfy the equation (0.1), it follows that the function v has almost everywhere an approximative second differential which satisfies the equation (1.3.1) with the coefficients satisfying (1.3.2).

Assume by contradiction that v is not a viscosity solution of (1.3.1). Then by Definition 1.3.2 it follows that either property (i) or (ii) fails to be true. We may assume without loss of generality that (ii) is not satisfied for the function v. That implies the existence of a point $y \in \Omega$, of a function $\phi \in C^2(\Omega)$, $\phi(y) = v(y)$, and a constant $\delta > 0$ such that for any $\varepsilon > 0$ there exists $r > 0$ such that in the ball $B = \{x : |x - y| < r\}$ the following inequalities hold: $\phi \leq v - \delta|x-y|^2$ and $L\phi \geq 0$ on $B \setminus E$ where E is a Borel subset of B such that $\text{meas}(E) < \varepsilon \,\text{meas}(B)$. Set $\psi = \phi - v + \delta r^2/2$. Let V be the convex envelope of ψ^+. By Theorem 1.2.3 one finds convergent sequences $u_n^- \to u_1$, $u_n^+ \to u_2$, $u_n^-, u_n^+ \in C^2$, such that the u_n^- are subelliptic, $F(D^2 u_n^-) > 0$, and the u_n^+ are superelliptic, $F(D^2 u_n^-) < 0$. Set $v_n = u_n^+ - u_n^-$. Then the v_n are superelliptic functions for a linear uniformly elliptic operator (1.3.1),
$$L_n v_n \leq 0.$$
Thus we can apply Lemma 1.3.3 to the function ψ. Denote by V the convex envelope of the function ψ. Let $J = a\,ds$ be the Jacobian of the Gauss map of the function V. Since the functions ψ_n are subelliptic, by Lemma 1.3.4, $a < C$, where the constant $C > 0$ depends on the C^2-norm of ϕ and the ellipticity constant of the operator F. Since the function v has almost everywhere a second differential satisfying (1.3.1), we conclude that $a = 0$ on $B \setminus E$. Thus by the APB maximum principle one has $\psi < Cr^2 \varepsilon^{1/n}, C > 0$. Since $\psi(y) = \delta r^2/2$, choosing a sufficiently small $\varepsilon > 0$, we get a contradiction. The theorem is proved. $\square$

COROLLARY 1.3.2. *Let u be a viscosity solution of the fully nonlinear equation (0.1). Then any partial derivative $v = u_{x_k}$ is a solution of a uniformly elliptic equation (1.3.1), with the coefficients a_{ij} defined almost everywhere by*
$$a_{ij} = F_{u_{ij}} := \frac{\partial F}{\partial u_{ij}}.$$

PROOF. We may assume that $k = 1$. Set
$$v_m = m\big(u(x_1, \ldots, x_n) - u(x_1 + \tfrac{1}{m}, \ldots, x_n)\big).$$
By Theorem 1.3.6 v_m is a viscosity solution of the equation (1.3.1) with the coefficients
$$a_{ij}^m = F_{u_{ij}}(\theta D^2 u_1 + (1-\theta) D^2 u_2),$$
$0 < \theta < 1$. For $\varepsilon > 0$ we denote
$$E(m, \varepsilon) = \{x \in \Omega : |D^2 u(x_1, \ldots, x_n) - D^2 u(x_1 + \tfrac{1}{m}, \ldots, x_n)| > \varepsilon\}.$$
Since $D^2 u$ is defined almost everywhere and is measurable on Ω, $\text{meas}(E(m, \varepsilon)) \to 0$ as $m \to \infty$ for any $\varepsilon > 0$ by Lusin's theorem. Thus $a_{ij}^m \to a_{ij}$ in $L_1(\Omega)$ as $m \to \infty$ and from Definition 1.3.1 it follows that v is a viscosity solution of (1.3.1). $\square$

COROLLARY 1.3.3. *Let u be a viscosity solution of the fully nonlinear equation of the form (0.1). Then the function u has almost everywhere the* approximative third differential, *i.e., for almost every point $y \in \Omega$ there exists a third order polynomial $p_y(x)$ such that $u(x) - p_y(x) = o(|x - y|^3)$.*

PROOF. By Corollary 1.3.2 the functions u_{x_i}, $i = 1, 2, \ldots$, are viscosity solutions of uniformly elliptic equations. Integrating the function u_{x_1} over dx_1 we get that the function u has almost everywhere an approximative third differential along the lines parallel to the x_1-axis. Consequently integrating functions u_{x_i} over dx_i, $i = 1, 2, 3, \ldots$, we get by induction that the function u has almost everywhere an approximative third differential along the planes parallel to $x_1 x_2$, along the subspaces parallel to $x_1 x_2 x_3$, etc. □

Finally we notice that no other a priori regularity for the viscosity solutions of the linear equation (1.3.1) except the Hölder regularity is known. It would be very important to understand whether a priori estimates in $W^{2,1}$ or in $W^{1,1}$ for solutions of equation (1.3.1) with measurable coefficients exist.

PROBLEM 1.3.1. Are viscosity solutions to a uniformly elliptic equation (1.3.1) locally in $W^{1,1}$?

1.4. Nonlinear equations with smooth solutions

In $n \geq 3$ dimensions the smoothness of the viscosity solutions is only known for a special but very important class of uniformly elliptic equations, namely the equations with a convex (or concave) function F defined on the space of symmetric matrices, or for F very close to that condition. The corresponding results are due to Evans and Krylov [83], [133] for the convex case and to Caffarelli-Cabre [43] and Caffarelli-Yuan [50] for some (rather slight) generalizations.

THEOREM 1.4.1. Assume that $F(D^2 u, Du, u; x)$ is a uniformly elliptic operator such that $F_u \leq 0$, F is a function convex in the variables $D^2 u$, and F is Hölder in the variable x. Then the Dirichlet problem (1.1.12) has a unique classical solution.

An important step in the proof of Theorem 1.4.1 is a maximum principle for the gradient of solutions of nonlinear elliptic equations with a convex function F.

THEOREM 1.4.2. Assume that u is a solution of a strictly elliptic equation (1.1.1) and F is a convex function of its arguments. Then the second derivatives u_{ii} are subelliptic solutions of the linear elliptic operator L (1.3.1).

SKETCH OF THE PROOF OF THEOREM 1.4.1. We follow [49]. Let $x_1, x_2 \in \Omega$. Then, since $u(x) - u(x - (x_2 - x_1))$ is a solution of a linear uniformly elliptic operator of type (1.3.1), we get

$$(1.4.1) \quad \operatorname{tr}[D^2 u(x_2) - D^2 u(x_1)]^+ \approx \operatorname{tr}[D^2 u(x_2) - D^2 u(x_1)]^-,$$

where $\approx$ denotes an equality up to a multiplicative constant depending on the ellipticity constant.

Let V be a subspace of $\mathbb{R}^n$ and let

$$w(x, V) = \Delta_V u(x) - \Delta_V u(0),$$

where $\Delta_V u(x)$ is the Laplacian of u restricted on the affine variety $x + V$.

From the convexity of F it follows that w is a supersolution of a linear elliptic equation on V and hence satisfies the estimate of Lemma 1.3.3. Note also that positive and negative parts of the Laplacian can be expressed as

$$\max_V w(x, V) = \operatorname{tr}[D^2 u(x_2) - D^2 u(x_1)]^+,$$

$$\min_V w(x, V) = -\operatorname{tr}[D^2 u(x_2) - D^2 u(x_1)]^-.$$

By iterating and dyadically rescaling, the theorem follows from the following claim: *For a positive constant k the condition $w(x, V) > -1$ for all subspaces V and all $x \in \Omega$ implies the inequality $w(x, V) > -1 + k$ for all $x \in B_{1/2}$. Here k depends on the ellipticity constant of F.*

Indeed, assume that $w(x_0, V_0) \leq -1 + k$ for some V_0 and $x_0 \in B_{1/2}$ for sufficiently small k which will be chosen later. By Lemma 1.3.3

$$w(x, V) + 1 \leq k^{1/2}$$

on a set G that covers almost the whole of $B_{1/4}$, in the sense that

$$|B_{1/4} \setminus G| \leq C k^{\varepsilon/2}.$$

Note that on G,

$$1 - k^{1/2} \leq -w(x, V) \leq \operatorname{tr}[D^2 u(x_2) - D^2 u(x_1)]^- \leq 1.$$

On the other hand,

$$\begin{aligned} w(x, V) + w(x, V^\perp) &= \Delta u(x) - \Delta u(0) \\ &= \operatorname{tr}[D^2 u(x_2) - D^2 u(x_1)]^+ - \operatorname{tr}[D^2 u(x_2) - D^2 u(x_1)]^-. \end{aligned}$$

Thus

$$0 \leq \operatorname{tr}[D^2 u(x_2) - D^2 u(x_1)]^+ - w(x, V^\perp) \leq k^{1/2}$$

for $x \in G$. Moreover, for k small, by (1.4.1) we have

$$\begin{aligned} -w(x, V) &\approx \operatorname{tr}[D^2 u(x_2) - D^2 u(x_1)]^+ \\ &\approx \operatorname{tr}[D^2 u(x_2) - D^2 u(x_1)]^- \\ &\approx w(x, V^\perp). \end{aligned}$$

Thus, for some constant $c > 0$ depending on the ellipticity constant of F we have

$$w(x, V^\perp) \geq c$$

on G. Set

$$v = (c - w(x, V^\perp))^+.$$

We have $0 \leq v \leq 2$ on $B_{1/4}$, $v(0) = c$, $v = 0$ on G. For a small k (i.e., when G is almost the whole of $B_{1/4}$) this contradicts Lemma 1.3.4 since c is a fixed positive constant. This completes the proof. □

As we mentioned above one can slightly relax the convexity condition. Namely, in [**43**] the authors proved the following generalization of Theorem 1.4.1.

THEOREM 1.4.3. *Let $F(A, x) = F = \min\{F_1, F_2\}$ be uniformly elliptic where the functional F_1 is convex and F_2 is concave. Then the Dirichlet problem (1.1.12) always has a classical solution.*

COROLLARY 1.4.1. *The conclusion holds for $F = \max\{\min\{L_1, L_2\}, L_3\}$ for linear uniformly elliptic operators $L_i, i = 1, 2, 3$.*

Another generalization was obtained in [**50**].

THEOREM 1.4.4. *Let $F(A) = F$ be uniformly elliptic and suppose that F satisfies the following conditions:*

(i) *Let $\Sigma := \{A : F(A) = 0\}$ be the zero level set of F. Then for any $t \in \mathbb{R}$ the set*
$$\Sigma_t := \Sigma \cap \{A : \operatorname{tr}(A) = t\}$$
is strictly convex.

(ii) *On the nonconvex part of Σ the angle between the identity matrix I and the normal to Σ is strictly positive.*

Then the Dirichlet problem (1.1.12) always has a classical solution.

Notice that it is not so easy to produce an example of nonconvex (and nonconcave) F verifying the conditions of this last theorem; however, nontrivial examples do exist [50].

In two dimensions much stronger results hold, thanks to the corresponding strong estimates in two dimensions for linear equations. The following result by Nirenberg [186] can be obtained with the help of Corollary 1.3.1

THEOREM 1.4.5. *The Dirichlet problem (1.1.12) such that F is uniformly elliptic and $F_u \leq 0$ in a 2-dimensional domain always has a classical solution.*

It is unclear whether the Dirichlet problem in a bounded convex 2-dimensional domain always has a classical solution for a strictly elliptic equation (0.1).

Nirenberg's theorem has a kind of extension to axially symmetric Dirichlet problems for fully nonlinear Hessian equations in $\mathbb{R}^3$ [181].

THEOREM 1.4.6. *Let $n = 3$ and let $F = F(X) \in C^1$ be a uniformly elliptic operator depending only on the eigenvalues of the Hessian of u. Let $\varphi \in C^{1,\varepsilon}(\partial\Omega)$ be an axially symmetric function, $0 < \varepsilon < \varepsilon_0$, where $\varepsilon_0 > 0$ depends on the ellipticity constant of F. Then the Drichlet problem (1.1.12) has a unique classical solution $u \in C^2(\Omega) \cap C^{1,\varepsilon}(\bar{\Omega})$.*

Before proving Theorem 1.4.6 we need some auxiliary results. Let $\Omega \subset \mathbb{R}^n$ and let

$$(1.4.2) \qquad Lw = \sum a_{ij}(x) \frac{\partial^2 w}{\partial x_i \partial x_j}$$

be a linear uniformly elliptic operator defined in a domain $\Omega \subset \mathbb{R}^n$,

$$C^{-1}|\xi|^2 \leq \sum a_{ij}\xi_i\xi_j \leq C|\xi|^2.$$

First we recall the following results [94], [135].

LEMMA 1.4.1. *Let $G \subset \mathbb{R}^n$ be a bounded domain with a smooth boundary, and let $u \in C^2(\bar{G})$ be a solution of*

$$Lu = 0 \quad \text{in } G, \ u_{|\partial G} = \phi.$$

Then

$$\|Du\|_{C^\alpha(\partial G)} \leq C\|\varphi\|_{C^{1,\alpha}(\partial G)},$$

where positive constants α and C depend on G and the ellipticity constant of the operator L.

LEMMA 1.4.2. *Assume that $F \in C^1$, $F(0) = 0$, $\partial\Omega \in C^2$, and the uniform ellipticity condition holds. Let $u \in C^2(\bar\Omega)$ be a solution of the Drichlet problem (1.1.12). Then*
$$\|Du\|_{C^\alpha(\partial\Omega)} \leq C\|\varphi\|_{C^{1,\alpha}(\partial\Omega)},$$
where positive constants α and C depend on Ω and on the ellipticity constant of F.

The following resuts are essentially 2-dimensional [**94**]:

PROPOSITION 1.4.1. *Let $u \in C^2(D_1)$, where $D_r \subset \mathbb{R}^2$ is the disk $|x| < r$, and let u solve $Lu = 0$ in D_1. Then*
$$\operatorname*{osc}_{D_1} u_{x_1} \leq (1+\xi) \operatorname*{osc}_{D_{1/2}} u_{x_1},$$
where $\xi > 0$ is a constant depending only on the ellipticity constant of operator L.

PROPOSITION 1.4.2. *Let $u \in C^2(D_1)$ be a solution of a C^1 fully nonlinear elliptic equation*
$$H(D^2u, Du, x) = 0$$
in D_1, and let $H(0,0,x) = 0$. Let $|u| < M$. Then
$$\|u\|_{C^{2,\alpha}(D_{1/2})} < CM,$$
where $\alpha \in]0,1[, C > 0$ are constants depending on the ellipticity constant and the C^1-norm of H.

As a corollary of Proposition 1.4.2 we have

LEMMA 1.4.3. *Let $u \in C^2(D_1)$ be a solution of the equation $Lu = 0$ in D_1 and let l be an affine linear function in D_1 with $|l - u| < M$. Then for any $\varepsilon > 0$ there are $\alpha \in]0,1[, r > 0$ depending only on ε and the ellipticity constant of L such that*
$$\|u - l\|_{C^{1,\alpha}(D_r)} < \varepsilon M.$$

Applying Lemma 1.4.3 to the derivative of solutions of a fully nonlinear elliptic equation we get

LEMMA 1.4.4. *Let $u \in C^2(D_1)$ be a solution of the fully nonlinear elliptic equation*
$$F(D^2u) = 0,$$
in D_1, and let $F(0) = 0$. Let q be a quadratic polynomial in D_1 such that $|q - u| < M$. Then for any $\varepsilon > 0$ there are $\alpha, \rho > 0$ depending only on ε and the ellipticity constant of F such that there exists
$$\|u - q\|_{C^{2,\alpha}(D_\rho)} < \varepsilon M.$$

PROOF OF THEOREM 1.4.4. We may assume without loss of generality that $F(0) = 0$.

Let x_1, x_2, x_3 be an orthonormal coordinate system in $\mathbb{R}^3$ such that the axis of symmetry of the domain Ω is the x_1-axis. Denote
$$\omega = \{x \in \Omega : x_3 = 0\}.$$
Let u be a classical axially symmetric solution of the Dirichlet problem (1.1.12). Denote
$$\|u\|_{C(\Omega)} = A.$$

Since u_{x_3} is a solution of the linear uniformly elliptic equation $Lu_{x_3} = 0$ and $u_{x_3} = 0$ on ω, by Proposition 1.4.2

$$\left\|\frac{\partial^2 u}{\partial x_3^2}\right\|_{C^\alpha(\omega')} \leq C\|\varphi\|_{C^{1,\alpha}(\partial\Omega)} \tag{1.4.3}$$

where $\omega' \Subset \omega$ and the positive constants $\alpha \in]0,1[$ and C depend on Ω, ω' and the ellipticity constant of the operator F.

We define the 2-dimensional Hessian elliptic operators $f_a, a \in \mathbb{R}$, by

$$f_a(\lambda_1, \lambda_2) = f(\lambda_1, \lambda_2, a).$$

Let $y \in \Omega$ be a point on the x_1-axis. Denote

$$h = \mathrm{dist}(y, \partial\Omega).$$

Let the function u_r on the unit disk $D_1 \subset \mathbb{R}^2$ be given for $r \in]0,h[$ by

$$u_r(x) = u_r(x_1, x_2) = \frac{u(r(x_1, x_2 - y)) - u(y)}{r^2}.$$

Set $a = u_{x_2 x_2}(y)$. Let v_r be a solution of the Dirichlet problem

$$\begin{cases} F(D^2 u) = 0 & \text{in } \Omega, \\ u = \varphi & \text{on } \partial\Omega. \end{cases} \tag{1.4.4}$$

By Nirenberg's theorem we get a classical solution of this problem.

Since the equation $F(D^2 u) = 0$ is homogeneous, we can assume without loss of generality that

$$1 < |\nabla F| < C$$

for a positive constant C. It follows from (1.4.3) and the last inequalities that for a sufficiently large constant C_0 the functions

$$u_r - C_0 r^\alpha (1 - |x|^2) \quad \text{and} \quad u_r + C_0 r^\alpha (1 - |x|^2)$$

are sub- and supersolutions of (1.4.4), respectively. Hence

$$|u_r - v_r| \leq C_0 r^\alpha.$$

Denote $w_r = u_r - v_r$ and let ρ be the constant of Lemma 1.4.3 for the elliptic operator f and $\varepsilon = 1/2$. Define the sequences of functions $u_n, v_n, w_n, n = 1, 2, \ldots$, in D_1, by

$$u_n = u_{h\rho^n}, \quad v_n = v_{h\rho^n}, \quad w_n = u_n - v_n.$$

From Lemma 1.4.3 we get that

$$\|v_{n+1} - q_{n+1}\|_{C(D_1)} \leq \frac{1}{2}\|v_n - q_n\|_{C(D_1)} + C_0 \rho^{\alpha n}$$

for quadratic polynomials q_n with $f(q_n) = 0$. Since $|u_1| < A/h^2$, we get quadratic polynomials q_n such that $f(q_n) = 0$ and

$$\|v_n - q_n\|_{C(D_1)} < 2AC_0 \rho^{\alpha n}/h^2,$$
$$\|u_n - q_n\|_{C(D_1)} < 2AC_0 \rho^{\alpha n}/h^2,$$
$$\|w_n\|_{C(D_1)} < C_0 \rho^{\alpha n}$$

for all n, and since the functions u_n are dilations of u, one obtains

$$\|q_{n+1} - q_n\|_{C(D_1)} < 2AC_0 \rho^{\alpha n - 2}/h^2. \tag{1.4.5}$$

It follows that
$$\|u_n\| < AC_1/h^2,$$
for all n, where $C_1 > 0$ is a constant depending only on the ellipticity constant of F.

Let
$$E = \{z = x + y : |x| < h/2,\ x_2/x_1 > 1/4\},$$
$$G = \{x \in D_1 : x_2 > 1/4,\ \text{dist}(x, \partial D_1) > 1/4\},$$
and define for any natural number n
$$G_n = \{x : x/h\rho^n \in G\}.$$
Set $g_n = u_n - q_n$. Then by Proposition 1.4.2 we have
$$\|g_n\|_{C^{1,\alpha}(G)} < AC_2/h^2,$$
where $C_2 > 0$ depends only on the ellipticity constant of F. Since
$$\|g_n\|_{C(D_1)} < 2AC_0 \rho^{\alpha n}/h^2,$$
interpolating between the last two inequalities, we get
$$\|g_n\|_{C^{1,\alpha^2}(G)} < AC_3 \rho^{\alpha^2 n}/h^2,$$
where $C_3 > 0$ depends on the ellipticity constant of the equation. Thus
$$\|u\|_{C^{1,\alpha^2}(G)} < AC_3/h^2, \quad n = 1, 2, \ldots.$$
Together with (1.4.5) the last inequality gives

(1.4.6) $$\|u\|_{C^{1,\alpha^2}(E)} < AC_4/h^2$$

where $C_4 > 0$ depends only on the ellipticity constant of the equation.

By the equation on the x_1-axis the second derivatives $u_{x_2 x_2} = u_{x_3 x_3}$ satisfy Hölder estimates. Since on the axis the mixed derivatives $u_{x_i x_j} = 0$ for $i \neq j$, we conclude from the equation that the second derivative $u_{x_1 x_1}$ satisfies Hölder estimates as well. These estimates together with (1.4.6) give the inequality
$$\|u\|_{C^{1,\alpha^2}(\omega')} < AC_5,$$
where $C_5 > 0$ depends on the ellipticity constant of the equation and the distance of ω' to the boundary $\partial \omega$.

Combining the last inequality with Proposition 1.4.2 we get the following a priori estimate for axially symmetric solutions:

LEMMA 1.4.5. *Let $u \in C^2(\Omega)$ be an axially symmetric solution of* (0.1). *Let Ω' be a compact subdomain of Ω. Then*
$$\|u\|_{C^{1,\alpha}(\Omega)} \leq C \|\varphi\|_{C^{1,\alpha}(\partial\Omega)}, \qquad \|u\|_{C^{2,\alpha}(\Omega')} \leq C' \|\varphi\|_{C^{1,\alpha}(\partial\Omega)},$$
where the positive constants C, C' and α depend on Ω and on the ellipticity constant of F. The constant C' also depends on the distance of Ω' to the boundary $\partial\Omega$.

The a priori estimate of this lemma and the standard method of continuation by parameter yield the classical solvability of the Dirichlet problem (1.4.4) for a uniformly elliptic equation and thus finish the proof of Theorem 1.4.4. $\square$

REMARK 1.4.1. The same result holds for the solutions of axially symmetric problems in $\mathbb{R}^n$ (i.e., for the solutions of the form $u(x) = u(x_1, x_2^2 + \cdots + x_n^2)$).

One could hope that in $n = 3$ dimensions the viscosity solutions of uniformly elliptic equations are always classical, but that is not clear. Maybe, we can at least suppose the following to hold.

CONJECTURE 1.4.1. *Let $u \in W^{2,\infty}(\Omega)$ be a viscosity solution of a uniformly elliptic Hessian equation (0.1) in $\Omega \subset \mathbb{R}^3$. Then $u \in C^2(\Omega)$.*

1.5. Degenerate elliptic equations

In this section we discuss degenerate elliptic equations. First we recall the "Dirichlet duality" theory by Harvey and Lawson [**104**], which establishes (under an appropriate explicit geometric assumption on the domain Ω) the existence and uniqueness of continuous solutions of the Dirichlet problem for fully nonlinear, degenerate elliptic equations, and then discuss the regularity problem for such equations.

1.5.1. Harvey-Lawson's Dirichlet duality. Consider fully nonlinear, degenerate elliptic equations

(1.5.1) $$\mathbf{F}(D^2 u) = 0.$$

Following the method of Krylov [**134**] Harvey-Lawson's theory takes a geometric approach to the equation which eliminates the operator $\mathbf{F}$ and replaces it with a closed subset F of the space $\mathrm{Sym}_n(\mathbb{R})$ with the property that ∂F is contained in $\{\mathbf{F} = 0\}$. We only need the case when Ω is a ball when the geometric assumption is automatically true and thus we do not discuss it below.

The general set-up of the theory is the following. Let F be a given closed subset of the space $\mathrm{Sym}_n(\mathbb{R})$. The theory formulates and solves the Dirichlet problem for the equation

$$D^2 u(x) \in \partial F \text{ for all } x \in \Omega$$

using the functions of "type F", i.e., which satisfy

$$D^2 u(x) \in F \text{ for all } x.$$

A priori these conditions make sense only for C^2 functions u. The theory extends the notion to functions which are only upper semicontinuous.

A closed subset $F \subset \mathrm{Sym}_n(\mathbb{R})$ is called a *Dirichlet set* if it satisfies the condition

$$F + \mathcal{P} \subset F$$

where

$$\mathcal{P} = \{A \in \mathrm{Sym}_n(\mathbb{R}) : A \geq 0\}$$

is the subset of nonnegative matrices. This condition corresponds to degenerate ellipticity in modern fully nonlinear theory; it implies that the maximum of two functions of type F is again of type F, which is the key requirement for solving the Dirichlet problem. Note that translates, unions (when closed), and intersections of Dirichlet sets are Dirichlet sets.

The *Dirichlet dual* set $\widetilde{F}$ is defined as

$$\widetilde{F} := \{\mathrm{Sym}_n(\mathbb{R}) \setminus \mathrm{Int}(F)\}.$$

By Lemma 4.3 in [**104**] this is equivalent to the condition

$$\widetilde{F} := \{A \in \mathrm{Sym}_n(\mathbb{R}) : \forall B \in F, A + B \in \widetilde{\mathcal{P}}\},$$

$\widetilde{\mathcal{P}}$ being the set of all quadratic forms except those that are negative definite.

An upper semicontinuous (USC) function u is called an *subaffine function* if it satisfies locally the following condition: *For each affine function a, if $u \leq a$ on the boundary of a ball B, then $u \leq a$ on B.*

Note that a C^2 function is subaffine if and only if $D^2 u$ has at least one nonnegative eigenvalue at each point. A USC function u is *of type F* if $u + v$ is subaffine for all C^2 functions v of type $\widetilde{F}$. In other words, u is of type F if for any "test function" $v \in C^2$ of the dual type $\widetilde{F}$, the sum $u+v$ satisfies the maximum principle. A function u on a domain is said to be *F-Dirichlet* if u is of type F and $-u$ is of type $\widetilde{F}$. Such a function u is automatically continuous, and at any point x where u is C^2, it satisfies the condition

$$D^2 u(x) \in \partial F \text{ for all } x \in \Omega.$$

The main result of the theory (in our restricted setting) is Theorem 6.2 in [**104**].

THEOREM 1.5.1 (The Dirichlet Problem). *Let $B \subset \mathbb{R}^n$ be a ball, and let F be a Dirichlet set. Then for each $\varphi \in C(\partial B)$, there exists a unique $u \in C(B)$ which is an F-Dirichlet function on B and equals φ on ∂B.*

Besides, one has the following (see Remark 4.9 in [**104**]).

PROPOSITION 1.5.1 (Viscosity Solutions). *In the conditions of Theorem 1.5.1 u is a viscosity solution of* (1.5.1).

Krylov's idea ([**134**, Theorem 3.2]) permits us to reconstruct from F a *canonical form* of the operator $\mathbf{F}$ such that
 1) $\partial F = \{\mathbf{F} = 0\}$,
 2) $F = \{\mathbf{F} \geq 0\}$.
It is sufficient to define

$$\mathbf{F}(A) := \operatorname{dist}(A, \partial F)(2\chi_F - 1),$$

χ_F being the characteristic function of F.

The operator $\mathbf{F}$ (in its canonical form) is *strictly elliptic* if for any $A \in F$ there exists $\delta(A) > 0$ such that $\mathbf{F}(A + P) \geq \delta(A) \cdot \|P\|$ for all $P \in \mathcal{P}$, and it is *uniformly elliptic* if $\mathbf{F}(A + P) \geq \delta \cdot \|P\|$ for all $P \in \mathcal{P}$, $A \in F$, and an absolute constant $\delta > 0$ (note that for $\mathbf{F}$ in its canonical form $\mathbf{F}(A+P) - \mathbf{F}(A) \leq \|P\|$ by definition). Moreover, the function $\mathbf{F}$ is concave iff F is concave, and it is convex iff $\widetilde{F}$ is convex.

Harvey-Lawson theory becomes especially elegant in the case of Hessian equations, i.e., when $\mathbf{F}(A)$ depends only on the eigenvalues $\lambda_1(A) \leq \lambda_2(A) \leq \ldots \leq \lambda_n(A)$ of A. Then the sets $\{\mathbf{F} = 0\}, F, \widetilde{F}$ are stable under the action of the orthogonal group $O_n(\mathbb{R})$ by conjugation. Consider the map $\operatorname{Sym}_n(\mathbb{R}) \longrightarrow D_n \subset \mathbb{R}^n$ defined as

$$A \longmapsto (\lambda_1(A), \lambda_2(A), \ldots, \lambda_n(A))$$

where

$$D_n := \{(\lambda_1, \lambda_2, \ldots, \lambda_n) \in \mathbb{R}^n : \lambda_1 \leq \cdots \leq \lambda_n\}.$$

The images $\{\mathbf{F}_\lambda = 0\}, F_\lambda, \widetilde{F}_\lambda$ of $\{\mathbf{F} = 0\}, F, \widetilde{F}$ determine completely their preimages. The sets

$$\{\mathbf{F}_\Lambda = 0\} := \bigcup_{\sigma \in \Sigma_n} \{\mathbf{F}_{\sigma(\lambda)} = 0\} \subset \mathbb{R}^n,$$

$$F_\Lambda := \bigcup_{\sigma \in \Sigma_n} F_{\sigma(\lambda)} \subset \mathbb{R}^n,$$

$$\widetilde{F}_\Lambda := \bigcup_{\sigma \in \Sigma_n} \widetilde{F}_{\sigma(\lambda)} \subset \mathbb{R}^n,$$

where $\sigma(\lambda) := (\lambda_{\sigma(1)}, \lambda_{\sigma(2)}, \ldots, \lambda_{\sigma(n)})$ are Σ_n-invariant subsets in $\mathbb{R}^n$ which determine $\{\mathbf{F} = 0\}$, F, and $\widetilde{F}$ as well. Here and below Σ_n denotes the permutation group on n symbols. Moreover, it is well known (see, e.g., [**21**], [**48**]) that the set F is convex iff F_Λ is convex.

1.5.2. Regularity for degenerate elliptic equations. The Dirichlet problem for a degenerate elliptic equation (0.1) in a bounded strictly convex domain $\Omega \subset \mathbb{R}^n$ with continuous Dirichlet data on $\partial\Omega$ always has a continuous viscosity solution (see [**71**]); compare with Section 1.5.1. We suppose in the present subsection that the second fundamental form of the hypersurface $\partial\Omega$ is everywhere strictly positive definite for a bounded strictly convex domain $\Omega \subset \mathbb{R}^n$.

The question of regularity for degenerate elliptic as well for strictly elliptic fully nonlinear equations is a rather subtle problem. We start our discussion with a classical result of Caffarelli, Nirenberg, and Spruck [**48**].

THEOREM 1.5.2. *Let $\Omega \subset \mathbb{R}^n$ be a smooth bounded strictly convex domain. The Dirichlet problem*

$$\begin{cases} \sigma_k(D^2 u) = f > 0 & in \ \Omega, \\ u = \varphi & on \ \partial\Omega, \end{cases}$$

σ_k *being the k-th symmetric function of the matrix $D^2 u$ and f, ϕ being smooth functions, has a smooth solution u.*

REMARK 1.5.1. The operators $\sigma_k(D^2 u)$, $k = 2, \ldots, n$, are not elliptic on the whole domain of definition. For each $k = 2, \ldots, n$ there is an open cone Γ_k in the functional space of solutions where the ellipticity of σ_k holds. We always consider solutions which are in Γ_k; the operators $\sigma_k(D^2 u)$ are strictly elliptic on Γ_k. Notice also that on Γ_k the function $\sigma_k^{\frac{1}{k}}$ becomes convex and this is essential for the proof.

The assumptions on f, ϕ, Ω in Theorem 1.5.2 are in a sense sharp. Pogorelov constructed examples [**205**] showing that convex generalized (= viscosity) solutions of the Monge-Ampère equation $\det D^2 u = f > 0$ for $n \geq 3$ need not be C^2 even if f is a real analytic function. Similar results for σ_k operators are given in [**259**]. Wang [**269**] has shown that the solution of the Dirichlet problem for the Monge-Ampère equation with a smooth $f > 0$ and φ only in $C^{2,1}$ fails to lie in the Sobolev space $W^{2,p}$ for sufficiently large p. For a degenerate Monge-Ampère equation with $f \equiv 0$ the best regularity of the solution u which one can expect is $C^{1,1}$ even if the boundary data φ is a real analytic function.

A sharp result on the regularity of solutions of the Dirichlet problem for the degenerate Monge-Ampère equation was proved by Guan, Trudinger, and Wang [**97**]:

THEOREM 1.5.3. *Let $\Omega \subset \mathbb{R}^n$ be a bounded strictly convex domain, $\partial\Omega \in C^{3,1}$, and let f be a nonnegative function such that $f^{1/(n-1)} \in C^{1,1}$. Then there exists a unique convex solution $u \in C^{1,1}$ of the Dirichlet problem*

(1.5.2) $$\begin{cases} \det(D^2 u) = f & \text{in } \Omega, \\ u = \varphi & \text{on } \partial\Omega. \end{cases}$$

Initially results somewhat weaker then Theorem 1.5.3 were proved by Krylov [**136**] and Ivochkina [**118**].

For a nonstrictly positive function f one cannot expect better regularity of u than in Theorem 1.5.3 even if the functions f, φ are smooth. For a positive f conditions of the existence of classical solutions of the Dirichlet problem are given by Caffarelli and Urbas [**38**], [**39**], [**40**], [**41**], [**261**]:

THEOREM 1.5.4. *Let u be a solution of the Dirichlet problem for the Monge-Ampère equation. Assume that $f > 0$, $f \in C^a$, $\partial\Omega \in C^{1,a}$, $\varphi \in C^{1,a}$, $a > 1 - \frac{2}{n}$ for $n > 2$ and $\varphi \in C^0$ for $n = 2$. Then u is a smooth solution in Ω.*

The Pogorelov examples [**205**] show that in dimensions $n \geq 3$ one cannot expect the existence of classical solutions for the Dirichlet problem with just continuous boundary data for the Monge-Ampère equation. Let us give one more optimal result for the Monge-Ampère equation [**258**].

THEOREM 1.5.5. *Assume that Ω is a uniformly convex domain with C^3 boundary, $\varphi \in C^3(\overline{\Omega})$, $f \in C^\alpha(\overline{\Omega})$ for some $\alpha \in (0,1)$, and let $f > 0$ (and thus $0 < c_0 \leq f \leq c_1$ for constants c_0, c_1). Then there is a unique convex solution $u \in C^{2,\alpha}(\overline{\Omega})$ to (1.5.2).*

Notice that for the Monge-Ampère equation $\det(D^2 u) = f$, $0 \leq f \leq C$, there is an alternative way to viscosity solutions to introduce weak solutions. We say that the function u is a weak solution if the Jacobian of the gradient map $X \to \nabla u$ coincides with f as a measure. By a result of Caffarelli [**38**], [**39**] weak solutions of the Monge-Ampère equation are just viscosity solutions.

One can also consider solutions of the Monge-Ampère equation in Sobolev spaces $W^{2,p}$. Remarkably, a theorem of Urbas [**259**] shows that if p is sufficiently large, then u is a smooth function:

THEOREM 1.5.6. *Let $p > \frac{n(n-1)}{2}$ and let $u \in W^{2,p}$ be a viscosity solution of $\det D^2 u = f$, $f > 0$, $f \in C^{0,1}$, in an n-dimensional domain. Then $u \in C^{2,a}$ for some $a \in (0,1)$.*

Now we consider the Dirichlet problem (1.4.4) for a general degenerate elliptic equation. We know that for a continuous φ the Dirichlet problem (1.4.4) has a continuous viscosity solution u. In general one cannot expect a better regularity of u, as is easy to see for the highly degenerate equation $\frac{\partial^2 u}{\partial x_1^2} = 0$. However, assuming some additional regularity of φ, we can get a better regularity of u:

THEOREM 1.5.7. *Let $\Omega \subset \mathbb{R}^n$ be a bounded strictly convex domain, let for $0 \leq a \leq 2$, $\varphi \in C^a(\partial\Omega)$, and let $u \in C(\Omega)$ be a viscosity solution of the Dirichlet problem (1.4.4). Then $u \in C^{0,a/2}(\Omega)$ and*

$$\|u\|_{C^{a/2}} \leq C\|\varphi\|_{C^a},$$

where $C > 0$ depends on a and Ω.

PROOF. Consider first an elliptic regularization of the operator F. Namely, set

$$F_\varepsilon(D^2 u) := F(D^2 u + (\varepsilon \Delta u) I_n),$$

for any $\varepsilon > 0$. Let u_ε be a viscosity solution of the Dirichlet problem (1.4.4) for the elliptic operator F_ε. Then $u_\varepsilon \to u$ as $\varepsilon \to 0$.

Indeed, let $v \in C^2(\Omega)$ be a subsolution of the operator F. Then for any $\delta > 0$ there is an $\varepsilon > 0$ such that $v + \delta|x|^2$ is a subsolution of the operator F_ε. Applying Perron's method to the definition of the solution u we immediately get the claim.

Moreover, since the constant C in the theorem is independent of the ellipticity constant of the equation $F_\varepsilon(D^u) = 0$, we can and will assume without loss of generality that F is a uniformly elliptic operator.

Let then $z \in \partial\Omega$. Denote by $n = n_z$ the inner normal to $\partial\Omega$ at z. By $G_z(h)$ we denote a "cap" which the plane $(x-z, n) = h$ cuts from Ω.

Let $y \in G_z(h)$. We are going to prove the inequality

(1.5.3) $$|u(y) - u(z)| \leq Ch^{a/2}.$$

Assume first that $a \leq 1$. Denote $g(h) = \partial G_z(h) \cap \partial\Omega$,

$$s(h) = \sup_{g(h)} |\varphi|.$$

Since by our assumptions $|u(x) - u(z)| \leq C_1 |x-z|^a$ and the domain Ω is strictly convex, we get the inequality $s(h) \leq C_2 h^{a/2}$. Set

$$p(h) = \sup_{t > h} s(t)/t.$$

We have

$$p(h) \leq C_3 \|\varphi\|_{C^a} h^{a/2}.$$

Define

$$L(x) := p(h)((x, n) - (z, n)).$$

Then $L(x) \geq |\varphi(x)|$ on $\partial\Omega \setminus g(h)$.

Set $\varphi_1(x) = \sup\{\varphi(x), L(x)\}$, $\varphi_2(x) = \inf\{\varphi(x), -L(x)\}$ on $g(h)$, $\varphi_1, \phi_2 = 0$ on $\partial\Omega \setminus g(t)$. Let u_1, u_2 be the solutions of the Dirichlet problem (1.4.4) with $\varphi = \varphi_1, \varphi_2$, respectively. Since $F(L) = 0$, we have the inequalities

$$u \leq L + u_1, \quad u \geq -L + u_2.$$

Since by the maximum principle $|u_1|, |u_2| \leq s(h)$, one gets (1.5.3) for $a \leq 1$.

Assume now that $1 < a \leq 2$. Denote $j = \nabla\varphi(z)$ on $\partial\Omega$. Set $l(x) = (x, j) - (z, j)$ and $\varphi_0 = \varphi - l(x)$. Let u_0 be a solution of the Dirichlet problem (1.4.4) with $\varphi = \varphi_0$. Then the above arguments are applicable to the function u_0 and we have the inequality

$$|u_0(y) - u_0(z)| \leq Ch^{a/2},$$

where $y, z \in G_z(h)$. Since the function l in $G_z(h)$ satisfies the inequality
$$|l(y) - l(z)| \leq Ch^{a/2},$$
we get (1.5.3).

Let then $e > 0$. Define $w_e(x) = u(x) - u(x+e)$ in $\Omega_e := \{x \in \Omega : x+e \in \Omega\}$. Then by Theorem 1.3.6 w_e is a viscosity solution of a linear elliptic equation and hence by the maximum principle we have
$$\sup_{\Omega_e} |w_e| \leq \sup_{\partial \Omega_e} |w_e|.$$

Since by (1.5.3)
$$\sup_{\partial \Omega_e} |w_e| \leq Ch^{a/2},$$
the theorem is proved. $\square$

1.6. Homogeneous solutions

In the present section we discuss homogeneous solutions of elliptic equations, first in the classical geometric setting of homogeneous order 1 functions in $\mathbb{R}^3$ and then in a more general setting, but supposing the uniform ellipticity.

1.6.1. Bernstein's and Alexandrov's theorems. In this subsection we discuss two classical results: Bernstein's geometric theorem and Alexanrov's theorem on uniqueness of closed surfaces.

Let $u \in C^2(\Omega)$, $\Omega \subset \mathbb{R}^n$, be a solution of a linear elliptic equation of the form
$$Lu = \sum a_{ij}(x) \frac{\partial^2 u}{\partial x_i \partial x_j} = 0,$$
where $a_{ij} \in L_\infty$, satisfying the inequality
$$0 < \sum a_{ij}(x) \xi_i \xi_j$$
at any point of the domain Ω. Then the graph of the function u has the following property: It is impossible to cut a "cap" off the graph of u by any hyperplane in $\mathbb{R}^{n+1}$. Axiomatizing this property we say that a function $f \in C(\Omega)$ is a *saddle function* if it is impossible to cut a "cap" off the graph of f by any hyperplane in $\mathbb{R}^{n+1}$. In other words, let a be any point on the graph of u, and let the function u_1 be (locally) defined on the corresponding tangent hyperplane T_a as having the same graph. Then u_1 does not admit a local extremum at the point a.

THEOREM 1.6.1. *Let $u \in C(\mathbb{R}^2)$ be a saddle function and let*
$$\lim_{r \to \infty} \frac{1}{r} \sup_{B_r} |u(x)| = 0.$$
Then the graph of u is a cylinder.

Theorem 1.6.1 was proved by Bernstein for bounded functions [25]. It was generalized to its present form by Adel'son-Vel'skii [3]; cf. [140].

We say that a continuous homogeneous order 1 function u in $\mathbb{R}^3$ is saddle if its restriction on any plane not passing through the origin is a saddle function. The following theorem of Alexandrov [6] might be regarded as an equivariant version of Bernstein's theorem.

THEOREM 1.6.2. *Let u be a saddle, homogeneous order 1, real analytic in $\mathbb{R}^3 \setminus \{0\}$ function. Then u is an affine linear function.*

SKETCH OF THE PROOF. The proof of Alexandrov's theorem is based on the following involution on the space of smooth homogeneous order 1 functions. Let f be a homogeneous order 1 function defined in a cone $K \subset \mathbb{R}^n$. Consider the map

$$h : \mathbb{S}^2 \cap K \to \nabla u.$$

Denote by $H(f) = \nabla f(K)$ the gradient surface of the function f (the surface $h(\mathbb{S}^2)$ called the "hérisson" of f; the notion of hérissons was introduced and studied in [**141**], [**155**]).

PROPOSITION 1.6.1. *At regular points of $H(f)$ the map h is an inverse to the Gauss map of $H(f)$.*

Proposition 1.6.1 is well known and can be regarded as a form of Minkowski duality; see, e.g., [**211**]. If $H(f)$ is the boundary of a convex body $K \subset \mathbb{R}^n$, then the function f is the support function of K:

$$f(y) = \sup_{x \in K} (y, x).$$

For $y \in \mathbb{S}^{n-1}$ the Hessian of the function f restricted on $y^\perp$ coincides with the second fundamental form of $H(f)$ at $h(y)$.

Let then $A \subset \mathbb{S}^2$ be the set where rank $h < 2$; A is an analytic subset of $\mathbb{S}^2$ of dimension at most 1. Therefore A is a union of a finite collection of points and of a finite collection of topological circles having a finite number of self-intersections. Thus if the dimension of A is 1, there is a simply connected domain $G \subset S^2$ such that $\partial G \subset A$ and $G \cap A$ is finite. From Proposition 1.6.1 it follows that if $x \in \mathbb{S}^2 \setminus A$, then $H(f)$ is negatively curved in $h(x)$. Thus if at a point $y \in H(f)$ the surface $H(f)$ has a supportive plane, then $H(f)$ is singular at y. It is possible to prove that $h(\partial G)$ is a single point and that $H(f)$ is nonsingular at $h(x)$ for any $x \in G$. Since the surface $h(G)$ has supportive planes, the theorem is proved. $\square$

The assumption on real analyticity of the function u in Theorem 1.6.2 cannot be dropped.

THEOREM 1.6.3. *There exists a homogeneous order 1 saddle nonlinear function u that is smooth in $\mathbb{R}^3 \setminus \{0\}$.*

A C^2-example was constructed by Martinez-Maure [**154**]. G. Panina [**195**] extended Theorem 1.6.3 to smooth saddle functions. Moreover, one can choose the function u in Theorem 1.6.3 in such a way that u will be a solution of an elliptic equation (1.4.2) in $\mathbb{R}^3 \setminus \{0\}$. These examples settle in the negative a conjecture of Alexandrov on the validity of Theorem 1.6.2 in the class of smooth functions. However, for certain special classes of nonanalytic homogeneous functions Theorem 1.6.2 remains true. First notice that Theorem 1.6.1 implies

THEOREM 1.6.4. *Let $u \in C^1(\mathbb{R}^3 \setminus \{0\})$ be a saddle, homogeneous order 1 function. Assume that the function u vanishes on the plane $\{x_1 = 0\}$. Then u is a linear function.*

PROOF. The restriction of u on the plane $\{x_1 = 1\}$ is a saddle function. Since $u \in C^1(\mathbb{R}^3 \setminus \{0\})$ and $u = 0$ on $\{x_1 = 0\}$, it follows that u is bounded on $\{x_1 = 1\}$. By Theorem 1.6.1 the graph of u over the plane $\{x_1 = 1\}$ is a cylinder. Without loss of generality we may assume that $u(x) = u(x_1, x_2)$ for $x_1 > 0$. Then since u has a differential at the point $(0, 0, 1)$ it follows that u is a linear function on $\{x_1 > 0\}$. A similar argument on the half-space $\{x_1 < 0\}$ finishes the proof. $\square$

Martinez-Maure [**154**] proved

THEOREM 1.6.5. *Let $u \in C^2(\mathbb{R}^3 \setminus \{0\})$ be a saddle, homogeneous order 1 function. Assume that u is odd, $u(x) = -u(-x)$. Then u is a linear function.*

Assuming uniform ellipticity of the operator L we have [**102**]

THEOREM 1.6.6. *Let $u \in W^{2,2}(\mathbb{R}^3 \setminus \{0\})$ be a homogeneous order 1 function. Assume that u is a solution of a uniformly elliptic equation (1.4.2). Then u is a linear function.*

In the class of real analytic homogeneous order α, $0 < \alpha < 1$, in $\mathbb{R}^3 \setminus \{0\}$ functions Safonov proved the existence of solutions of uniformly elliptic equations [**234**].

THEOREM 1.6.7. *For any α, $0 < \alpha < 1$, there exists a uniformly elliptic operator L of the form (1.4.2) defined in $\mathbb{R}^3$ and a homogeneous order α function u that is real analytic in $\mathbb{R}^3 \setminus \{0\}$ such that u is a viscosity solution of the equation (1.4.2) in $\mathbb{R}^3$. The ellipticity constant C of the operator L depends on α.*

As a corollary of Theorem 1.6.2 Alexandrov proved a uniqueness theorem for closed surfaces.

THEOREM 1.6.8. *Let $f(R_1, R_2, n)$ be a function defined for $n \in \mathbb{S}^2$ and $R_1, R_2 \geq 0$. Assume that for each $n \in \mathbb{S}^2$ the function f is monotone as a function of R_1 and R_2. Assume that there are two real analytic closed surfaces M_1, M_2 of positive Gauss curvature such that for any $n \in \mathbb{S}^2$ we have*

$$f(R_1', R_2', n) = f(R_1'', R_2'', n),$$

R_1', R_2' (resp. R_1'', R_2'') being the principal curvatures of the surface M_1 (resp. M_2) at a point with the normal n. Then M_1 and M_2 coincide up to a translation.

On can get the proof by applying Theorem 1.6.2 to the difference of the support functions of the surfaces M_1 and M_2.

In Theorem 1.6.8 one can assume that f is a C^1 function satisfying the inequalities $\frac{\partial f}{\partial R_1} > 0$, $\frac{\partial f}{\partial R_2} > 0$ on the whole of $\mathbb{S}^2$ instead of the real analyticity of M_1 and M_2. Then the uniqueness result for closed C^2 surfaces will follow from Theorem 1.6.6. For $f \in C^3$ such a result was proved by Pogorelov [**203**].

Theorem 1.6.8 implies the following characterization of the round sphere [**6**], [**169**].

THEOREM 1.6.9. *Let M be a 2-dimensional closed surface of genus zero analytically immersed in $\mathbb{R}^3$ so that its principal curvatures R_1, R_2 satisfy the Weingarten inequality: $(R_1 - c)(R_2 - c) \leq 0$. Then M is a sphere.*

Theorem 1.6.4 gives a smooth counterexample to Theorem 1.6.9; see [**154**]. For certain special classes of nonanalytic surfaces Theorem 1.6.9 remains true. For instance, as a corollary of Theorem 1.6.6 one gets immediately that if M is a convex C^2 surface with principal curvatures of M satisfying the Weingarten inequality and if the orthogonal projection of M on the $\{x_1, x_2\}$-plane is a round disk, then M is a sphere (for the case of surfaces of strictly positive Gauss curvature see [**131**]). Theorem 1.6.6 implies that Alexandrov's conjecture holds for convex C^2 surfaces of a constant width.

One can expect that Alexandrov's conjecture holds in the following lopped form:

CONJECTURE 1.6.1. *If M is a convex C^2 surface such that principal curvatures of M satisfy the Weingarten inequality, then there is a domain $G \subset M$ which is congruent to a spherical cap.*

As consequences of Theorem 1.6.1 and Theorem 1.6.8 one can also deduce the two following classical results [**25**], [**8**]:

COROLLARY 1.6.1. *A minimal graph in $\mathbb{R}^3$ is a plane.*

PROOF. Let u be a solution of the minimal surface equation:
$$\left(1 + \left(\frac{\partial u}{\partial x_2}\right)^2\right)\frac{\partial^2 u}{\partial x_1^2} - 2\left(\frac{\partial u}{\partial x_1}\frac{\partial u}{\partial x_2}\right)\frac{\partial^2 u}{\partial x_1 \partial x_2} + \left(1 + \left(\frac{\partial u}{\partial x_1}\right)^2\right)\frac{\partial^2 u}{\partial x_2^2} = 0.$$
Substituting $v = \arctan(\partial u/\partial x_1)$ in the equation, we obtain that v is a saddle function. □

COROLLARY 1.6.2. *A convex constant mean curvature surface in $\mathbb{R}^3$ is a round sphere.*

Note that by a result of Chern a constant mean curvature hypersurface in Euclidean space which is a graph of an entire function is necessarily a minimal surface [**64**].

A counterpart to the last results in the hyperbolic space H^3 is given by a result of do Carmo and Lawson [**53**]:

THEOREM 1.6.10. *Let S be a surface of constant curvature in H^3. Suppose that S admits a one-to-one projection onto a geodesic plane H^2 (i.e., S is a graph over H^2). Then S is a plane.*

1.6.2. Homogeneous solutions in four dimensions. Principal results of the present volume can be regarded as a step to a definite classification of homogeneous solutions of fully nonlinear uniformly elliptic equations. Here we summarize our achievements and formulate some conjectures permitting us to complete such a classification. We place ourselves in the simplest setting,
$$F(D^2 u) = 0.$$
The results of Chapters 4 and 5 give us that even if we suppose F to be Hessian, that is, depending only on the eigenvalues of $D^2 u$, then for any $\varepsilon \in]0,1]$ there exist homogeneous order $1 + \varepsilon$ solutions in all dimensions starting from 5, thanks to the following result (see Section 5.3 and [**183**]):

THEOREM 1.6.11. *The function $w_{5,\delta} = P_5(x)/|x|^{1+\delta}$ is a viscosity solution in the unit ball of a smooth uniformly elliptic equation $F(D^2 w) = 0$ with smooth Hessian F for any $\delta \in]0,1]$.*

Here P_5 is some minimal cubic polynomial in five variables (Cartan's cubic).

This theorem establishes the optimality of $C^{1,\varepsilon}$-regularity for viscosity solutions in $\mathbb{R}^n, n \geq 5$.

On the other hand, we believe the following conjecture to be true.

CONJECTURE 1.6.2. *Let $\varepsilon \in]0,1]$ and let $u \neq 0$ be a homogeneous analytic in $\mathbb{R}^n \setminus \{0\}$ order $1+\varepsilon$ solution to a fully nonlinear uniformly elliptic equation $F(D^2 u) = 0$ in $\mathbb{R}^n, n \leq 4$. Then $\varepsilon = 1$ and u is a quadratic polynomial.*

Therefore, nonclassical homogeneous solutions do not exist up to four dimensions and do exist, even in the most restrictive setting, starting from five dimensions.

Note that for $n \leq 2$ this is a consequence of the classical regularity in that setting. For $n = 3$ and 4 the conjecture is not known; however there are some partial results in its direction. Namely, it holds for $\varepsilon = 1$, i.e., for homogeneous order 2 solutions.

First of all, Theorem 1.6.2 implies the absence of homogeneous order 2 real analytic in $\mathbb{R}^3 \setminus \{0\}$ solutions to fully nonlinear equations different from quadratic forms (in the $C^{2,\alpha}$ setting it is proved in [**102**]). Unfortunately, this argument does not work in four dimensions, the analogue of Alexandrov's theorem in four dimensions being false ($u = (x_1^2 + x_2^2 - x_3^2 - x_4^2)/|x|$ gives a counterexample; cf. [**144**]).

However, the result still holds in four dimensions at least for analytic functions:

THEOREM 1.6.12. *Let u be a homogeneous order 2 real analytic function in $\mathbb{R}^4 \setminus \{0\}$. If u is a solution of the uniformly elliptic equation $F(D^2 u) = 0$ in $\mathbb{R}^4 \setminus \{0\}$, then u is a quadratic polynomial.*

PROOF. We begin with some auxiliary results.

LEMMA 1.6.1. *Let v be a smooth homogeneous order 1 function in $\mathbb{R}^3 \setminus \{0\}$. Assume that $y \in \mathbb{S}^2$ and that the quadratic form $D^2 v(y)$ changes sign. Let $e \in \mathbb{S}^2, e \neq \pm y$, and let $G \subset \mathbb{R}^3$ be an open domain, $y \in G$. Then*

$$\sup_G v_e(x) > v_e(y).$$

PROOF. Let $L \subset \mathbb{R}^3$ be an affine 2-dimensional plane transversal to the vector y such that $y \in L$ and e is parallel to L. Denote by v' the restriction of the function v on L. Since v is a homogeneous order 1 function the quadratic form, $D^2 v'(y)$ changes sign. Thus there is a neighborhood D of the point y where v' satisfies a uniformly elliptic equation on L of the form

$$\sum a_{ij}(x)\frac{\partial^2 v'}{\partial x_i \partial x_j} = 0.$$

Thus by Theorem 1.3.3, giving the maximum principle for the gradient of a solution of elliptic equations in dimension 2, v'_e cannot attain the supremum at the point y. The lemma is proved. □

LEMMA 1.6.2. *Let v be a real analytic homogeneous order 1 function in $\mathbb{R}^n \setminus \{0\}$. Assume that v is a solution of a linear uniformly elliptic equation*

$$Pv = \sum a_{ij}(x/|x|)\frac{\partial^2 v}{\partial x_i \partial x_j} = 0,$$

where the coefficients a_{ij} are smooth functions on $\mathbb{S}^{n-1}$. Let $e_1, \ldots, e_n \in \mathbb{S}^{n-1}$ be linearly independent unit vectors. Assume that the functions v_{e_i}, $i = 1, \ldots, n$, attain a local supremum at $a \in \mathbb{S}^{n-1}, a \neq \pm e_i, i = 1, \ldots, n$. Then v is a linear function.

PROOF. Denote by L an affine hyperplane in $\mathbb{R}^n$ orthogonal to a, $a \in L$. Then the restriction v' of the function v on L satisfies a linear uniformly elliptic equation of the type

$$(1.6.1) \qquad P(v') = \sum a'_{ij}(y) \frac{\partial^2 v'}{\partial x_i \partial x_j} = 0,$$

where $y \in L$ and the a'_{ij} are smooth functions on L. Indeed, $D^2 v(a) = 0$ since v is order 1 homogeneous; thus the partial derivatives of v' coincide with ones of v in an appropriate coordinate system which gives the uniform ellipticity of (1.6.1). We consider then a coordinate system on L such that the point a becomes the origin, assuming without loss of generalty that $v'(0) = 0, \nabla v'(0) = 0$. After a linear transformation of $\mathbb{R}^n$ we can assume that $P(0)$ is the Laplacian, i.e., $a'_{ij}(0) = \delta_i^j$. Suppose that v' is not a linear function and let p, $\deg p = k \geq 2$, be the first nonzero homogeneous polynomial of the Taylor expansion of v' at 0; clearly p is harmonic. Let $B \subset L$ be a small ball centered at 0, let g be the gradient map

$$g : L \to \mathbb{R}^{n-1}, \ g := \nabla v',$$

and let $\Gamma = g(B)$. Then $\Gamma \subset K := \bigcap_{i=1}^n \{e_i \leq 0\}$, K being a strictly convex cone in $\mathbb{R}^n$ since the e_i are linearly independent. Denote $K_0 = \{K + a\} \cap L$; if K_0 is nonempty, then K_0 is a strictly convex cone in L. Since p is harmonic, $\nabla p(L)$ intersects the complement of K_0 and thus $\nabla p(l^+) \cap K_0 = \emptyset$ for a line $l \subset L$ and a ray $l^+ \subset l$. The curve $g(l^+) \subset \mathbb{R}^n$ is tangent to $\nabla p(l^+)$ at the point $\{a\}$ of order at least k since $g(x) - \nabla p(x) = O(|a - x|^k)$. Therefore $g(l^+ \cap B)$ intersects the complement of K, and the lemma follows. $\square$

LEMMA 1.6.3. *Let v be a real analytic homogeneous order* 1 *function in $\mathbb{R}^4 \setminus \{0\}$. Assume that v is a solution of a linear uniformly elliptic equation*

$$(1.6.2) \qquad Pv = \sum a_{ij}(x/|x|) \frac{\partial^2 v}{\partial x_i \partial x_j} = 0$$

and that the rank of the gradient map $\nabla v : \mathbb{S}^3 \to \mathbb{R}^4$ is ≤ 2. Then v is a linear function.

PROOF. Notice that if the rank of the gradient map ∇v is equal to 1, then by the *Hartman-Nirenberg cylinder theorem* [**101**] v is a cylindrical function and the proof immediately follows. If it equals 2, the function v is not necessarily cylindrical (for some exceptions see [**4**]) and we need to do a more careful analysis of v involving the global structure of v in the whole space.

Let $y \in \mathbb{S}^3$, let $m = y^\perp \subset \mathbb{R}^4$ be its perpendicular hyperplane, and let $M \subset \mathbb{R}^4$ be the affine hyperplane parallel to m with $y \in M$. Denote by f the restriction of v on M. Then f is a real analytic function on M such that for any $x \in M$ the Hessian $D^2 f(x)$ is degenerate and either the quadratic form $D^2 f(x)$ changes sign or $D^2 f(x) = 0$. Let

$$H := \{x \in \mathbb{R}^3 : \operatorname{rank}(D^2 f(x)) = 2\}.$$

We assume without loss of generalty that $\dim(\mathbb{R}^3 \setminus H) \leq 2$. Indeed, if not, then f is constant or a cylindrical function by the Hartman-Nirenberg cylinder theorem [**101**], which easily implies the conclusion. For $x \in H$ let $z(x)$ be the zero eigenspace of $D^2 f(x)$. By assumption of the lemma $z(x)$ is a line analytically depending on the point $x \in H$. By Chern-Lashof's lemma, [**65**, Lemma 2], [**234**, Lemma VI 5.1], in the neighborhood of any point $x \in M$ the plane M is foliated by a 2-dimensional family of straight lines L such that for any line $l \in L$ the restriction of f on l is an affine function. Moreover l is parallel to the line $z(x)$ at any point $x \in l$; see the proof of Lemma 2 in [**65**]. By the analyticity of f the family L foliates the whole space M without intersection. Let $l \in L$ and let p be a 2-dimensional subspace spanned by l in $\mathbb{R}^4$. Then by homogeneity v is linear on a half-plane of p and by analyticity it is linear on the whole plane p. Denote the whole set of these planes p by P. Then any two planes of P intersect only at $\{0\}$ and foliate $\mathbb{R}^4 \setminus m$.

Let $y' \in \mathbb{S}^3$, $m' = (y')^\perp \subset \mathbb{R}^4$ and let P' be the foliation of $\mathbb{R}^4 \setminus m'$ by 2-dimensional planes corresponding to y'. We will prove that P and P' coincide on $\mathbb{R}^4 \setminus (m \cup m')$. Assume not. Then there is a 4-dimensional subset $X \subset \mathbb{R}^4$ such that for any $x \in X$ one has $x \in p \cap p'$ for some $p \in P$, $p' \in P'$, $p \neq p'$. Since the planes p and p' are zero eigenspaces of $D^2 v$, it follows that the zero eigenvalue has multiplicity at least 3 at x, and hence $D^2 v(x) = 0$. Thus $D^2 v$ vanishes on X and hence by analyticity v is linear. Thus choosing different $y \in \mathbb{S}^3$ we get a foliation P of $\mathbb{R}^4 \setminus \{0\}$ by 2-dimensional planes which are zero eigenspaces of $D^2 v$.

Notice that any 3-dimensional subspace of $\mathbb{R}^4$ contains at most one plane of P, since any two different planes in 3-dimensional space have nontrivial intersection.

Let $m \subset \mathbb{R}^4$ be a 3-dimensional subspace such that $m \supset p$, $p \in P$. Denote by v' the restriction of the function of v to m; subtracting a linear function we can assume that $v' = 0$ on p. Let $x \in m \setminus p$, $x \in p'$ for some $p' \in P$. Then p' is trasversal to m. Since p' is a zero eigenspace of $D^2 v$, we get that either $D^2 v'(x)$ changes sign or $D^2 v'(x) = 0$ on m. Thus the function v' is a solution of an elliptic equation at x and satisfies (1.6.2) on $m \setminus p$. Let $e \in m$ be a vector parallel to p. Let $z \in \mathbb{S}^2 \subset m$ be a point at which v'_e attains its maximum on $\mathbb{S}^2$. If $v'_e(z) > 0$, then $z \in \mathbb{S}^2 \setminus p$ since by our assumption $v' = 0$ on p. Since in a neighborhood of z the function v' is a solution of (1.6.2), this contradicts Lemma 1.6.2. Thus $v'_e(z) \leq 0$ and thus $v'_e \leq 0$ everywhere since $v'_e(z)$ is maximal. Applying the same argument to the function $-v'$ we get $v'_e \geq 0$ everywhere and thus $v'_e \equiv 0$ for any vector e parallel to p. Hence v' is a function which depends only on the coordinate orthogonal to p and therefore v' is a linear function. Thus we get that for any 3-dimensional subspace m of $\mathbb{R}^4$ the restriction of v on m is a linear function. Hence v is a linear function on $\mathbb{R}^4$ and the lemma is proved. $\square$

REMARK 1.6.1. The proof of the lemma can seem somewhat involved, but the main reason for this is that the Hartman-Nirenberg cylinder theorem does not hold for rank = 2, which complicates the argument in that case.

LEMMA 1.6.4. *Let* $Q(x, y, z) \in \mathbb{R}[x, y, z]$ *be a cubic form such that for any* $e \in \mathbb{S}^2$ *the quadratic form* Q_e *is degenerate. Then* Q *is a function of two variables in some coordinate system.*

PROOF. First of all, the conditions as well as the conclusion of the lemma are invariant under nonsingular linear transformations. Considering $Q(x, y, z) = 0$ as an equation of a plane projective cubic curve E_Q and applying the usual argument giving its Weierstrass form (see, e.g., pp. 45–46 in the proof of Proposition 1.2 of Chapter 2 in [**161**]) one gets the following:

(1) E_Q is elliptic or irreducible possessing a singular point with $y \neq 0$; in this case Q is equivalent under a linear transfomation to the Weierstrass form
$$Q^{\text{Weier}} = y^2 z + x^3 + px^2 z + qz^3.$$

(2) E_Q is irreducible possessing a singular point with $y = 0$; then
$$Q^{\text{sing}} = x^3 + axyz + bxz^2 + cyz^2 + dz^3$$
after a suitable nonsingular linear transformation.

(3) E_Q is reducible; then either
$$Q^{\text{reduc}} = z(x^2 + ay^2 + bz^2 + cxz + dyz)$$
modulo such a transformation or Q verifies the conclusion.

In the first case, if $Q = Q^{\text{Weier}}$ and $e = (k, l, m)$, then
$$r = r(k, l, m) := \det(D^2 Q_e) = 3k^2 mp + 9km^2 q - 3kl^2 - m^3 p^2$$
should be identically zero; in particular,
$$-6 = r_{kll} = \frac{\partial^3 r}{\partial k \partial l^2} = 0,$$
which is clearly not the case.

If $Q = Q^{\text{sing}}$, then
$$r/2 = a^3 klm + a^2 bkm^2 + a^2 clm^2 - 3a^2 dm^3 + 4abcm^3 - 3a^2 k^3 - 12ack^2 m - 12c^2 km^2;$$
hence
$$r_{klm} = 2a^3 = 0,$$
$$r_{kmm} = 4(a^2 b - 12c^2) = 0,$$
implying $c = a = 0$ and the conclusion.

Similarly, if $Q = Q^{\text{reduc}}$, then
$$r/8 = 3abm^3 - ac^2 m^3 - a^2 l^2 m - ackm^2 - adlm^2 - d^2 m^3 - ak^2 m;$$
hence
$$r_{llm} = -16a^2 = 0,$$
$$r_{mmm} = 48(3ab - ac^2 - d^2) = 0.$$
Thus $a = d = 0$ as necessary and the proof is finished. □

LEMMA 1.6.5. *Let $Q(x, y, z) \in \mathbb{R}[x, y, z]$ be a cubic form such that for any $a \neq b \in C \subset \mathbb{S}^2$ the partial derivative Q_{ab} vanishes as a linear form, C being a curve on $\mathbb{S}^2$. Then Q is a function of two variables in some coordinate system.*

PROOF. The proof is very similar to that of the previous lemma, but sightly more combersome. We consider the same three main cases, each of them being divided into subcases depending on the curve $C \subset \mathbb{S}^2$.

1) The Weierstrass case. There are two subcases:

1a) The curve C is not in $\mathbb{S}^2 \cap \{yz = 0\}$.

1b) The curve $C \subset \mathbb{S}^2 \cap \{yz = 0\}$.

In the subcase 1a) we can suppose without loss of generalty that $a = (a_1, b_1, c_1)$, $b = (a_2, b_2, c_2)$ with $c_1b_2 + c_2b_1 \neq 0$. A brute force calculation gives $Q_{aby}/2 = c_1b_2 + c_2b_1 \neq 0$ and thus we get a contradiction.

In the subcase 1b) we suppose without loss of generalty that $a = (a_1, b_1, 0)$, $b = (a_2, b_2, 0)$ with $a_1a_2 \neq 0$, but then $Q_{abx}/6 = a_1a_2 \neq 0$.

2) The singular case (singularity at $y = 0$), $Q = x^3 + pxyz + qxz^2 + ryz^2 + sz^3$. There are two subcases:

2a) The curve C is not in $\mathbb{S}^2 \cap \{z = 0\}$.

2b) The curve $C \subset \mathbb{S}^2 \cap \{z = 0\}$.

Suppose subcase 2a), $a = (a_1, b_1, c_1)$, $b = (a_2, b_2, c_2)$, $c_1c_2 \neq 0$. Then the condition $Q_{abx} = 0$ implies $2c_1c_2r = -(a_2c_1 + c_2a_1)p$. If there exists $c = (a_3, b_3, c_3) \in C$ such that $c_3a_2 \neq a_3c_2$, then $0 = Q_{acy} = -c_1p(c_3a_2 - a_3c_2)/c_2$ gives $p = 0, r = 0$, which proves the lemma. If $a_3c_2 = a_2c_3$, we can suppose that $b_3c_2 \neq c_3b_2$, and the condition $0 = Q_{acx} = c_1p(c_2b_3 - b_2c_3)/c_2$ gives $r = p = 0$.

In the subcase 2b) we get $a = (a_1, b_1, 0), b = (a_2, b_2, 0), a_1a_2 \neq 0$, and hence $Q_{abx} = 3a_1a_2 \neq 0$.

3) The reducible case, $Q = z(x^2 + py^2 + qz^2 + rxz + syz)$. There are two subcases:

3a) The curve C is not in $\mathbb{S}^2 \cap \{z = 0\}$.

3b) The curve $C \subset \mathbb{S}^2 \cap \{z = 0\}$.

Suppose subcase 3a) and $a = (a_1, b_1, c_1)$, $b = (a_2, b_2, c_2)$, $c_1c_2 \neq 0$. Then the condition $Q_{aby} = 0$ implies that $c_1c_2s = -(b_2c_1 + c_2b_1)p$. For any $c = (a_3, b_3, c_3)$ one gets $0 = Q_{acy} = p(b_3c_2 - c_3b_2)c_1/c_2$ with $b_3c_2 \neq c_3b_2$ since $b_3c_2 = c_3b_2$ gives $Q_{acx} = (a_3c_2 - c_3a_2)c_1/c_2 \neq 0$. Hence $s = p = 0$.

Suppose subcase 3b), $a = (a_1, b_1, 0), b = (a_2, b_2, 0), c = (a_3, b_3, 0), a_1a_2b_1b_2 \neq 0$, $a_2b_3 \neq a_3b_2$. Then

$$0 = Q_{abz} = pb_1b_2 + a_1a_2, \quad p = -\frac{a_1a_2}{b_1b_2}, \quad Q_{acz} = a_1(a_3b_2 - b_3a_2)/b_2 \neq 0,$$

a contradiction and the proof is finished. $\square$

The proof of Theorem 1.6.12 is based on the following construction. Let $x \in \mathbb{S}^3$. Set

$$A(x) = \left\{ (a, b) \in \mathbb{S}^3 \times \mathbb{S}^3, a \neq b : u_{ab}(x) = \sup_{y \in \mathbb{S}^3} u_{ab}(y) \right\};$$

note that $A(x)$ is a semianalytic subset of $\mathbb{S}^3 \times \mathbb{S}^3$ and that $(a, b) \in A(x)$ implies $(b, a) \in A(x)$. The semianalyticity of $A(x)$ implies the subanalyticity of all the sets below in the proof. In particular they satisfy *Whitney's stratification theorem* [**273**] as was showed by Hironaka [**107**]; i.e., each such set M is stratified in a finite union of open k-dimensional smooth submanifolds, $k = 0, 1, \ldots, m = \dim M$.

Let then $\mathfrak{C}(x)$ for $x \in \mathbb{R}^4 \setminus \{0\}$ be the cubic form of the Taylor expansion of the function u at the point x; i.e., $D^3\mathfrak{C}(x) = D^3u(x)$. Let us notice first that for any vector $e \in \mathbb{R}^4$ the function u_e is homogeneous order 1 and hence x is a zero eigenvector of the quadratic form $\mathfrak{C}(x)_e$. We need the following two simple properties of this form.

LEMMA 1.6.6. *Let* $(a,b) \in A(x)$. *Then* b *is a zero eigenvector of the quadratic form* $\mathfrak{C}(x)_a$.

PROOF. From our assumptions it follows that for any vector $e \in \mathbb{R}^4$ one has $u_{abe}(x) = 0$. Hence $\mathfrak{C}(x)_{abe} = 0$. This implies that b is a zero eigenvector of $\mathfrak{C}(x)_a$. □

LEMMA 1.6.7. *Let* $a, x, b_1, b_2, b_3 \in \mathbb{S}^3$ *with linearly independent* b_1, b_2, b_3. *Suppose that for all* $i = 1, 2, 3$ *one has* $(a, b_i) \in A(x)$. *Then* $\mathfrak{C}(x)_a = 0$.

PROOF. By Lemma 1.6.6 the vectors b_i are zero eigenvectors of the quadratic form $\mathfrak{C}(x)_a$; i.e., the quadratic form $\mathfrak{C}(x)_a$ has the zero eigenvalue with multiplicity at least 3. Since $\mathfrak{C}(x)_a$ should change sign or be equal to zero, the lemma follows. □

We can now finish the proof of the theorem. Let
$$X := \{x \in \mathbb{S}^3 : \dim A(x) \geq 3\}.$$
Then $X \neq \emptyset$ since
$$\bigcup_{x \in \mathbb{S}^3} A(x) = \mathbb{S}^3 \times \mathbb{S}^3 \setminus \Delta, \quad \dim(\mathbb{S}^3 \times \mathbb{S}^3 \setminus \Delta) = 6,$$
$\Delta \subset \mathbb{S}^3 \times \mathbb{S}^3$ being the diagonal. We denote by $d \in [0, 3]$ the dimension of X.

Let $\Gamma = \bigcup_{x \in X} A(x)$; then $\dim((\mathbb{S}^3 \times \mathbb{S}^3 \setminus \Delta) \setminus \Gamma) \leq 5$, $\dim(\Gamma) = 6$. We have four possibilities for d, namely, $d = 0, 1, 2$, or 3.

1. Let $d = 0$. Then $\dim A(y) = 6$ for some $y \in X$, and
$$\dim((\mathbb{S}^3 \times \{e\}) \cap A(y)) \geq 3$$
for $e \in \mathbb{S}^3$. In this case one can find linearly independent vectors $e_1, \ldots, e_4$, $e_i \neq y$, such that $(e, e_i) \in A(y)$. Applying Lemma 1.6.7 to the function u_e we get the required conclusion.

2. Let $d = 1$. Then we can suppose without loss of generality that $\dim A(y) = 5$ for any $y \in X$ and
$$\dim((\mathbb{S}^3 \times \{e\}) \cap A(y)) \geq 2, \quad \dim((\{e\} \times \mathbb{S}^3) \cap A(y)) \geq 2;$$
thus
$$E_1 \times E_2 \subset A(y)$$
$E_1, E_2 \subset \mathbb{S}^3, \dim(E_1) = \dim(E_2) = 2$. Denote the set of all $y \in \mathbb{S}^3$ satisfying $E_1 \times E_2 \subset A(y)$ by Y. Let $y \in Y$, $a \in E_1$. Then by Lemma 1.6.2 $\mathfrak{C}(y)_a = 0$. Since E_1 is a 2-dimensional set, the cubic form $\mathfrak{C}(y)$ depends at most on one coordinate. Since its derivatives change sign, it follows that $\mathfrak{C}(y) = 0$. Thus if Y_1 is a connected component of Y, then $D^2 u$ is constant on Y_1. Since Y is a real analytic set, it contains only a finite number of connected components, $Y_1, \ldots, Y_n$. At each Y_i function u has a fixed Hessian and thus one returns to the previous case of $d = 0$.

3. Let $d \geq 2$, let $y \in X$, and let A be a connected component of maximal dimension in $A(y)$. Denote by $d_1 = d_1(A)$, $d_2 = d_2(A)$ the dimensions of the projections of A to the first and the second factor in the product $\mathbb{S}^3 \times \mathbb{S}^3$, respectively. By symmetry one can suppose $d_1 \geq d_2$. We study first the case when $d_1 = 3, d_2 \geq 1$.

Set $Z(x) := pr_1(A(x)) \subset \mathbb{S}^3$, $\dim Z(x) = 3$. Then
$$\forall a \in Z(x), \ a \times h(a) \in A(x),$$
where $h(a) \in \mathbb{S}^3$.

Let $y \in X$ and let $L = y^\perp \subset \mathbb{R}^4$. Since u is homogeneous order 2, the function $\mathfrak{C}(y)$ depends only on the coordinates of L. Thus there exists a 2-dimensional set $E \subset \mathbb{S}^2 \subset L$ such that $\mathfrak{C}(y)_e$ is degenerate for any $e \in E$ and hence for any $e \in \mathbb{S}^2$. Thus by Lemma 1.6.4 the cubic form $\mathfrak{C}(y)$ depends only on two variables and for any $e \in \mathbb{S}^3$ the rank of the gradient map $\nabla \mathfrak{C}(y)_e : \mathbb{R}^4 \to \mathbb{R}^4$ is at most 2 at any point $y \in Z(x)$. Therefore since u_e is a homogeneous order 1 function, the rank of the gradient map $\nabla_x u_e : \mathbb{S}^3 \to \mathbb{R}^4$ is at most 2 at any point $y \in Z(x)$. For an affine hyperplane $L \subset \mathbb{R}^4, 0 \notin L$, let Z' be the spherical projection of $Z(x)$ on L, and let $s = u_e|_L$. Since u_e is a homogeneous order 1 function, the gradient map of $u_e(x)$ depends only on the spherical coordinate of x and hence $\det D^2 s = 0$ on Z'. Since s is real analytic and Z' is 3-dimensional, we get $\det D^2 s = 0$ on the whole plane L and by Lemma 1.6.3 u_e is linear.

4. Now we have to consider all other possible combinations of d, d_1, d_2, i.e., $d \geq 2 \geq d_1 \geq d_2$, or $d = d_1 = 3, d_2 = 0$.

Let first $d = 2$. We suppose without loss of generality that $\dim A(y) = 4$ for any $y \in X$.

Since $d_1 + d_2 \geq \dim A = 4$, we have the unique possibility of $d_1 = 2, d_2 = 2$ with $d_1 + d_2 = \dim A$, the manifold A being itself a product, and we return to the proof of case 2.

Let $d = 3$. We suppose without loss of generality that $\dim A(y) = 3$ for any $y \in X$. For a connected component A of $A(y)$ with $\dim A = 3$ we get $d_1 + d_2 \geq 3$. We have the following possibilities:

4a) $d_1 = 2, d_2 = 1$.

4b) $d_1 = d_2 = 2$.

4c) $d_1 = 3, d_2 = 0$.

In the cases 4a) and 4c) one has $d = d_1 + d_2$, A is a product, and the proofs above remain valid.

Assume finally case 4b). Let $y \in X$ and let $L = y^\perp \subset \mathbb{R}^4$, where as earlier $\mathfrak{C}(y)$ depends only on the coordinates of L. Then by Lemma 1.6.5 the cubic form $\mathfrak{C}(y)$ depends only on two coordinates, which we denote by z_1, z_2; let l be parallel to L and orthogonal to the linear span of z_1, z_2. Thus l is a zero eigenspace of $\mathfrak{C}(y)_e$ for any $e \in \mathbb{S}^3$. By our assumption one finds $(a, b) \in A(y)$, $b \notin l$. Therefore the multiplicity of the zero eigenvalue of $\mathfrak{C}(y)_a$ is at least 3. Again, since its derivatives change sign, it follows that $\mathfrak{C}(y) = 0$ and one finishes the proof as before. $\square$

Let us then say something about homogeneous solutions of degenerate elliptic equations. First, Alexandrov's theorem implies that a homogeneous order 2 real analytic solution in $\mathbb{R}^3 \setminus \{0\}$ of a degenerate elliptic equation $F(D^2 u) = 0$ is a quadratic polynomial. Moreover, we guess

CONJECTURE 1.6.3. *The same holds for smooth solutions of a degenerate elliptic equation $F(D^2 u) = 0$ in $\mathbb{R}^3$.*

It is not true in $\mathbb{R}^4$: Let f be a homogeneous order 1 concave function defined in the first quadrant of $\mathbb{R}^2$. Then a straightforward and simple calculation gives

PROPOSITION 1.6.2. *The function $u = f(x_1^2 + x_2^2, x_3^2 + x_4^2)$ is a solution of a degenerate elliptic equation $F(D^2 u) = 0$ in $\mathbb{R}^4 \setminus \{0\}$.*

For homogeneous solutions of fully nonlinear equations in $\mathbb{R}^n \setminus \{0\}$ one has [**184**]

THEOREM 1.6.13. *Let u be a homogeneous order $d \neq 2$ in $\mathbb{R}^n \setminus \{0\}$ solution to an elliptic equation $F(D^2 u) = 0$ with $F \in C^1$. Then u is harmonic in a possible new coordinate system in $\mathbb{R}^n$, namely*

$$\sum_{i,j=1}^{n} F_{ij}(0) D_{ij} u(x) = 0.$$

Consequently, $u \equiv 0$ if $-(n-2) < d < 0$ or d is not an integer; otherwise, u is a homogeneous harmonic polynomial of degree d.

REMARK 1.6.2. For the validity of the result it is essential that the solution u is defined in the whole punctured space $\mathbb{R}^n \setminus \{0\}$ rather than in the punctured ball $B_1 \setminus \{0\}$.

REMARK 1.6.3. Theorem 1.6.13 is not true for Lipschitz functionals F. In [**13**] the authors have proven that for a positively homogeneous functional F of degree 1 there exists a homogeneous solution of $F(D^2 u) = 0$ in $\mathbb{R}^n \setminus \{0\}$, where u is positive near $\{0\}$. Existence of similar homogeneous solutions changing sign near $\{0\}$ follows from [**14**].

1.7. Liouville type theorems. Removable singularities

In this section we consider two particular but rather important subjects, namely Liouville type theorems and removable singularities for fully nonlinear elliptic equations.

1.7.1. Liouville theorems for fully nonlinear elliptic equations.
We consider here solutions of fully nonlinear elliptic equations defined in the whole space $\mathbb{R}^n$. Properties of solutions defined in the whole space are in a sense related to local properties of solutions of the same equations. The classical *Liouville theorem* says that if u is a bounded harmonic function defined in $\mathbb{R}^n$, then u is a constant, or, in a stronger form, any positive harmonic function in $\mathbb{R}^n$ is a constant. This fact admits a direct generalization to uniformly elliptic fully nonlinear equations $F(D^2 u) = 0$.

THEOREM 1.7.1. *Let u be a positive viscosity solution of a uniformly elliptic equation $F(D^2 u) = 0$ defined in $\mathbb{R}^n$. Assume that $F(0) = 0$. Then $u \equiv const$.*

PROOF. Since $F(0) = 0$, constants are solutions of the equation $F(D^2 u) = 0$. Therefore by Theorem 1.3.5 the function u is a viscosity solution of a linear uniformly elliptic equation $Lu = 0$. Considering u in balls B_R and taking $R \to \infty$ we get from the Harnack inequality, Theorem 1.3.5, that u is a constant. □

From the results of Chapter 5 it follows that the assumption $F(0) = 0$ cannot be dropped in $n \geq 5$ dimensions. On the other hand for lower dimensions that remains unclear.

CONJECTURE 1.7.1. *Let u be a positive viscosity solution of a uniformly elliptic equation $F(D^2 u) = 0$ defined in $\mathbb{R}^n$, $n \leq 4$. Then $u \equiv const$.*

Liouville's theorem is known for certain classes of strictly elliptic Hessian equations. For the Monge-Ampère equation the Liouville property is given by the celebrated *Jörgen-Calabi-Pogorelov theorem*, [**125**], [**51**], and [**204**].

THEOREM 1.7.2. *Any convex solution u of the Monge-Ampère equation*
$$\det(D^2 u) = 1$$
defined on $\mathbb{R}^n$ is a quadratic polynomial.

Similar results are known for the special Lagrangian equation and for the σ_2 equation [**281**], [**60**]:

THEOREM 1.7.3. *Any convex solution u of a special Lagrangian equation defined on $\mathbb{R}^n$ is a quadratic polynomial.*

THEOREM 1.7.4. *Any convex solution u of the equation $\sigma_2 = 1$ defined on $\mathbb{R}^n$ is a quadratic polynomial.*

Theorem 1.7.2 has important consequences in the affine geometry. Applying Theorem 1.7.2 to earlier results of Calabi, Cheng and Yau proved [**63**] that every properly immersed parabolic affine sphere is a paraboloid. We recall that a C^3-hypersurface is an affine sphere if the lines formed by affine normals all meet at a point, called the center. A convex affine sphere is called elliptic, parabolic, or hyperbolic according to whether the affine normals point to the center, are parallel, or point away from the center, respectively.

The existence of the hyperbolic affine sphere is equivalent to the existence of the solution to the Monge-Ampère equation
$$\det(D^2 u) = (-1/n)^{n+2},$$
where u is defined in a bounded convex domain and has zero Dirichlet data. For a function u of two variables, this equation was solved by Loewner and Nirenberg [**151**].

A kind of Liouville theorem for special Lagrangian equations was proved in [**126**]. Another Liouville type theorem for special Lagrangian equations can be found in [**270**].

1.7.2. A generalization of Jörgens-Calabi-Pogorelov's theorem. In fact, Jörgens-Calabi-Pogorelov's theorem admits a generalization to equations linear in σ_k, namely for the equations

(1.7.1) $$L[f] \equiv \sum_{i=1}^{n} a_i(x) \sigma_i(\lambda) = 0$$

in the notation of Examples 1.1.11 and 1.1.12 above. This generalization uses the following condition:

(Q) Either $a_k(x) \equiv 0$ on $\mathbb{R}^n$ or there exist two positive constants μ_1, μ_2 such that
$$\mu_1 \leq |a_k(x)| \leq \mu_2.$$

Let us denote by $J = J(L)$ the set of indices i, $1 \leq i \leq n$, such that $a_i(x) \not\equiv 0$. The following generalization of Jörgens-Calabi-Pogorelov's theorem was obtained in [**248**].

THEOREM 1.7.5. *Let $f(x)$ be an entire convex C^2 solution of (1.7.1) and let the structural condition (Q) be satisfied. If*

$$\limsup_{|x|\to\infty} \frac{|f(x)|}{|x|^2} = 0, \tag{1.7.2}$$

then $\sigma_i(\lambda) \equiv 0$ for any $i \in J$; in particular, $\det D^2 f(x) = 0$. If additionally $a_1(x) \not\equiv 0$, then $f(x)$ is an affine function.

Notice that a paricular case of this result concerning special Lagrangian equations was given in [**30**].

PROOF. We begin with the following preliminary result:

LEMMA 1.7.1. *Let $A(x) \geq 0$ be a continuous $n \times n$ symmetric matrix solution of*

$$L(A(x)) \equiv \sum_{i=1}^{n} a_i(x)\sigma_i(A(x)) = 0, \qquad x \in \mathbb{R}^n, \tag{1.7.3}$$

where L is subject to condition (Q) and $\sigma_i(A(x))$ denotes the i-th symmetric function of the spectrum of $A(x)$. Then either $\sigma_i(A(x)) \equiv 0$ for any $i \in J$ or there exist $k \in J$ and a constant c_0 depending on μ_1 and μ_2 such that for all $x \in \mathbb{R}^n$ the following inequality holds:

$$\sigma_k(A(x)) \geq c_0 > 0.$$

PROOF OF LEMMA 1.7.1. Note that $\sigma_k(A(x)) \geq 0$ since $A(x) \geq 0$. Then, if all (nonidentically zero) coefficients a_i have the same sign, then $\sigma_k(A(x)) \equiv 0$ for any $i \in J$. Now suppose that there exists $x_0 \in \mathbb{R}^n$ and $k \in J$ such that $\sigma_k(A(x_0)) > 0$. In that case, there exist two coefficients a_i having different signs. Observe that by condition (Q) this is true in the whole of $\mathbb{R}^n$. Rewrite (1.7.3) as

$$|a_{i_1}(x_0)|\sigma_{i_1}(A(x_0)) + \cdots + |a_{i_m}(x_0)|\sigma_{i_m}(A(x_0))$$
$$= |a_{j_1}(x_0)|\sigma_{j_1}(A(x_0)) + \cdots + |a_{j_p}(x_0)|\sigma_{j_p}(A(x_0)),$$

where $i_1 < \cdots < i_m$, $j_1 < \cdots < j_p$, and also $i_1 < j_1$. We claim that $k = i_1$ satisfies the conclusion of the lemma. Indeed, we have

$$\sigma_{i_1}(A(x_0)) \leq b_1 \sigma_{j_1}(A(x_0)) + \cdots + b_p \sigma_{j_p}(A(x_0)), \tag{1.7.4}$$

where $b_k = |a_{j_k}(x_0)|/|a_{i_1}(x_0)| \leq \mu_2/\mu_1$. Now, using Proposition 3.2.2 in [**153**], we have

$$\left(\frac{\sigma_k(A(x_0))}{\binom{n}{k}}\right)^m \leq \left(\frac{\sigma_m(A(x_0))}{\binom{n}{m}}\right)^k,$$

for any $1 \leq m \leq k \leq n$; therefore by (1.7.4)

$$\sigma_{i_1}(A(x_0)) \leq \frac{\mu_2}{\mu_1} \sum_{k=1}^{p} \alpha_k \cdot (\sigma_{i_1}(A(x_0)))^{\nu_k},$$

where $\nu_k = j_k/i_1 > 1$ and $\alpha_k = \binom{n}{j_k} \cdot \binom{n}{i_1}^{-\nu_k}$. Observe that the left-hand side of the equation

$$\frac{\mu_2}{\mu_1} \sum_{k=1}^{p} \alpha_k \cdot c^{\nu_k - 1} = 1$$

is an increasing function of $c \geq 0$, and let $c = c_0$ denote its unique positive root. Then $\sigma_{i_1}(A(x_0)) \geq 0$ implies $\sigma_{i_1}(A(x_0)) \geq c_0$. By the continuity assumption on $A(x)$, the latter inequality also holds in the whole of $\mathbb{R}^n$, which proves the lemma. $\square$

COROLLARY 1.7.1. *Let $f(x) \in C^2(\mathbb{R}^n)$ be a convex solution of (1.7.1) under condition (Q). Then either $\det D^2 f \equiv 0$ in $\mathbb{R}^n$ or there exists $k \in J$ such that the inequality*

(1.7.5) $$\sigma_k(D^2 f(x)) \geq c_0 > 0$$

holds for all $x \in \mathbb{R}^n$ with k, c_0 chosen as in Lemma 1.7.1.

Let us then finish the proof of the theorem. First, we claim $\sigma_i(D^2 f(x)) \equiv 0$ for any $i \in J$. Indeed, if not, we have (1.7.5) in the whole of $\mathbb{R}^n$ for some $k \in J$. One can assume without loss of generality, replacing if needed $f(x)$ by $f(x) + c + \langle a, x \rangle$, that $f(x) \geq 0$ in $\mathbb{R}^n$. Given an arbitrary $\varepsilon > 0$, condition (1.7.2) yields the existence of a constant $p \in \mathbb{R}$ such that $f(x) \leq \frac{\varepsilon}{2}|x|^2 + p$ for any $x \in \mathbb{R}^n$. But $g(x) = \frac{\varepsilon}{2}|x|^2 - f(x) \to \infty$ uniformly as $x \to \infty$; hence it attains its minimum value at some point, say $x_0 \in \mathbb{R}^n$, and

$$D^2 g(x_0) = D^2\left(\frac{\varepsilon}{2}\|x\|^2 - f(x)\right)|_{x_0} \geq 0$$

holds, which yields $D^2 f(x_0) \leq \varepsilon I$ with I being the unit matrix.

Since $D^2 f(x_0) \geq 0$, we obtain applying the monotonicity theorem for matrices (see, for instance, Corollary 4.3.3 in [**108**]) that

$$\sigma_k(D^2 f(x_0)) \leq \sigma_k(\varepsilon I) = \varepsilon^k \binom{n}{k}.$$

But $\sigma_k(D^2 f(x)) \geq c_0$ is impossible since ε is arbitrary. This proves our claim. In particular, by convexity of f we also have $D^2 f(x) \geq 0$; hence $D^2 f(x)$ has zero eigenvalues for any $x \in \mathbb{R}^n$, implying $\det D^2 f(x) \equiv 0$ in $\mathbb{R}^n$ (see also Corollary 1.7.1). If, additionally, $a_1(x) \not\equiv 0$, then $1 \in J$ and the claim implies $\sigma_1(D^2 f(x)) = \Delta f(x) \equiv 0$. The convexity of $f(x)$ gives that $D^2 f(x) \equiv 0$ in $\mathbb{R}^n$; hence $f(x)$ is an affine function and the proof of the theorem is finished. $\square$

The following example shows that (1.7.2) is optimal in the sense that there exist operators L satisfying condition (Q) and possessing solutions $f(x) \sim \|x\|^2$ as $x \to \infty$ growing quadratically and such that $\operatorname{Hess} f(x) \not\equiv 0$.

EXAMPLE 1.7.1. Let $\alpha(t)$ be a positive function, nonidentically constant and such that for some $0 < q < 1$ one has $q \leq \alpha(t) \leq q^{-1}$. Let us consider the function

$$f(x_1, \ldots, x_n) = \sum_{i=1}^n \int_0^{x_i} (x_i - t)\alpha(t)dt.$$

Then $D^2 f = (\alpha(x_i)\delta_{ij})_{1 \leq i,j \leq n}$; hence $f(x)$ is convex and satisfies

$$\det D^2 f - \omega(x)\Delta f(x) = 0$$

with $\omega(x) = \alpha(x_1) \cdots \alpha(x_n)/\sum_{i=1}^n \alpha(x_i)$. We have $q^{n+1}/n \leq a_1(x) \leq q^{-n-1}/n$, which establishes that L satisfies condition (Q). On the other hand, $q\|x\|^2/2 \leq f(x) \leq \|x\|^2/2q$; thus $f(x)$ has quadratic growth at infinity.

1.7.3. Removable singularities.

Let $u \in C^2(B_1 \setminus \{0\})$ be a viscosity solution of a fully nonlinear equation (0.1). We are concerned then with the behavior of $u(x)$ when x approaches the origin. If u is a bounded harmonic function, the classical result says that u is a harmonic function in the whole ball B_1; i.e., the singularity at 0 is *removable*. For solutions of general fully nonlinear elliptic equations the removability of the singularity does not hold. Indeed, it is easy to check that the function $u = |x|^a$ is a solution of a uniformly elliptic equation in $B_1 \setminus \{0\}$, $B_1 \subset \mathbb{R}^n$ for any $n \geq 2$, $0 < a < 1$.

However, Caffarelli, Li, and Nirenberg [44], [45], see also [187], recently obtained a general result on removable singularities for solutions of strictly elliptic fully nonlinear equations.

THEOREM 1.7.6. *Let $u \in C^2(B_1 \setminus \{0\})$, $B_1 \subset \mathbb{R}^n$, $n \geq 2$, be a viscosity solution of a strictly elliptic equation (0.1). Assume that for any small $r > 0$,*

$$\inf_{0 < |x| \leq r} \quad \text{and} \quad \sup_{0 < |x| \leq r} \ (u + \text{any linear function}) \text{ both occur on } \{|x| = r\}.$$

Then u is a viscosity solution of $F(D^2 u) = 0$ in B_1.

COROLLARY 1.7.2. *Let $u \in C^2(B_1 \setminus \{0\})$, $B_1 \subset \mathbb{R}^n$, $n \geq 2$, be a viscosity solution of a strictly elliptic equation $F(D^2 u) = 0$. Assume that $u \in C^{1+\varepsilon}(B_1)$, $\varepsilon > 0$. Then u is a viscosity solution of $F(D^2 u) = 0$ in B_1.*

The paper [45] also contains some generalizations of Theorem 1.7.6.

Singularities of solutions of uniformly elliptic Hessian equations defined in a punctured ball admit a more detailed description [179]:

THEOREM 1.7.7. *Let u be a viscosity solution of a uniformly elliptic Hessian equation in a punctured ball $B_1 \setminus \{0\}$, $B_1 \subset \mathbb{R}^n$. Assume that $u \in C^0(B_1)$. Then $u = v(x) + l(x) + o(|x|^{1+\varepsilon})$, where the constant $\varepsilon > 0$ depends only on the ellipticity constant of the equation, v is a monotone function of the radius, $v(x) = v(|x|)$, $v \in C^\varepsilon(B_1)$, and l is a linear function.*

PROOF. Let P be a linear elliptic operator of the form

$$P = \sum_{i,j} a_{ij}(x) \frac{\partial^2}{\partial x_i \partial x_j},$$

defined in the half-ball $B_+ = \{x \in B \subset \mathbb{R}^n, x_1 > 0\}$, $a_{ij} \in L_\infty(B_+)$, and satisfying the inequalities

$$C^{-1}|\xi|^2 \leq \sum a_{ij}(x)\xi_i \xi_j \leq C|\xi|^2, \ \forall \xi \in \mathbb{R}^n, \forall x \in B_+.$$

Let $z \in C^2(B_+)$ and let $Pz = 0$ in B_+, $z = 0$ on L, where $L = \{x \in B, x_1 = 0\}$. Assume that $z < 1$ in B_+. Then it is well known [94] that

$$|\nabla z(0)| \leq K,$$

where the constant K depends on the ellipticity constant C.

LEMMA 1.7.2. *The following inequality holds with positive constants K, ε depending on the ellipticity constant C:*

$$|z(x) - dz(0)| \leq K|x|^{1+\varepsilon},$$

where dz is the differential of the function z.

1.7. LIOUVILLE TYPE THEOREMS. REMOVABLE SINGULARITIES

This follows directly from P. Bauman's boundary Harnack inequality [23].

We may assume without loss of generality that $F(0) = 0$; otherwise we consider the function $u + c|x|^2$ with a suitable constant c instead of u.

Set
$$v(r) = \sup_{|x|=r} u(x),$$
$$u_i = u(x_1, \ldots, -x_i, \ldots, x_n),$$
$$z_i = u - u_i.$$

Since u is a solution of a Hessian equation, the functions u_i are solutions of the same equation as well. Hence the functions z_i given as the difference of two solutions of the fully nonlinear elliptic equation are solutions to a linear elliptic equation $Pz_i = 0$ in B. Define a linear function l as
$$l = \tfrac{1}{2} \sum dz_i(0).$$

Set
$$u_0 = u - l.$$

Let $|y| = |y'| = r < 1$. Choose in $\mathbb{R}^n$ an orthonormal coordinate system $y_1, \ldots, y_n$ such that $y_1 = (y - y')/|y - y'|$. Set
$$\widetilde{u}(y_1, \ldots, y_n) = u_0(-y_1, \ldots, y_n),$$
$$v = u_0 - \widetilde{u}.$$

Since $F(u') = 0$, we get $Pv = 0$ in B. Moreover
$$\nabla v(0) = 0.$$

Hence by Lemma 1.7.1 $v(x) = o(|x|^{1+\varepsilon})$ and therefore
$$u_0(y) - u_0(y') = o(|y|^{1+\varepsilon}).$$

Set
$$h(r) = \inf_{|x|=r} u_0(x), \quad h_0(r) = \sup_{|x|=r} u_0(x).$$

Then
(1.7.6) $$h(|x|) - h_0(|x|) = o(|x|^{1+\varepsilon}).$$

Since $F(0) = 0$, we may assume without loss of generality that $u(0) = 0$, $h'_0(1) > 0$. Then by the maximum principle $h'_0(r)$ is a monotone function of r. If $h(r) = o(|x|^{1+\varepsilon/2})$, we may set $h \equiv 0$ and the theorem is proved. Assume that $h(r) > \varepsilon |x|^{1+\varepsilon/2}$. Then from (1.7.6) it follows that $|h(r)|$ is a positive function for sufficiently small r.

By a direct computation one gets
$$\lambda(D^2 h(|x|)) = (h'', h'/|x|, \ldots, h'/|x|).$$

Hence h has no local minimums, and since $h > 0$, we get $h' > 0, h'' < 0$ for sufficiently small r. Therefore h is a monotone, concave function for small r.

For any $0 < r < 1$ there exists a point $x_0, |x_0| = r$, such that $u_0(x_0) = h(r)$ and since $h - u_0 \leq 0$ the quadratic part of the function $u_0 - h$ is nonnegatively defined. Hence from the uniform ellipticity condition for F we get the inequality
$$-|x|h''/h' > \delta$$

on an interval $(0, a)$ for some $a > 0$, where δ depends on the ellipticity constant. From the last inequality it follows that
$$h(r) > r^{1-\delta}$$
on $(0, a)$. Since we can redefine h on $(a, 1)$ as a monotone, concave function, the theorem is proved. □

COROLLARY 1.7.3. *Let u be a homogeneous order α, $0 < \alpha < 1$, solution of a uniformly elliptic Hessian equation in a punctured ball $B_1 \setminus \{0\}$, $B_1 \subset \mathbb{R}^n$. Then $u = c|x|^\alpha$.*

Removable singularities for special classes of fully nonlinear equations were studied in [138] and [139]. The results of these two papers were vastly generalized in [13] where a complete characterization of isolated singularities for Pucci operators and other fully nonlinear operators was given.

CHAPTER 2

Division Algebras, Exceptional Lie Groups, and Calibrations

The present chapter is completely classical and does not contain recent results; we give here a brief introduction to real division algebras, some exceptional Lie groups, and calibrations generated by them. We restrict our attention mainly to their properties used in Chapter 4 to construct singular solutions of certain fully nonlinear elliptic equations and used in Chapter 7 to treat calibrated geometries. It means essentially that we need only exceptional groups arising in the study of trialities and calibrations, i.e., Spin(7), G_2, and the like. The chapter is divided into two parts. The first one, containing Sections 2.1 to 2.3, is concentrated around the Cayley-Dickson construction and trialities; these results are used in Chapter 4. The second part gives a treatment of the same objects from a (multi-)linear algebra point of view which naturally leads to calibrations used in Chapter 7.

More precisely, in Section 2.1 we give the principal properties of the real division algebras, especially of octonions using the Cayley-Dickson construction. In Section 2.2 we define Clifford algebras and spinor groups, and in Section 2.3 we describe the exceptional group G_2, as well as trialities. In Sections 2.4 and 2.5 the arguments are purely (multi-)linear. More precisely, Section 2.4 contains constructions of associative and coassociative calibrations based on double cross products. In Section 2.5 we show that the notions of a cross product and of a normed algebra are essentially equivalent. We introduce triple and quadruple cross products and associated Cayley calibrations. This section also contains a multi-linear algebra description of the groups G_2 and Spin(7).

Our exposition is elementary and does not depend on the representation theory of Lie groups or their classification. A complete exposition of the constructions in Sections 2.1, 2.2, and 2.3 (and many more) can be found in [**2**] and [**19**]; for Sections 2.4 and 2.5 see the appendices in [**103**], the preprint [**217**] (in these sections we follow this paper rather closely), [**160**], and the short note [**80**].

Note that the principle result of the first three sections forming the first half of the chapter is a construction of certain trilinear polynomials (trialities) and the calculation of their automorphism groups. However, these (poly-)linear objects are constructed in terms of algebraic multiplicative structures, such as the Cayley-Dickson construction and Clifford algebras. The main result of the second half of the chapter is the constructions of three calibrations, associative, coassociative, and Cayley, which are also polylinear. In this second part we choose to construct them by means of (poly-)linear algebra, which is in some respects a more direct and "elementary" approach than that in the first part of the chapter. However, one notes that this more direct approach (cf. [**217**]) needs in fact much more lengthy calculations than the algebraic approach adopted in, say, the appendices in [**103**].

We do not give these calculations since this would require a very considerable space (see [**217**]). In some sense, the logic of the second part of the chapter is inverse to that of the first part: there the division algebras are the result of linear algebra costructions. However, the two approaches are equivalent and each of them has its own merit.

2.1. The Cayley-Dickson construction. Octonions

All vector spaces that we consider here are finite-dimensional over the field $\mathbb{R}$ of real numbers. In particular any algebra A below will be a vector space with a bilinear multiplication map $(a,b) \mapsto ab$; A will be unitary, i.e., possessing an element $1 \in A$ called the unit such that $1a = a1 = a$. These algebras are not supposed to be commutative or associative. We identify $\mathbb{R}$ with its image

$$1 \cdot \mathbb{R} = \{a \in A : a = 1r, r \in \mathbb{R}\}$$

in A. We are mainly interested in *division algebras* for which the operations of left and right multiplication by any nonzero element are invertible.

An algebra A is a *normed division algebra* if it is also a normed vector space with a multiplicative norm $|\cdot| : A \to A$:

$$|ab| = |a| \cdot |b|.$$

Thus A is a division algebra with $|1| = 1$.

An algebra is called *alternative* if the subalgebra generated by any two elements is associative. In fact, an algebra A is alternative if for all $a, b \in A$ any two of the following conditions are satisfied:

(i) $(aa)b = a(ab)$;   (ii) $(ab)a = a(ba)$;   (iii) $(ba)a = b(aa)$.

It is also equivalent to the fact that the *associator* defined by

(2.1.1) $$[a,b,c] := (ab)c - a(bc)$$

is an alternating bilinear map.

The main result on real division algebras is

THEOREM 2.1.1. (i) *All division algebras have dimension* $1, 2, 4,$ *or* 8.

(ii) *The fields* $\mathbb{R}, \mathbb{C}$ *(dimensions* 1 *and* 2*), the algebra of* quaternions $\mathbb{H}$ *(dimension* 4*), and the algebra of* octonions $\mathbb{O}$ *(dimesion* 8*) are the only (up to isomorphism) normed division algebras.*

(iii) $\mathbb{R}, \mathbb{C}, \mathbb{H},$ *and* $\mathbb{O}$ *are the only (up to isomorphism) alternative division algebras.*

Note that it is not true that $\mathbb{H}$ and $\mathbb{O}$ are the only division algebras in their respective dimensions. However, we do not use other real division algebras.

Let us give the classical *Cayley-Dickson* construction for our algebras, which partially explains their properties. As is well known, the complex number $a + bi$ can be thought of as a pair (a, b) of real numbers with componentwise addition and with multiplication given by

$$(a,b)(c,d) = (ac - db, ad + cb),$$

the conjugate being $\overline{(a,b)} = (a, -b)$. Notice also that a normed algebra carries a canonical inner product $\langle a, b \rangle := \text{Re}(a\bar{b})$.

One can define the *quaternions* in a similar way. A quaternion can be thought of as a pair of complex numbers added componentwise and multiplied in the following way:
$$(a,b)(c,d) = (ac - d\bar{b}, \bar{a}d + cb).$$
The conjugate of a quaternion is given by $\overline{(a,b)} = (\bar{a}, -b)$.

If we use the same formulas for quaternions, we get the algebra of *octonions*, and one can repeat the procedure to get new agebras of dimensions $2^k, k = 0, 1, \ldots$. It follows that all these algebras have multiplicative inverses since one can check that $(a,b)\overline{(a,b)} = \overline{(a,b)}(a,b) = p(1,0)$ for a positive real number (the square of the norm of (a,b)). Hence whenever $(a,b) \neq 0$, $(a,b)^{-1} = \overline{(a,b)}/p$. However, they are division algebras only for $k \leq 3$.

It can be explained as follows. Let A be an algebra equipped with a conjugation, that is, a real linear antihomomorphism
$$\overline{(\cdot)} : A \longrightarrow A, \ a \mapsto \bar{a}, \ \overline{ab} = \bar{b}\bar{a}.$$
We also suppose that A is *nicely normed*, i.e., $a + \bar{a} \in \mathbb{R}$, $a\bar{a} = \bar{a}a > 0$ for all nonzero $a \in A$. For such an A, we set
$$\mathrm{Re}(a) = (a + \bar{a})/2 \in \mathbb{R}, \quad \mathrm{Im}(a) = (a - \bar{a})/2 = a - \mathrm{Re}(a)$$
and define a norm on A by $|a|^2 = a\bar{a}$; A has multiplicative inverses given by $a^{-1} = \bar{a}/|a|^2$.

Any nicely normed alternative A is a normed division algebra. Indeed, for any $a, b \in A$, the elements $a, b, \bar{a}, \bar{b} \in \mathbb{R}[\mathrm{Im}(a), \mathrm{Im}(b)]$, this last algebra being associative, we get
$$|ab|^2 = (ab)\overline{ab} = ab(\bar{b}\bar{a}) = a(b\bar{b})\bar{a} = |a|^2|b|^2.$$

Starting from a nicely normed algebra A, the Cayley-Dickson construction gives a new algebra A'. Elements of A' are pairs $(a,b) \in A^2$, with multiplication and conjugation as above.

Brute force calculations show the effect of the Cayley-Dickson construction:

PROPOSITION 2.1.1. (i) *A is real (and thus commutative) iff A' is commutative.*
(ii) *A is commutative and associative iff A' is associative.*
(iii) *A is associative and nicely normed iff A' is alternative and nicely normed.*
(iv) *A is nicely normed iff A' is nicely normed.*

Therefore, $\mathbb{R}, \mathbb{C}, \mathbb{H}$, and $\mathbb{O}$ are normed division algebras. It also follows that the octonions are neither real nor commutative nor associative. The 16-dimensional algebra $\mathbb{O}'$ of *sedenions* has zero divisors, e.g., $xy = 0$ for $x = (e_1, 0) + (0, e_2), y = (0, e_7) - (e_4, 0) \in \mathbb{O}'$, where $(e_0 = 1, e_1, \ldots, e_7)$ is the canonical basis of $\mathbb{O}$ arising from the Cayley-Dickson construction. In fact, the zero divisors of $\mathbb{O}'$ form a 14-dimensional submanifold of $\mathbb{O}'$ homeomorphic to the exceptional Lie group G_2. As all the other algebras obtained by the construction contain the sedenions, they all have zero divisors.

For the canonical basis $(e_0 = 1, e_1, \ldots, e_7)$ of $\mathbb{O}$ one can deduce the following:
1) $e_0 = 1$, $e_i^2 = -1$, $i = 1, \ldots, 7$.
2) e_i and e_j anticommute for $i \neq j$:
$$e_i e_j = -e_j e_i = \pm e_k.$$

TABLE 2.1.1. Octonion multiplication table.

	e_1	e_2	e_3	e_4	e_5	e_6	e_7
e_1	-1	e_4	e_7	$-e_2$	e_6	$-e_5$	$-e_3$
e_2	$-e_4$	-1	e_5	e_1	$-e_3$	e_7	$-e_6$
e_3	$-e_7$	$-e_5$	-1	e_6	e_2	$-e_4$	e_1
e_4	e_2	$-e_1$	$-e_6$	-1	e_7	e_3	$-e_5$
e_5	$-e_6$	e_3	$-e_2$	$-e_7$	-1	e_7	e_4
e_6	e_5	$-e_7$	e_4	$-e_3$	$-e_1$	-1	e_2
e_7	e_2	e_6	$-e_1$	e_5	$-e_4$	$-e_2$	-1

3) The *index cycling* identity holds:
$$e_i e_j = e_k \text{ implies } e_{i+1} e_{j+1} = e_{k+1}$$
with indices modulo 7.

4) The *index doubling* identity holds:
$$e_i e_j = e_k \text{ implies } e_{2i} e_{2j} = e_{2k}.$$

This permits us to recover the whole multiplication table from $e_1 e_2 = e_4$: If we denote $i := e_1$, $j := e_2$, $e := e_3$, $k := e_4$, then $e_5 = je, e_6 = ie_5 = -ke, e_7 = ie$; the subspace H generated by e_0, e_1, e_2, and e_4 is a subalgebra isomorphic to $\mathbb{H}$ while e_3, e_5, e_5, and e_7 generate its orthogonal complement $\mathbb{H}^\perp$.

2.2. Clifford algebras and spinor groups

We consider only compact real Lie groups. The group $\mathrm{Spin}(n)$ is the universal covering group of the special orthogonal group $SO(n)$. The groups $\mathrm{Spin}(n)$ can be constructed by the Clifford algebras which we recall in our (very restricted) setting.

Let n be an integer ≥ 0. By the real *Clifford algebra* C_n we mean the associative algebra over $\mathbb{R}$ with n generators x_i ($1 \leq i \leq n$) such that
$$x_i^2 = -1, \quad x_i x_j + x_j x_i = 0 \ (i \neq j).$$
It is of dimension 2^n and the elements
$$1, \ x_{i_1} x_{i_2} \cdots x_{i_k} \ (1 \leq i_1 < i_2 < \cdots < i_k \leq n)$$
form a basis of C_n. The elements $x_i x_j$ generate the even subalgebra C_n^+ in C_n which is linearly spanned by
$$1, \ x_{i_1} x_{i_2} \cdots x_{i_{2r}} \ (1 \leq i_1 < i_2 < \cdots < i_{2r} \leq n).$$
Note also that there exists a natural isomorphism
$$C_{n-1} \longrightarrow C_n^+, \ e_i \mapsto x_i x_n.$$
Thus, it is easy to see that
$$C_0 = \mathbb{R}, \ C_1 = \mathbb{C}, \ C_2 = \mathbb{H};$$

note, however, that $C_3 \neq \mathbb{O}$ since $\mathbb{O}$ is not associative (and C_3 is not a division algebra). In fact, it is not difficult to show that

$$C_3 \simeq \mathbb{H} \oplus \mathbb{H},$$
$$C_4 \simeq \mathbb{H}[2],$$
$$C_5 \simeq \mathbb{C}[4],$$
$$C_6 \simeq \mathbb{R}[8],$$
$$C_7 \simeq \mathbb{R}[8] \oplus \mathbb{R}[8],$$

$A[n]$ being the algebra of $n \times n$-matrices over an associative algebra A.

Moreover, one has the Bott periodicity isomorphism

$$C_{n+8} \simeq C_n[16].$$

The algebra C_n has a unique antiautomorphism β which fixes all x_i, $\beta(x_i) = x_i$, $i = 1, \ldots, n$. One defines $\mathrm{Spin}(n)$ as the group of all elements $u \in C_n^+$ such that

(i) u is invertible;
(ii) $u V_n u^{-1} = V_n$, V_n being the linear span of all $x_i, i = 1, \ldots, n$; it is considered with the scalar product for which the basis $x_i, i = 1, \ldots, n$, is orthonormal;
(iii) $\beta(u)u = 1$.

The group $\mathrm{Spin}(n)$ is a connected compact Lie group, and the action $p(u)$ of any $u \in \mathrm{Spin}(n)$ on V_n given by $p(u)v = uvu^{-1}$ is an orthogonal transformation defining a surjective homomorphism $\mathrm{Spin}(n) \longrightarrow SO(n)$ with kernel $\{\pm 1\}$ which gives the universal covering of $SO(n)$.

Another description of $\mathrm{Spin}(n)$ is the group of all elements of C_n^+ of the form $u_1 u_2 \cdots u_{2k}$ for $u_i \in V_n, |u_i| = 1, \forall i = 1, \ldots, 2k$.

From the above explicit description of $C_n, n = 1, \ldots, 5$, and the isomorphism $C_{n-1} \simeq C_n^+$ one can deduce that

$$\mathrm{Spin}(2) \simeq U(1) \simeq SO(2),$$
$$\mathrm{Spin}(3) \simeq Sp(1) \simeq SU(2),$$
$$\mathrm{Spin}(4) \simeq Sp(1) \times Sp(1),$$
$$\mathrm{Spin}(5) \simeq Sp(2),$$
$$\mathrm{Spin}(6) \simeq SU(4).$$

Here, as usual,

$$U(n) := \{M \in GL(n) : M^* M = I_n\},$$
$$SU(n) := \{M \in U(n) : \det M = 1\},$$
$$Sp(n) := \{M \in GL(2n) : M^t J_{2n} M = I_{2n}\}$$

are the unitary group, the special unitary group, and the symplectic group, respectively, and J_{2n} denotes the block diagonal matrix with the standard 2×2-blocks

$$\begin{pmatrix} 0 & -1 \\ 1 & 0 \end{pmatrix}$$

on the diagonal.

Note, however, that this chain of *exceptional isomorphisms* stops at $\mathrm{Spin}(6)$; there is no similar isomorphisms for $\mathrm{Spin}(n), n \geq 7$.

Let us say a few words about the representations of the spinor groups. It is easiest to analyze them via representations of the algebras C_n.

Since Clifford algebras are built from matrix algebras over $\mathbb{R}, \mathbb{C}$, and $\mathbb{H}$, it is easy to determine their irreducible representations (that is, not possesing nontrivial subrepresentations; note that the algebras are semisimple and thus every representation is a direct sum of irreducible ones). The only irreducible representation of $\mathbb{R}[n]$ is its obvious one via matrix multiplication on $\mathbb{R}^n$. The same is true for $\mathbb{C}[n]$ acting on $\mathbb{C}^n$, and for $\mathbb{H}[n]$ acting on $\mathbb{H}^n$.

Since C_n is a matrix algebra unless n equals 3 or 7 modulo 8, it has a unique irreducible representation. It is called the space of pinors and is denoted P_n. For n equal to 3 or 7 modulo 8, the algebra C_n is a direct sum of two matrix algebras, so it has two irreducible representations, the positive pinors P_n^+ and negative pinors P_n^-.

The irreducible representations of $C_n, n \leq 7$, can be described as follows:

n	C_n	irreps
0	$\mathbb{R}$	$P_0 = \mathbb{R}$
1	$\mathbb{C}$	$P_1 = \mathbb{C}$
2	$\mathbb{H}$	$P_2 = \mathbb{H}$
3	$\mathbb{H} \oplus \mathbb{H}$	$P_3^+ = \mathbb{H}, P_3^- = \mathbb{H}$
4	$\mathbb{H}[2]$	$P_4^+ = \mathbb{H}^2$
5	$\mathbb{C}[4]$	$P_5 = \mathbb{C}^4$
6	$\mathbb{R}[8]$	$P_6 = \mathbb{R}^8$
7	$\mathbb{R}[8] \oplus \mathbb{R}[8]$	$P_7^+ = \mathbb{R}^8, P_7^- = \mathbb{R}^8$

Let us turn then to irreducible representations of $\mathrm{Spin}(n)$. First of all we have an n-dimensional irreducible representation V_n of $\mathrm{Spin}(n)$ which comes from the canonical surjection $\mathrm{Spin}(n) \longrightarrow SO(n)$. It is called a *vector representation*. All other irreducible representations of $\mathrm{Spin}(n)$ are said to be of *spin type*; they are nontrivial on the kernel of $\mathrm{Spin}(n) \longrightarrow SO(n)$. The *spin representations* are the irreducible representations that have the smallest dimension among the spin type representations of $\mathrm{Spin}(n)$.

To get irreducible representations of $\mathrm{Spin}(n), n \leq 8$, we can restrict representations of $C_{n-1} \simeq C_n^+$ to $\mathrm{Spin}(n)$. The irreducible representations of C_{n-1} remain irreducible and thus we get the following spin representations of $\mathrm{Spin}(n), n \leq 8$:

n	$\mathrm{Spin}(n)$	$\dim(\mathrm{Spin}(n))$	spinreps
2	$U(1)$	1	$S_2 = \mathbb{C}$
3	$Sp(1)$	3	$S_3 = \mathbb{H}$
4	$Sp(1) \times Sp(1)$	6	$S_4^+ = S_4^- = \mathbb{H}$
5	$Sp(2)$	10	$S_5 = \mathbb{H}^2$
6	$SU(4)$	15	$S_6 = \mathbb{C}^4$
7	$\mathrm{Spin}(7)$	21	$S_7 = \mathbb{R}^8$
8	$\mathrm{Spin}(8)$	28	$S_8^+ = S_8^- = \mathbb{R}^8$

One can also verify that the representations S_n^+ and S_n^- are self-dual for $n = 4, 8$.

We can then pass to constructing principal objects of the first part of the chapter, namely, the exceptional group G_2 and trialities.

2.3. The group G_2. Trialities

2.3.1. The simple group G_2. We begin with the exceptional isomorphisms

$$\mathrm{Spin}(5) \simeq Sp(2), \quad \mathrm{Spin}(6) \simeq SU(4).$$

First of all we get that the action of $\mathrm{Spin}(5)$ on $\mathbb{S}^7 \subset S_5 = \mathbb{H}^2$ is transitive since it is true for $Sp(2)$. Next, the group $\mathrm{Spin}(6)$ acts transitively on pairs $(x, z), x \in \mathbb{S}^6, z \in \mathbb{S}^7 \subset S_6 = \mathbb{C}^4$. Indeed, since $\mathrm{Spin}(6) \longrightarrow SO(6)$ is surjective, we get the transitivity on $\mathbb{S}^6$. For $x = (0, \ldots, 0, 1) \in \mathbb{S}^6$ the stabilizer is $\mathrm{Spin}(5)$ and we apply the previous remark. That in turn implies the transitivity of the action of $\mathrm{Spin}(7)$ on triples

$$(x, y, z) \in \mathbb{S}^6 \times \mathbb{S}^6 \times \mathbb{S}^7 \subset V_7 \times V_7 \times S_7 = \mathbb{R}^7 \times \mathbb{R}^7 \times \mathbb{R}^8, \ x \perp y,$$

since $\mathrm{Spin}(7) \longrightarrow SO(7)$ is surjective and the stabilizer of $y = (0, \ldots, 0, 1) \in \mathbb{S}^6$ is $\mathrm{Spin}(6)$.

Let us define now the group $G_2 \subset \mathrm{Spin}(7)$ as the stabilizer of $z = (0, \ldots, 0, 1) \in S_7 = \mathbb{R}^8$. It is a compact connected Lie group of dimension

$$\dim G_2 = \dim \mathrm{Spin}(7) - \dim \mathbb{S}^7 = 21 - 7 = 14$$

which acts transitively on pairs $(x, y) \in \mathbb{S}^6 \times \mathbb{S}^6, x \perp y$. One verifies that this definition coincides with the usual definition of G_2 via Dynkin diagrams.

2.3.2. Trialities. Duality is ubiquous in algebra; triality is similar, but much rarer. For two real vector spaces V_1 and V_2, a duality is simply a nondegenerate bilinear map

$$f : V_1 \times V_2 \longrightarrow \mathbb{R}.$$

Similarly, for three real vector spaces V_1, V_2, and V_3, a *triality* is a trilinear map

$$t : V_1 \times V_2 \times V_3 \longrightarrow \mathbb{R}$$

that is nondegenerate in the sense that if we fix any two arguments to any nonzero values, the linear functional induced on the third vector space is nonzero. Each vector spaces V_1 has the dual vector space $V_2 = V_1^*$. Trialities are in fact exceptional and eventually come from division algebras. Indeed, let

$$t : V_1 \times V_2 \times V_3 \longrightarrow \mathbb{R}$$

be a triality. By dualizing one gets a bilinear map

$$m : V_1 \times V_2 \longrightarrow V_3^*.$$

By the nondegeneracy of t, the three spaces V_1, V_2, and V_3^* can be identified with a single vector space, say V, which gives a product

$$m : V \times V \longrightarrow V.$$

Applying the nondegeneracy once more one sees that V is actually a division algebra. It follows from (i) in Theorem 2.1.1 that trialities only occur in dimensions

1, 2, 4, or 8. The 1-dimensional case is trivial. Examples of trialities in dimensions 2, 4, and 8 are given by

$$t_2 : \mathbb{C} \times \mathbb{C} \times \mathbb{C} \longrightarrow \mathbb{R}, \quad t_2(z_1, z_2, z_3) = \mathrm{Re}(z_1 z_2 z_3),$$
$$t_4 : \mathbb{H} \times \mathbb{H} \times \mathbb{H} \longrightarrow \mathbb{R}, \quad t_4(q_1, q_2, q_3) = \mathrm{Re}(q_1 q_2 q_3),$$
$$t_8 : \mathbb{O} \times \mathbb{O} \times \mathbb{O} \longrightarrow \mathbb{R}, \quad t_8(o_1, o_2, o_3) = \mathrm{Re}((o_1 o_2) o_3) = \mathrm{Re}(o_1 (o_2 o_3)).$$

Note that the identity $\mathrm{Re}((xy)z) = \mathrm{Re}(x(yz))$ for any $x, y, z \in \mathbb{O}$ follows from the fact the associator is alternative:

$$\overline{[x, y, z]} = \overline{(xy)z} - \overline{x(yz)} = \bar{z}(\bar{y}\bar{x}) - (\bar{z}\bar{y})\bar{x} = -[\bar{x}, \bar{y}, \bar{z}] = [z, y, x] = -[x, y, z];$$

the explicit formula below proves this identity as well.

A choice of $\mathbb{R}$-bases in $\mathbb{C}, \mathbb{H}$, and $\mathbb{O}$ transforms t_2, t_4, and t_8 into the following cubic harmonic (since they are polylinear) forms in 6, 12, and 24 variables, respectively:

$$P_6 = x_0 y_0 z_0 - x_0 y_1 z_1 - x_1 y_1 z_0 - x_1 y_0 z_1,$$

$$P_{12} = (y_0 z_0 - y_1 z_1 - y_2 z_2 - y_3 z_3)x_0 + (y_3 z_2 - y_0 z_1 - y_1 z_0 - y_2 z_3)x_1$$
$$+ (y_1 z_3 - y_0 z_2 - y_2 z_0 - y_3 z_1)x_2 + (y_2 z_1 - y_0 z_3 - y_1 z_2 - y_3 z_0)x_3,$$

$$P_{24} = (z_0 y_0 - z_1 y_1 - z_2 y_2 - z_3 y_3 - z_4 y_4 - z_5 y_5 - z_6 y_6 - z_7 y_7)x_0$$
$$+ (-z_1 y_0 - z_0 y_1 - z_4 y_2 - z_7 y_3 + z_2 y_4 - z_6 y_5 + z_5 y_6 + z_3 y_7)x_1$$
$$+ (-z_2 y_0 + z_4 y_1 - z_0 y_2 - z_5 y_3 - z_1 y_4 + z_3 y_5 - z_7 y_6 + z_6 y_7)x_2$$
$$+ (-z_3 y_0 + z_7 y_1 + z_5 y_2 - z_0 y_3 - z_6 y_4 - z_2 y_5 + z_4 y_6 - z_1 y_7)x_3$$
$$+ (-z_4 y_0 - z_2 y_1 + z_1 y_2 + z_6 y_3 - z_0 y_4 - z_7 y_5 - z_3 y_6 + z_5 y_7)x_4$$
$$+ (-z_5 y_0 + z_6 y_1 - z_3 y_2 + z_2 y_3 + z_7 y_4 - z_0 y_5 - z_1 y_6 - z_4 y_7)x_5$$
$$+ (-z_6 y_0 - z_5 y_1 + z_7 y_2 - z_4 y_3 + z_3 y_4 + z_1 y_5 - z_0 y_6 - z_2 y_7)x_6$$
$$+ (-z_7 y_0 - z_3 y_1 - z_6 y_2 + z_1 y_3 - z_5 y_4 + z_4 y_5 + z_2 y_6 - z_0 y_7)x_7.$$

These cubic polynomials will serve to construct nonclassical solutions to elliptic equations in Chapter 4.

Note that the above trialities t_2, t_4, and t_8 are *normed*, i.e.,

$$|t_j(v_1, v_2, v_3)| \leq |v_1| |v_2| |v_3|, \quad \forall (v_1, v_2, v_3) \in V^3, \forall j \in \{2, 4, 8\}.$$

Given a normed triality, picking unit vectors in any two of the spaces V_i allows us to identify all three spaces and get a normed division algebra, and this shows that the above normed trialities are essentially unique.

For purposes of Chapter 4 we need automorphisms of the normed trialities, which are interesting in themselves. An automorphism of the normed triality

$$t : V_1 \times V_2 \times V_3 \longrightarrow \mathbb{R}$$

is a triple of norm-preserving maps $f_i : V_i \longrightarrow V_i$, $i = 1, 2, 3$, such that

$$t(f_1(v_1), f_2(v_2), f_3(v_3)) = t(v_1, v_2, v_3), \quad \forall (v_1, v_2, v_3) \in V_1 \times V_2 \times V_3.$$

The automorphism groups of our normed trialities are
$$Aut(t_2) = \{(g_1, g_2, g_3) \in U(1)^3 : g_1 g_2 g_3 = 1\} \times \{\pm 1\},$$
$$Aut(t_4) = Sp(1)^3/\{\pm(1,1,1)\},$$
$$Aut(t_8) = \mathrm{Spin}(8),$$
where
$$U(1) = \mathbb{S}^1 = \{u \in \mathbb{C} : |u| = 1\},\ Sp(1) = SU(2) = \{v \in \mathbb{H} : |v| = 1\}.$$

For t_2 and t_4 the automorphism group can be calculated directly since a norm-preserving $\mathbb{R}$-linear endomorphism of $\mathbb{C}$ is simply a multiplication by an element of $U(1)$, and a norm-preserving $\mathbb{R}$-linear endomorphism of $\mathbb{H}$ can be written as $q \mapsto uqv$ for some $u, v \in Sp(1)$ since $\mathrm{Spin}(4) = Sp(1) \times Sp(1)$ is a double cover of $SO(4)$.

In fact, the normed trialities come from spinors. We explain this only in the two nontrivial cases $n = 4$ and $n = 8$ (the only other possible values $n = 1, 2$ being completely obvious).

Since $V_n \subset C_n$, we can restrict the action of the C_n on pinors to get a map
$$m_n : V_n \times P_n^\pm \longrightarrow P_n^\mp,\ n = 3, 7.$$

If we restrict the vector representation to the subgroup $\mathrm{Spin}(n+1)$, we get a map
$$m_n : V_n \times S_n^\pm \longrightarrow S_n^\mp,\ n = 4, 8.$$

Dualizing the above maps we get trilinear maps:
$$t_n : V_n \times S_n^\pm \times S_n^\mp \longrightarrow \mathbb{R},\ n = 4, 8.$$

Note that for $n = 4, 8$ the dimension of the vector representation equals that of spinor representations (it occurs also for $n = 1$ and $n = 2$), which leads to normed trialities t_4 and t_8. In particular, we get an action of $\mathrm{Spin}(8)$ on t_8. Its action on three copies of the algebra $\mathbb{O}$ is given by three possible projections $\mathrm{Spin}(8) \longrightarrow SO(8)$; the center of $\mathrm{Spin}(8)$ is isomorphic to $(\mathbb{Z}/2\mathbb{Z})^2$ and thus has exactly three subgroups of order 2 which are kernels of the vector and the two spinor representations.

We also need the following fact:

THEOREM 2.3.1. *The group G_2 is the automorphism group of the algebra $\mathbb{O}$.*

Indeed, the product $V_8 \times S_8^+ \times S_8^-$ gets identified with $\mathbb{O} \times \mathbb{O} \times \mathbb{O}$ by choosing a triple of unit vectors $x_0 \in V_8, y_0 \in S_8^+, z_0 \in S_8^-$. The automorphism group $Aut(\mathbb{O})$ is the subgroup of $\mathrm{Spin}(8)$ preserving (x_0, y_0, z_0) (it also preserves the norm and the triality). The stabilizer of z_0 is $\mathrm{Spin}(7)$ and the stabilizer of x_0 is G_2 (by the definition given above in this section); it automatically preserves y_0 as well.

The restrictions of the three projections $\mathrm{Spin}(8) \longrightarrow SO(8)$ on G_2 coincide since the action of G_2 on $\mathbb{O}$ is determined by the structure of $\mathbb{O}$ as an algebra.

2.4. Cross products, associative and coassociative calibrations

We assume then until the end of Chapter 2 that V is a finite-dimensional real Hilbert space. In this and the next section we give essentially a (multi-)linear treatment of the objects in Sections 2.1– 2.3. The beginning of the method is a simple remark that the product in a real division algebra restricted to the orthogonal complement of the algebra's unity defines a cross product (for noncommutative algebras). In fact, we are going to show that, conversely, a cross product with its usual properties is completely equivalent to a noncommutative real division algebra.

2.4.1. Cross products and associative calibrations.

DEFINITION 2.4.1. *A skew-symmetric bilinear map*
$$V \times V \longrightarrow V : (u,v) \mapsto u \times v$$
is called a cross product *if it satisfies*

(2.4.1) $$\langle u \times v, u \rangle = \langle u \times v, v \rangle = 0,$$

(2.4.2) $$|u \times v|^2 = |u|^2 |v|^2 - \langle u, v \rangle^2$$

for all $u, v \in V$.

PROPOSITION 2.4.1. *V admits a nontrivial cross product if and only if its dimension is 3 or 7. In dimension 3 it is unique up to sign and determined by an orientation of V, and in dimension 7 it is unique up to an orthogonal transformation.*

This result corresponds to the existence and unicity of quaternions and octonions; it can be deduced from the following lemmas, which are verified by direct calculations.

LEMMA 2.4.1. *For a skew-symmetric bilinear map the following conditions are equivalent:*

  (i) *(2.4.2) holds for all $u,v \in V$.*
  (ii) *For all $u,v,w \in V$, $\langle u \times v, w \rangle = \langle u, v \times w \rangle$.*
  (iii) *The map $\phi : V^3 \longrightarrow \mathbb{R}$ defined by*

(2.4.3) $$\varphi(u,v,w) := \langle u \times v, w \rangle$$

  is an alternating 3-form (called the associative calibration *of V).*

LEMMA 2.4.2. *For a skew-symmetric bilinear map the following conditions are equivalent:*

  (i) *(2.4.2) holds for all $u,v \in V$.*
  (ii) *If u and w are orthonormal, then $|u \times w| = 1$.*
  (iii) *If $|u| = 1$ and $u \perp w$, then $u \times (u \times w) = w$.*
  (iv) *For all $u, w \in V$, $u \times (u \times w) = \langle u \times w, u \rangle - |u|^2 w$ for all $u, w \in V$.*
  (v) *For all $u, v, w \in V$,*

(2.4.4) $$u \times (v \times w) + v \times (u \times w) = \langle u, w \rangle v + \langle v, w \rangle u - 2 \langle u, v \rangle w \ .$$

LEMMA 2.4.3. *Let $\dim V = 3$.*

  (i) *A cross product on V determines a unique orientation such that $u, v, u \times v$ form a positive basis for any pair of linearly independent vectors $u, v \in V$.*
  (ii) *For a cross product on V the 3-form (2.4.3) is the volume form associated to the inner product and the orientation in (i).*
  (iii) *$(u \times v) \times w = \langle u, w \rangle v - \langle v, w \rangle u$, for all $u, v, w \in V$.*
  (iv) *For an orientation let φ be the associated volume form; then (2.4.3) determines a cross product on V.*

2.4. CROSS PRODUCTS, ASSOCIATIVE AND COASSOCIATIVE CALIBRATIONS

In three dimensions $u \times v$ is given by the familiar vector product formula equivalent to (2.4.3) with the standard volume form. In $\mathbb{R}^7$ the standard cross product corresponds to

$$\varphi_0 = e^{123} - e^{145} - e^{167} - e^{246} - e^{275} - e^{347} - e^{356} \tag{2.4.5}$$

for $e^{ijk} := dx_i \wedge dx_j \wedge dx_k$.

Notice that if one identifies $V = \mathbb{R}^7$ with the space

$$\operatorname{Im} \mathbb{O} := \{u \in \mathbb{O} : \operatorname{Re}(u) = 0\}$$

of purely imaginary octonions, then

$$u \times v = \operatorname{Im}(v\bar{u})$$

for the standard cross product and any $u, v \in \operatorname{Im} \mathbb{O}$.

Let then $\dim V = 7$.

DEFINITION 2.4.2. A 3-form $\varphi \in \Lambda^3 V^*$ is called *nondegenerate* if, for every pair of linearly independent vectors $u, v \in V$, there is a vector $w \in V$ such that $\varphi(u, v, w) \neq 0$. An inner product on V is called *compatible* with φ if (2.4.3) is a cross product.

Nondegenerate forms produce associative calibrations using the following results:

PROPOSITION 2.4.2. *Let $\varphi, \varphi' \in \Lambda^3 V^*$, then:*

(i) *φ is nondegenerate if and only if it admits a compatible inner product.*
(ii) *The inner product in (i), if it exists, is uniquely determined by φ.*
(iii) *If φ, φ' are nondegenerate, then $g^*\varphi = \varphi'$ for some $g \in \operatorname{Aut}(V)$.*

This result can be proved using the following lemma, which is useful in itself.

LEMMA 2.4.4. *Let $\varphi \in \Lambda^3 V^*$. Then the following conditions are equivalent:*

(i) *φ is compatible with the inner product.*
(ii) *There is an orientation on V such that for all $u, v \in V$*

$$i(u)\varphi \wedge i(v)\varphi \wedge \varphi = 6\langle u, v \rangle \omega_0,$$

with $\omega_0 \in \Lambda^7 V^*$ being the associated volume form.

These conditions imply that φ is nondegenerate and uniquely determines the orientation in (ii).

Notice that (2.4.4) immediately implies that for a cross product $V \times V \longrightarrow V$: $(u, v) \mapsto u \times v$ and orthonormal $u, v, w := u \times v$ one has $v \times w = u$ and $w \times u = v$.

DEFINITION 2.4.3. Any 3-form $\varphi \in \Lambda^3 V^*$ satisfying the conditions of the last lemma is called an associative calibration on V.

Let us then construct the associator bracket, which is useful in studing associative calibration.

The formula (2.4.4) implies that the expression $(u \times v) \times w$ is alternating on any triple of pairwise orthogonal vectors $u, v, w \in V$. Hence it extends uniquely to an alternating 3-form

$$V^3 \to V : (u, v, w) \mapsto [u, v, w]$$

called *the associator bracket*, which can be given by

$$[u, v, w] = (u \times v) \times w + \langle v, w \rangle u - \langle u, w \rangle v$$

or, equivalently, by
$$[u,v,w] = \frac{(u \times v) \times w + (v \times w) \times u + (w \times u) \times v}{3}.$$

Again, for $V = \text{Im}\,\mathbb{O}$ it coincides with the associator (2.1.1) up to a scalar factor. More precisely, $2[u,v,w] = (uv)w - u(vw)$ for all $u,v,w \in \text{Im}\,\mathbb{O}$.

The following property relating φ and the associator bracket is of fundamental impotance since it is the base of applications to calibrated geometries; see Chapter 7.

PROPOSITION 2.4.3. *For all $u,v,w \in V$ one has*
$$\varphi(u,v,w)^2 + |[u,v,w]|^2 = |u \wedge v \wedge w|^2 = \det \text{Gram}(u,v,w),$$
$\text{Gram}(u,v,w)$ *being the Gram matrix of the vectors (u,v,w).*

PROOF. By Gram-Schmidt one supposes that w is orthogonal to u and v; then $|[u,v,w]|^2 = |(u \times v) \times w|^2 = |u \times v|^2|w|^2 - \langle u, v \times w\rangle^2 = |u \wedge v \wedge w|^2 - \varphi(u,v,w)^2$ by (2.4.1) and (2.4.3). □

The following notion is important in calibrated geometry since it corresponds to tangent spaces to an associative manifold.

DEFINITION 2.4.4. A 3-dimensional subspace $\Lambda \subset V$ is called *associative* if $[u,v,w] = 0$ for all $u,v,w \in \Lambda$.

Here are some other characterizations of associative spaces.

LEMMA 2.4.5. *Let $\Lambda \subset V$ be a 3-dimensional vector subspace. Then the following are equivalent:*
  (i) Λ *is associative.*
  (ii) *If $\{u,v,w\}$ is an orthonormal basis of Λ, then $\varphi(u,v,w) = \pm 1$.*
  (iii) *If $u,v \in \Lambda$, then $u \times v \in \Lambda$.*
  (iv) *If $u \in \Lambda^\perp, v \in \Lambda$, then $u \times v \in \Lambda^\perp$.*
  (v) *If $u,v \in \Lambda^\perp$, then $u \times v \in \Lambda$.*

For linearly independent u,v the subspace spanned by u, v, and $u \times v$ is associative.

2.4.2. Coassociative calibrations. Let us then characterize (Hodge) duals of associative calibrations.

PROPOSITION 2.4.4. *Let $\psi: V^4 \to \mathbb{R}$ be defined by*
$$\psi(u,v,w,x) := \langle [u,v,w], x\rangle = \frac{\varphi(u \times v, w, x) + \varphi(v \times w, u, x) + \varphi(w \times u, v, x)}{3};$$
then $\psi \in \Lambda^4 V$. Moreover, $\psi = *\varphi$ for the Hodge operator associated to the inner product and the corresponding orientation; ψ is called the coassociative calibration *dual to φ.*

The coassociative calibration of $\mathbb{R}^7 = \text{Im}\,\mathbb{O}$ dual to φ_0 is

(2.4.6) $$\psi_0 = e^{4567} - e^{1247} - e^{1256} + e^{1346} - e^{1357} - e^{2345} - e^{2367}.$$

As before for associative calibration, it is useful to define the corresponding 4-fold bracket in the following way.

LEMMA 2.4.6. *For all $u, v, w, x \in V$ one has*
$$[u, v, w, x] := \varphi(u, v, w)x - \varphi(x, u, v)w + \varphi(w, x, u)v - \varphi(v, w, x)u$$
$$= \tfrac{1}{3}(-[u, v, w] \times x + [x, u, v] \times w - [w, x, u] \times v + [v, w, x] \times u).$$

The resulting multilinear map
$$V^4 \longrightarrow V : (u, v, w, x) \mapsto [u, v, w, x]$$
is alternating and is called the coassociator bracket *on V*.

As above for φ, the principal property of ψ is

PROPOSITION 2.4.5. *For all $u, v, w, x \in V$ one has*
$$\psi(u, v, w, x)^2 + |[u, v, w, x]|^2 = |u \wedge v \wedge w \wedge x|^2 = \det(\mathrm{Gram}(u, v, w, x))$$
for the corresponding Gram matrix.

Tangent spaces to coassociative manifolds in calibrated geometry are given by

DEFINITION 2.4.5. *A 4-dimensional subspace $H \subset V$ is called coassociative if $[u, v, w, x] = 0$ for all $u, v, w, x \in H$.*

Their alternative descriptions are

LEMMA 2.4.7. *Let $H \subset V$ be a 4-dimensional linear subspace. Then the following are equivalent:*
  (i) *H is coassociative.*
  (ii) *If $\{u, v, w, x\}$ is an orthonormal basis of H, then $\psi(u, v, w, x) = \pm 1$.*
  (iii) *For all $u, v, w \in H$ we have $\varphi(u, v, w) = 0$.*
  (iv) *If $u, v \in H$, then $u \times v \in H^\perp$.*
  (v) *If $u \in H$ and $v \in H^\perp$, then $u \times v \in H$.*
  (vi) *If $u \in H^\perp$ and $v \in H^\perp$, then $u \times v \in H^\perp$.*
  (vii) *$H^\perp$ is associative.*

2.4.3. Automorphism group. We are going then to revisit the group G_2 as the automorphism group of both associative and coassociative calibrations.

Indeed, let $G(V, \varphi)$ be the group of automorphisms of φ:
$$G(V, \varphi) := \{g \in GL(V) : g^*\varphi = \varphi\}.$$

One has $G(V, \varphi) \subset SO(V)$ and hence
$$G(V, \varphi) := \{g \in SO(V) : gu \times gv = g(u \times v), \ \forall u, v \in V\}.$$

For the standard structure φ_0 on $\mathbb{R}^7$ we denote the group by $G_2 := G(\mathbb{R}^7, \varphi_0)$; the group $G(V, \varphi)$ is isomorphic to G_2 for every nondegenerate 3-form on a 7-dimensional vector space. Indeed the condition $gu \times gv = g(u \times v)$ is equivalent to the commuting of g with the product in $\mathbb{O}$, which implies the isomorphism with G_2 by Theorem 2.3.1.

THEOREM 2.4.1. *The group $G(V, \varphi)$ is a 14-dimensional simple, connected, simply connected Lie group. It acts transitively on the unit sphere and, for any unit vector $u \in V$, the isotropy subgroup $G_u := \{g \in G(V, \varphi) : gu = u\}$ is isomorphic to $SU(3)$. Thus there is a fibration $SU(3) \longrightarrow G_2 \longrightarrow \mathbb{S}^6$ (that is, with the base $\mathbb{S}^6$ and the fiber $SU(3)$).*

Let us then consider the action of the group $G(V, \varphi)$ on the space

$$\mathcal{S} := \{(u, v, w) \in V : |u| = |v| = |w| = 1, u \perp v, u \perp w, v \perp w, u \times v \perp w\}.$$

The space $\mathcal{S}$ is a bundle of 3-spheres over a bundle of 5-spheres over a 6-sphere and thus is a compact connected simply connected 14-dimensional manifold. Indeed, the space $\mathcal{S}$ is a bundle over $\mathbb{S} \subset V$ whose fiber over u is the space of Hermitian orthonormal pairs in the tangent space $T_u\mathbb{S}$ with its natural complex structure $v \mapsto u \times v$.

THEOREM 2.4.2. *The group $G(V, \varphi)$ acts freely and transitively on $\mathcal{S}$.*

COROLLARY 2.4.1. *The group $G(V, \varphi)$ acts transitively on the space of associative subspaces of V and on the space of coassociative subspaces of V.*

2.5. Cross products, normed algebras, and Cayley calibrations

In the present section we show how to reconstitute the normed division algebra structure from cross products and how to use normed division algebras to define Cayley calibrations, which in a sense generalize both associative and coassociative calibrations.

2.5.1. Cross products and normed division algebras.
Let us begin with the relation between vector spaces with (double) cross products and normed division algebras.

THEOREM 2.5.1. (i) *If W is a normed division algebra, then $V := 1^\perp = \mathrm{Im}(W)$ is equipped with a cross product $V \times V \to V : (u, v) \mapsto u \times v$ defined by*

$$(2.5.1) \qquad u \times v := uv + \langle u, v \rangle$$

for $u, v \in V$.

(ii) *If V is a finite-dimensional Hilbert space equipped with a cross product, then $W := \mathbb{R} \oplus V$ is a normed division algebras with the product*

$$uv := u_0 v_0 - \langle u_1, v_1 \rangle + u_0 v_1 + v_0 u_1 + u_1 \times v_1$$

for $u = u_0 + u_1, v = v_0 + v_1 \in \mathbb{R} \oplus V$ (a real number λ is identified with the pair $(\lambda, 0) \in \mathbb{R} \oplus V$ and a vector $v \in V$ with the pair $(0, v) \in \mathbb{R} \oplus V$).

These constructions are inverses of each other. In particular, a normed division algebra has dimension 1, 2, 4, or 8 and is isomorphic to $\mathbb{R}, \mathbb{C}, \mathbb{H}$, or $\mathbb{O}$.

Thus constructed normed division algebras enjoy the usual properties:

PROPOSITION 2.5.1. *In a normed division algebras W one has the following for all $u, v, w \in W$:*
  (i) $\langle uv, w \rangle = \langle v, \bar{u}w \rangle$, $\langle uv, w \rangle = \langle u, w\bar{v} \rangle$.
  (ii) $u\bar{u} = |u|^2$, $u\bar{v} + v\bar{u} = 2\langle u, v \rangle$.
  (iii) $\langle u, v \rangle = \langle \bar{u}, \bar{v} \rangle$, $\overline{uv} = \overline{v}\overline{u}$.
  (iv) $u(\bar{v}w) + v(\bar{u}w) = 2\langle u, v \rangle w$, $(u\bar{v})w + (u\bar{w})v = 2\langle v, w \rangle u$.

Let us now show how to construct Cayley calibrations using our normed division algebras in eight dimensions.

2.5. CROSS PRODUCTS, NORMED ALGEBRAS, AND CAYLEY CALIBRATIONS

THEOREM 2.5.2. *Let W be an 8-dimensional normed division algebra.*

(i) *The map*
$$W^3 \longrightarrow W : (u,v,w) \mapsto u \times v \times w := \frac{(u\bar{v})w - (w\bar{v})u}{2}$$
(the triple cross product of W) is alternating and satisfies
$$\langle x, u \times v \times w \rangle = -\langle u \times v \times x, w \rangle,$$
$$|u \times v \times w| = |u \wedge v \wedge w|,$$
for all $u, v, w, x \in W$ and
$$\langle e \times u \times v, e \times w \times x \rangle = |e|^2 \langle u \times v \times w, x \rangle$$
whenever $e, u, v, w, x \in W$ are orthonormal.

(ii) *The map*
$$\Phi : W^4 \longrightarrow \mathbb{R}, \ \Phi(x, u, v, w) := \langle x, u \times v \times w \rangle$$
(the Cayley calibration of W) is an alternating 4-form. Moreover, Φ is self-dual,
$$\Phi = *\Phi$$
for the Hodge $*$-operator associated to the inner product and the natural orientation (as $\mathbb{R} \oplus V$).

(iii) *Let $V := 1^\perp$ with the cross product and the associator bracket defined as above, and let*
$$\varphi \in \Lambda^3 V^*, \ \psi \in \Lambda^4 V^*$$
be the respective associative and coassociative calibrations of V. Then the Cayley calibration can be written as
$$\Phi = 1^* \wedge \varphi + \psi.$$

As before for associative and coassociative calibrations, Cayley calibrations are intimately related to a cross product, but this time with a 4-fold one.

DEFINITION 2.5.1. Let W be an 8-dimensional normed division algebra. The *4-fold cross product* on W is the alternating multi-linear map
$$W^4 \longrightarrow W : (x, u, v, w) \mapsto x \times u \times v \times w$$
defined by
$$(2.5.2) \quad x \times u \times v \times w := (u \times v \times w)\bar{x} - (v \times w \times x)\bar{u} + (w \times x \times u)\bar{v} - (x \times u \times v)\bar{w}.$$

The basic property of Cayley calibrations is, as before, the following relation (2.5.3) below:

THEOREM 2.5.3. *Let W be an 8-dimensional normed division algebra with triple cross product, Cayley calibration, and 4-fold cross product defined as above. Then, for all $x, u, v, w \in W$, one has*
$$|x \times u \times v \times w| = |x \wedge u \wedge v \wedge w|,$$
$$\mathrm{Re}(x \times u \times v \times w) = \Phi(x, u, v, w),$$
$$\mathrm{Im}(x \times u \times v \times w) = [x_1, u_1, v_1, w_1] - x_0[u_1, v_1, w_1] + u_0[v_1, w_1, x_1]$$
$$- v_0[w_1, x_1, u_1] + w_0[x_1, u_1, v_1].$$

In particular,

(2.5.3) $$\Phi(x,u,v,w)^2 + |\mathrm{Im}(x \times u \times v \times w)|^2 = |x \wedge u \wedge v \wedge w|^2.$$

2.5.2. Triple cross products and Cayley calibrations. Now we show how to recover the Cayley calibration and the normed division algebra structure on W from the triple cross product. In fact we shall see that every unit vector in W can be used as a multiplicative unity for the normed algebra structure. We continue to assume that W is a finite-dimensional real Hilbert space.

DEFINITION 2.5.2. *An alternating multi-linear map*
$$W \times W \times W \longrightarrow W : (u,v,w) \mapsto u \times v \times w$$
is called a triple cross product *if it satisfes*

(2.5.4) $$\langle u \times v \times w, u \rangle = \langle u \times v \times w, v \rangle = \langle u \times v \times w, w \rangle = 0,$$

(2.5.5) $$|u \times v \times w| = |u \wedge v \wedge w|$$

for all $u,v,w \in W$.

Note that condition (2.5.5) alone implies that the map is necessarily alternating.

LEMMA 2.5.1. *For an alternating multi-linear map, (2.5.4) is equvalent to the condition*
$$\langle x, u \times v \times w \rangle + \langle u \times v \times x, w \rangle = 0$$
for all $x,u,v,w \in W$.

DEFINITION 2.5.3. *Let $\dim W = 8$ and let*
$$\Phi(x,u,v,w) := \langle x, u \times v \times w \rangle.$$
Then $\Phi \in \Lambda^4(W^*)$ is called the Cayley calibration *of W.*

The following construction restores the normed division algebra by Cayley calibration.

THEOREM 2.5.4. *Let Φ be the Cayley calibration corresponding to a triple cross product on W, $\dim W = 8$, and let $e \in W$ be a unit vector.*

(i) *Define the map $\psi_e : W^4 \longrightarrow \mathbb{R}$ by*
$$\psi_e(u,v,w,x) := \langle e \times u \times v, e \times w \times x \rangle - (\langle u,w \rangle - \langle u,e \rangle \langle e,w \rangle)(\langle v,x \rangle - \langle v,e \rangle \langle e,x \rangle)$$
$$+ (\langle u,x \rangle - \langle u,e \rangle \langle e,x \rangle)(\langle v,w \rangle - \langle v,e \rangle \langle e,w \rangle).$$
Then $\psi_e \in \Lambda^4(W^*)$ and
$$\Phi = e^* \wedge \varphi_e + \varepsilon \psi_e, \quad \varphi_e := i(e)\Phi \in \Lambda^3(W^*)$$
where $\varepsilon \in \{\pm 1\}$ is the sign of the triple cross product, which can be defined as follows. Let $e,u,v \in W$ be an orthonormal triple and let $w \in W$ be orthogonal to e,u,v, and to $e \times u \times v$. Then
$$e \times u \times (e \times v \times w) = \varepsilon(u \times v \times w).$$

(ii) *The subspace $V_e := e^\perp$ carries a cross product*
$$V_e \times V_e \longrightarrow V_e : (u,v) \mapsto u \times_e v := u \times e \times v,$$
the restriction of φ_e to V_e being its associative calibration, and the restriction of ψ_e to V_e being its coassociative calibration.

2.5. CROSS PRODUCTS, NORMED ALGEBRAS, AND CAYLEY CALIBRATIONS

(iii) *The space W is a normed division algebra with unit e and with multiplication and conjugation given by*
$$uv := u \times e \times v + \langle u, e \rangle v + \langle v, e \rangle u \langle u, v \rangle e, \quad \bar{u} := 2\langle u, e \rangle e - u.$$
If the triple cross product is positive ($\varepsilon = 1$), then $(u\bar{v})w - (w\bar{v})u = 2u \times v \times w$.

The following result characterizes Cayley subspaces, i.e., tangent spaces to Cayley manifolds in calibrated geometry.

LEMMA 2.5.2. *Let Φ be the Cayley calibration corresponding to a triple cross product on W with $\dim W = 8$ and let $\Lambda \subset W$ be a 4-dimensional vector subspace. Then the following conditions are equivalent:*

(i) *If $u, v, w \in \Lambda$, then $u \times v \times w \in \Lambda$.*

(ii) *If $U, v \in \Lambda$ and $w \in \Lambda^\perp$, then $u \times v \times w \in \Lambda^\perp$.*

(iii) *if $u \in \Lambda$ and $v, w \in \Lambda^\perp$, then $u \times v \times w \in \Lambda$.*

(iv) *If $u, v, w \in \Lambda^\perp$, then $u \times v \times w \in \Lambda^\perp$.*

(v) *If $u, v, w \in \Lambda$ and $x \in \Lambda^\perp$, then $\Phi(x, u, v, w) = 0$.*

(vi) *If $\{x, u, v, w\}$ is an orthonormal basis of Λ, then $\Phi(x, u, v, w) = \pm 1$.*

A 4-dimensional subspace that satisfies these equivalent conditions is called a Cayley subspace of W. If $u, v, w \in W$ are linearly independent, then
$$\Lambda := \mathrm{span}\{u, v, w, u \times v \times w\}$$
is a Cayley subspace of W.

One can also give an intrinsic characterization of Cayley calibrations in terms of so-called Cayley-forms.

DEFINITION 2.5.4. *A 4-form $\Phi \in \Lambda^4(W^*)$ is nondegenerate if, for every triple u, v, w of linearly independent vectors in W, there is a vector $x \in W$ such that $\Phi(u, v, w, x) \neq 0$. An inner product on W is called compatible with Φ if the map*
$$W^3 \longrightarrow W : (u, v, w) \mapsto u \times v \times w, \quad \langle x, u \times v \times w \rangle := \Phi(x, u, v, w)$$
is a triple cross product; Φ is a Cayley-form if it admits a compatible inner product.

PROPOSITION 2.5.2. *The standard Cayley-form Φ_0 on $W = \mathbb{R}^8 = \mathbb{O}$ with coordinates $x_0, x_1, \ldots, x_7$ is given by*

(2.5.6)
$$\begin{aligned}\Phi_0 = {} & e^{0123} - e^{0145} - e^{0167} - e^{02456} + e^{0257} - e^{0347} - e^{0356} \\ & + e^{4567} - e^{2367} - e^{2345} - e^{1357} + e^{1346} - e^{1256} - e^{1247}.\end{aligned}$$

It is compatible with the standard inner product and induces the standard triple cross product on $\mathbb{R}^8$. Note that $\Phi_0 \wedge \Phi_0 = 14\omega_0$.

REMARK 2.5.1. The constructions of the algebra $\mathbb{O}$ given in the last proposition and in Section 2.1 give different natural bases in $\mathbb{O}$; more precisely, the basis $(e'_1, e'_2, \ldots, e'_8)$ in the proposition is given by
$$1 = e'_1 = e_0, \ e'_2 = e_1, \ e'_3 = e_2, \ e'_4 = e_4, \ e'_5 = e_3, \ e'_6 = e_7, \ e'_7 = e_5, \ e'_8 = -e_6$$

in terms of the basis $(e_0, e_1, \ldots, e_7)$ of Section 2.1. However, we use the basis $(e_0, e_1, \ldots, e_7)$ in Chapters 4 and 7; it implies only some changes of signs; cf. for instance the formula pairs (2.4.5) and (7.1.5), (2.4.6) and (7.1.7), and (2.5.6) and (7.1.8).

2.5.3. The automorphism group. First one notes that all Cayley-forms are essentially equivalent under the automorphism group of W (i.e., G_2):

THEOREM 2.5.5. *If $\Phi, \Psi \in \Lambda^4(W^*)$ are two Cayley-forms, then there is an automorphism $g \in Aut(W)$ such that either $g^*\Phi = \Psi$ or $g^*\Phi = -\Psi$.*

Let us turn then to the automorphism group of a Cayley-form Φ corresponding to an 8-dimensional real Hilbert space W equipped with a positive triple cross product. Let W be oriented so that $\Phi \wedge \Phi > 0$; then Φ is self-dual, relative to the associated Hodge operator. Further, the orientation of W is compatible with the decomposition $W = \langle e \rangle \oplus V_e = \langle e \rangle \oplus e^\perp$ for any unit $u \in W$. Let $G(W, \Phi)$ be the group of automorphisms of Φ:

$$G(W, \Phi) := \{g \in GL(W) : g^*\Phi = \Phi\}.$$

As before one has

$$G(W, \Phi) := \{g \in SO(W) : gu \times gv \times gw = g(u \times v \times w), \forall u, v, w \in W\},$$

and for the standard form Φ_0 on $\mathbb{R}^8$ the group is just

$$\mathrm{Spin}(7) = G(\mathbb{R}^8, \Phi_0).$$

The group $G(W, \Phi)$ is isomorphic to $\mathrm{Spin}(7)$ for any Cayley-form on an 8-dimensional vector space.

THEOREM 2.5.6. *The group $G(W, \Phi)$ is a 21-dimensional simple, connected, simply connected Lie group. It acts transitively on the unit tangent bundle of the unit sphere and, for every unit vector $e \in W$, the isotropy subgroup $G_e := \{g \in G(W, \Phi) : ge = e\}$ is isomorphic to G_2. Thus there is a fibration $G_2 \longrightarrow \mathrm{Spin}(7) \longrightarrow \mathbb{S}^7$.*

Let us examine now the action of the group $G(W, \Phi)$ on the space

$$\mathcal{S}' := \{(u, v, w, x) \in W : \{u, v, w, x, u \times v \times w\} \text{ are orthonormal}\}.$$

The space $\mathcal{S}'$ is a bundle of 3-spheres over a bundle of 5-spheres over a bundle of 6-spheres over a 7-sphere. Hence it is a compact connected simply connected 21-dimensional manifold.

THEOREM 2.5.7. *The group $G(W, \Phi)$ acts freely and transitively on $\mathcal{S}'$.*

PROOF. Since $\mathrm{Spin}(7)$ acts transitively on $\mathbb{S}^7$ with isotropy subgroup G_2, the result follows immediately from Theorem 2.4.2. □

COROLLARY 2.5.1. *The group $G(W, \Phi)$ acts transitively on the space of Cayley subspaces of W.*

CHAPTER 3

Jordan Algebras and the Cartan Isoparametric Cubics

In the present chapter we give a review of Jordan algebras regarding their connections to Cartan's isoparametric cubics. In Section 3.1 we recall the definition of these cubics and their connection to the so-called eiconal equation, which is characteristic for the cubics (together with harmonicity). Section 3.2 is devoted to the definition and principal properties of Jordan algebras, including a brief historic introduction. Finally, Section 3.3 gives a more thorough study of the relation between the eiconal equation and cubic Jordan algebras. Most of the results given in the chapter are classical, but Theorem 3.1.1 is new and can be applied to classify the so-called radial eigencubics; see Chapter 6.

Notice that throughout the present chapter we use the notation $\mathbb{F}_d$ for the classical normed division algebra of real dimension $d = 1, 2, 4, 8$. Therefore, here $\mathbb{F}_1 = \mathbb{R}$, $\mathbb{F}_2 = \mathbb{C}$, $\mathbb{F}_4 = \mathbb{H}$, and $\mathbb{F}_8 = \mathbb{O}$.

3.1. Isoparametric cubics and the eiconal equation

We begin with the general Cartan isoparametric functions.

3.1.1. Isoparametric functions. According to È. Cartan, a smooth function $f : M \to \mathbb{R}$ on a Riemannian manifold M is called *isoparametric* if the squared norm of the gradient $|\nabla f|^2$ and the Laplacian $\Delta = \operatorname{div} \nabla f(x)$ are smooth functions of f.

A hypersurface of M is called *isoparametric* if it has constant principal curvatures.

If f is an isoparametric function, then any element of the family of regular level hypersurfaces $\{f^{-1}(c), c \in \mathbb{R}\}$, called an isoparametric foliation, is itself an isoparametric hypersurface. Conversely, any connected hypersurface of M with constant principal curvatures can be obtained as a level set of an isoparametric function.

The most interesting case is isoparametric hypersurfaces in the real space forms. Isoparametric hypersurfaces in Euclidean space $M = \mathbb{R}^n$ were classified by Levi-Civita [146] for $n = 3$ and Segre [220] for all n, in the late 1930s. At the same time, È. Cartan solved the problem in the hyperbolic case $M = H_n$. In both cases the number g of distinct principal curvatures is at most 2, and the hypersurfaces are essentially tubes over a totally geodesic subspace.

In the spherical case $M = \mathbb{S}^{n-1} \subset \mathbb{R}^n$, however, Cartan found the situation quite different [54], [55], [56]. The orbits of a group of isometries acting on the sphere obviously have constant principal curvatures if they happen to be hypersurfaces, and Cartan was able to construct such homogeneous examples with up to $g = 4$ distinct principal curvatures. He also established that all isoparametric hypersurfaces with $g \leq 3$ distinct principal curvatures are homogeneous and algebraic. If $g = 3$, Cartan showed that all the principal curvatures must have the same multiplicity $d = 1, 2, 4$, or 8 and that the isoparametric hypersurface must be a tube of constant radius over a standard Veronese embedding of a projective plane $\mathbb{F}_d P^2$ into the standard sphere $\mathbb{S}^{3d+1}$ ($\mathbb{F}_d$ denotes as mentioned above the real normed division algebra of dimension d). More explicitly, the defining polynomials of these hypersurfaces are given by

$$C_d(x) = x_{3d+2}^3 + \frac{3}{2}x_{3d+2}(|z_1|^2 + |z_2|^2 - 2|z_3|^2 - 2x_{3d+1}^2)$$
(3.1.1)
$$+ \frac{3\sqrt{3}}{2}x_{3d+1}(|z_2|^2 - |z_1|^2) + 3\sqrt{3}\,\mathrm{Re}(z_1 z_2 z_3),$$

where $z_k = (x_{kd-d+1}, \ldots, x_{kd}) \in \mathbb{R}^d = \mathbb{F}_d$. The real part in (3.1.1) is well-defined even in the nonassociative case of the algebra of octonions; see Section 2.3.

We shall refer to (3.1.1) as to a *Cartan isoparametric cubic*. Any Cartan isoparametric cubic u satisfies the so-called *Cartan-Münzner system*

(3.1.2) $\qquad |\nabla u(x)|^2 = 9|x|^4, \qquad \Delta u(x) = 0,$

and, conversely, any irreducible cubic solution $u(x)$ of the above system is congruent to some $C_d(x)$; see [**250**]. In general, a celebrated result due to H. F. Münzner [**170**], [**171**] asserts that any isoparametric hypersurface with g distinct principal curvatures in the standard sphere arises as a solution of the system

(3.1.3) $\qquad |\nabla u(x)|^2 = g^2|x|^{2g-2}, \qquad \Delta u(x) = \frac{m_2 - m_1}{2} g^2 |x|^{g-2}$

(which obviously is equivalent to (3.1.2) for $g = 3$ and $m_1 = m_2 = d$) and the number g of distinct principal curvatures must be $1, 2, 3, 4$, or 6. If a homogeneous degree g polynomial $u(x)$ satisfies (3.1.3) with $m_2 \geq m_1 \geq 1$, then any level set $f^{-1}(c)$, $c \in (-1,1)$, is an isoparametric hypersurface in $\mathbb{S}^{n-1} \subset \mathbb{R}^n$ with g distinct principal curvatures, and m_1 and m_2 are the multiplicities of the maximal and minimal principal curvatures of M ($m_1 = m_2$ when g is odd). The converse is also true. We refer the interested reader to [**58**], [**245**], and [**67**] for a detailed account concerning $g = 4$ and $g = 6$. See also Section 5.3. In the present chapter we are mostly interested in the cubic case, $g = 3$.

The theory of Riemannian isoparametric submanifolds has a natural generalization to the pseudo-Riemannian (sub)manifolds which has been suggested by Nomizu [**190**], Magid [**152**], and Hahn [**100**] and then considered by many authors; see for instance [**57**] and the references therein. Despite many similarities between the Riemannian and pseudo-Riemannian cases, there are also many differences. In particular, even in the simplest case of the pseudo-Euclidean space $\mathbb{R}_p^n$, there are isoparametric hypersurfaces which are not level sets of isoparametric functions.

3.1.2. Transnormal functions and the eiconal equation. On the other hand, the gradient equation in (3.1.2) alone turned out to have many applications in analysis and geometry. We abuse the terminology and simply call it the *eiconal equation*. As we shall see later, the cubic polynomial solutions of the eiconal equation have remarkable connections with Jordan algebras. We begin by fixing some notation.

A function $f: M \to \mathbb{R}$ is called *transnormal* on a complete Riemannian manifold M if it satisfies the eiconal type equation

$$|\nabla f(x)|^2 = b(x), \tag{3.1.4}$$

where $b: M \to \mathbb{R}$ is a smooth function. It is seen from the definition that any isoparametric function is transnormal. On the other hand, a striking result of Wang [**267**] (see also [**166**]) states that the *level hypersurfaces* of any transnormal function on $M = \mathbb{R}^n$ or $M = \mathbb{S}^n$ are isoparametric. Notice that, however, this result does not hold anymore for the hyperbolic space $M = H_n$.

Geometrically, the definition of a transnormal function means that the level hypersurfaces $f^{-1}(c)$ for regular c are parallel in M, which originates from geometrical optics and wave propagation [**213**], [**24**]. Transnormal functions appear also in many contexts, e.g., in the theory of harmonic morphisms [**20**], entire solutions of the eiconal equation [**149**], transnormal foliations [**210**], [**26**], and classification of cubic minimal cones [**249**].

Now let us return to the eiconal equation

$$|\nabla u(x)|^2 = 9|x|^4, \quad x \in \mathbb{R}^n. \tag{3.1.5}$$

If $u(x)$ is a cubic solution of (3.1.5), then its restriction $f(x) = u(x)|_{\mathbb{S}^{n-1}}$ on the unit sphere is a transnormal function. By homogeneity, the covariant gradient ∇f can then be found as the tangent projection of the gradient of u in $\mathbb{R}^n$,

$$\nabla f(x) = \overline{\nabla} u(x) - \partial_x u \cdot x = \overline{\nabla} u(x) - 3u(x)x, \quad x \in \mathbb{S}^{n-1}.$$

Since $|x| = 1$ on $\mathbb{S}^{n-1}$, we have by (3.1.5)

$$|\nabla f(x)|^2 = |\overline{\nabla} u(x)|^2 - 9u^2(x) = 9(|x|^4 - f^2(x)) \equiv 9(1 - f^2(x)),$$

which shows that f satisfies (3.1.4) with $b(f) = g^2(1 - f^2)$.

In particular, Wang's result [**267**] yields that level sets of $f(x)$ are isoparametric submanifolds of the unit sphere (observe that we did not require u to be harmonic). In [**250**] it is established that any cubic solution of the eiconal equation (up to an orthogonal transformation in $\mathbb{R}^n$) is one of the following: either a Cartan isoparametric cubic (3.1.1) or a reducible cubic form

$$r_n(x) = x_n^3 - 3x_n(x_1^2 + \cdots + x_{n-1}^2), \quad x \in \mathbb{R}^n, n \geq 2. \tag{3.1.6}$$

On the other hand, it is well known that the Cartan cubic $C_d(x)$ can be represented by the determinant of a trace free element in a formally real Jordan algebra of 3×3 Hermitian matrices with entries in the division algebra $\mathbb{F}_d$. A closer look at the reducible solution $r_n(x)$ reveals that it also satisfies a similar determinant representation for another Jordan algebra, the so-called reduced cubic factor, which is a direct sum of the algebra of reals and a spin factor. By the celebrated Jordan-von Neumann-Wigner classification, these two classes of Jordan algebras are the only possible formal real Jordan algebras of rank *three*.

This observation suggests characterizing explicitly the correspondence between cubic solutions of (3.1.5) and formally real cubic Jordan algebras. As we shall see later, such a correspondence really does exist even in a more general setting. We outline the main idea of our construction below and then discuss it in more detail in the remaining part of this chapter.

We start with an intrinsic examination of the eiconal equation. Let us consider an arbitrary quadratic space, i.e., a pair (W, Q), where W is a vector space over the field $\mathbb{F}$ and $Q: W \to \mathbb{F}$ is a quadratic form on W which will always be assumed to be nonsingular. Recall that Q is called nonsingular if the kernel of its associated bilinear form

$$Q(x;y) = \frac{1}{2}(Q(x+y) - Q(x) - Q(y)), \quad Q(x;x) = Q(x),$$

is $\{0\}$. Two evident examples of quadratic spaces are the Euclidean space $\mathbb{R}^n$ with $Q(x) = |x|^2$ and the Minkowski space $\mathbb{R}_1^n$ with $Q(x) = M(x) := x_n^2 - \sum_{i=1}^{n-1} x_i^2$.

The bilinear form $Q(x;y)$ is symmetric and induces an inner product on W. Now, considering (W, Q) as a (pseudo-)Riemannian manifold equipped with the metric tensor Q, one can define as usual the notion of the covariant derivative. In particular, given a smooth function $u : W \to \mathbb{F}$ on a nonsingular quadratic space (W, Q), one defines the covariant gradient $\nabla u(x)$ to be the unique vector field satisfying

$$(3.1.7) \qquad Q(\nabla u(x); y) = \partial_y u|_x \equiv \lim_{t \to 0} \frac{u(x+ty) - u(x)}{t}, \qquad \forall y \in W.$$

The uniqueness obviously follows from the nonsingularity of Q.

DEFINITION 3.1.1. Given a nonsingular quadratic space (W, Q) over a field $\mathbb{F}$, a cubic form $u : W \to \mathbb{F}$ is said to be a *cubic eiconal* on (W, Q) if

$$(3.1.8) \qquad Q(\nabla u(x)) = 9Q(x)^2, \quad x \in W,$$

and the latter is said to be the (cubic) eiconal equation in (W, Q).

REMARK 3.1.1. Note that if $u(x)$ is a cubic eiconal on (W, Q), then $f(x) = u(x)|_{\mathbb{S}_Q(W)}$ is a transnormal function on the unit Q-sphere $\{x \in W : Q(x) = 1\}$.

Two cubic forms $u(x)$ and $\widetilde{u}(\widetilde{x})$ on quadratic spaces (W, Q) and $(\widetilde{W}, \widetilde{Q})$, respectively, are called *congruent* if there exists an isometry $O : (W, Q) \to (\widetilde{W}, \widetilde{Q})$ (i.e., O is invertible and $\widetilde{Q}(Ox) = Q(x)$ for all $x \in W$) such that $u(x) = \widetilde{u}(Ox)$. It is easily derived from (3.1.8) that a cubic form congruent to a cubic eiconal is a cubic eiconal too.

Now we are ready to formulate the main new result of this chapter.

THEOREM 3.1.1. *There is a one-to-one correspondence between the congruence classes of cubic eiconals on quadratic spaces, on one hand, and the isomorphic classes of cubic Jordan algebras, on the other hand.*

In this correspondence, any cubic eiconal $u(x)$ on (W, Q) gives rise to a cubic Jordan algebra $J(u)$ on $\mathbb{F} \times W$ with unit $e = (1, 0)$, the generic norm

$$N_u(\mathbf{x}) = x_0^3 - \frac{3}{2} x_0 Q(x) + \frac{1}{\sqrt{2}} u(x), \quad \mathbf{x} = (x_0, x) \in J(u),$$

and the trace bilinear form T such that $T|_{e^\perp} = Q$, and the eiconal is recovered from $J(u)$ by the following determinant-like representation:

$$(3.1.9) \qquad u(x) = \sqrt{2} \, N|_{e^\perp}(x),$$

where $e^\perp = \{\mathbf{x} \in J(u) : T(\mathbf{x}; e) = 0\}$ is the trace free subspace of $J(u)$. In the converse direction, any cubic Jordan algebra J gives rise to a cubic eiconal u defined by virtue of (3.1.9) on the quadratic space (W, Q), where $W = e^\perp$ and $Q = T|_{e^\perp}$.

This correspondence is natural in the sense that two cubic Jordan algebras are isomorphic if and only if the associated cubic eiconals are congruent.

We explain the above terminology in more detail when we review the basic Jordan algebra theory in Section 3.2 below. Then we shall prove Theorem 3.1.1 in Section 3.3 and discuss some applications in Section 5.1.

REMARK 3.1.2. It is also convenient to provide a "local" version of the equivariant eiconal equation. Let (W, Q) be a nonsingular quadratic space and let $e_1, \ldots, e_k$, $k = \dim W$, be a basis of W and let $x_i = Q(x, e_i)$ be the associated local coordinates of $x \in W$. Let $Q_{ij} = Q(e_i; e_j)$ denote the matrix representation of Q in the local coordinates and let Q^{ij} denote its inverse. Decomposing the gradient in the basis as $\nabla u(x) = \sum_{i=1}^{k} u_i e_i$, we obtain from (3.1.7)

$$u_i = \sum_{j=1}^{k} Q^{ij} \frac{\partial u(x)}{\partial x_j};$$

thus (3.1.8) takes the following form in the local coordinates:

$$\sum_{i,j=1}^{k} Q^{ij} \frac{\partial u(x)}{\partial x_i} \frac{\partial u(x)}{\partial x_j} = 9 \left(\sum_{i,j=1}^{k} Q_{ij} x_i x_j \right)^2.$$

Observe that one obtains the Euclidean eiconal equation if $Q_{ij} = \delta_{ij}$, the Kronecker symbol.

3.2. Jordan algebras

3.2.1. Historical remarks. Introduced at the beginning of the 1930s in an attempt to discover a nonassociative algebraic setting for quantum mechanics, Jordan algebras found later applications in many different areas of mathematics, spanning from Lie algebras and group theory to real and complex differential geometry. The usual interpretation of quantum mechanics makes use of the concept of an *observable* which is essentially a Hermitian matrix (or a Hermitian operator on Hilbert space), and the physically significant information is encoded by the spectrum of the observable. On the other hand, the underlying algebraic structure is not completely satisfactory because it is not compatible with the Hermitian structure. Indeed, the scalar multiple λx is not again Hermitian unless the scalar λ is real, and the matrix product xy is non-Hermitian unless x and y commute.

The program proposed by Jordan [**123**] was to find a new algebraic setting for quantum mechanics, in particular, to capture intrinsic algebraic properties of Hermitian matrices, and then to see what other possible nonmatrix systems satisfied these axioms. By linearizing the quadratic squaring operation, Jordan replaced the usual matrix multiplication by the *anticommutator* product (also called the *Jordan product*)

(3.2.1) $$x \bullet y = \frac{1}{2}(xy + yx)$$

satisfying the following two identities: the commutativity condition

(3.2.2) $$x \bullet y = y \bullet x$$

and the *Jordan identity*

(3.2.3) $$x^{\bullet 2} \bullet (x \bullet y) = x \bullet (x^{\bullet 2} \bullet y),$$

where $x^{\bullet 2} := x \bullet x$. The second identity can be thought of as a substitute for the associative law. The new bullet product was no longer associative but instead commutative.

According to Jordan, any algebra satisfying (3.2.2)–(3.2.3) and which is additionally *formally real* (i.e., where $x_1^{\bullet 2} + \cdots + x_k^{\bullet 2} = 0$ implies $x_1 = \cdots = x_k = 0$) was called an *r-system*. The name *"Jordan algebras"* was given later by A. A. Albert.

In the 1934 fundamental paper [**124**], Pascual Jordan, John von Neumann, and Eugene Wigner classified all finite-dimensional r-systems and showed that any such system is a direct sum of five simple building blocks, namely,

(1) the symmetric matrices $\mathrm{Sym}_n(\mathbb{R})$, $n \geq 1$,
(2) the Hermitian matrices over the complexes, $\mathrm{Herm}_n(\mathbb{C})$, $n \geq 3$,
(3) the Hermitian matrices over quaternions, $\mathrm{Herm}_n(\mathbb{H})$, $n \geq 3$,
(4) the Albert algebra of Hermitian matrices over octonions, $\mathrm{Herm}_3(\mathbb{O})$,
(5) the spin factors.

In cases (1)–(4), the multiplication is given by (3.2.1). See Examples 3.2.3 and 3.2.5 below for the above notation.

3.2.2. Definitions and basic properties. For a general panorama of Jordan algebra theory the reader is referred to the excellent books [**121**] and [**158**] and to [**85**] and [**130**] for Euclidean Jordan algebras especially. A comprehensive account of the modern theory and its applications can be found in [**115**].

Let us write $\mathbb{F}$ for $\mathbb{F}_d$ with $d = 1$ or $d = 2$, i.e., in the commutative case.

DEFINITION 3.2.1. A vector space V over a field $\mathbb{F}$ is called a (linear) Jordan algebra if there exists an $\mathbb{F}$-bilinear map $\bullet : V \times V \to V$ satisfying (3.2.2) and (3.2.3).

We shall assume that every Jordan algebra is finite dimensional and has an identity element, denoted by e. Let us denote by $L(x)$ the multiplication operator by x given by

$$L(x)y = x \bullet y, \qquad x, y \in V,$$

which is obviously an endomorphism of V. Then the property (3.2.3) says that for every element $x \in V$ the endomorphisms $L(x)$ and $L(x^{\bullet 2})$ commute:

(3.2.4) $$[L(x), L(x^{\bullet 2})] = 0.$$

Defining recursively $x^{\bullet k} = x \bullet x^{\bullet k-1}$, one can show that (3.2.4) implies that

$$[L(x^{\bullet p}), L(x^{\bullet q})] = 0, \qquad \forall p, q \geq 0,$$

which implies that the subalgebra $\mathbb{F}[x]$ generated by a single element $x \in V$ is associative or, equivalently,

$$x^{\bullet p} \bullet x^{\bullet q} = x^{\bullet(p+q)}$$

for any $x \in V$ and all $p, q \geq 1$; see, for instance, [**130**, Chapter 3]. We shall also use the following corollary of (3.2.4):

(3.2.5) $$L(x^{\bullet 3}) = 3L(x^{\bullet 2})L(x) - 2L(x)^3;$$

see, for instance, [**85**, p. 27].

Since V is finite dimensional, one can define the order $\nu(x)$ of any $x \in V$ as the minimal $k > 0$ such that the system $(e, x, \ldots, x^k)$ is linearly independent. Observe that $\nu(x) \leq \dim_{\mathbb{F}} V$. The number
$$r := \max\{\nu(x) : x \in V\}$$
is called the *rank* of the Jordan algebra V. An element x is said to be *regular* if $\nu(x) = r$. For a regular $x \in V$, the elements $e, x, \ldots, x^{\bullet(r-1)}$ are linearly independent, so that
$$x^r = \sigma_1(x) x^{r-1} - \sigma_2(x) x^{r-2} + \cdots + (-1)^{r-1} \sigma_r(x) e,$$
where the $\sigma_i(x) \in \mathbb{F}$ are uniquely defined by x. The polynomial
$$(3.2.6) \qquad m_x(\lambda) = \lambda^r - \sigma_1(x)\lambda^{r-1} + \cdots + (-1)^r \sigma_r(x) \in \mathbb{F}[\lambda]$$
is called the *minimum polynomial* of the regular element $x \in V$ [120]. In fact, it can be readily shown that the coefficients $\sigma_i(x)$ actually are polynomial functions of x homogeneous of degree i in the following sense. If $\{e_i\}_{1 \leq i \leq n}$ is an arbitrary basis of V and
$$x_\xi := \sum_{i=1}^n \xi_i e_i, \qquad \xi \in \mathbb{F}^n,$$
is a generic element of the algebra V, then
$$x_\xi^r - \sigma_1(\xi) x_\xi^{r-1} + \cdots + (-1)^r \sigma_r(\xi) e = 0,$$
where $\sigma(\xi)$ is a polynomial function of x homogeneous of degree i.

DEFINITION 3.2.2. The coefficient $\sigma_1(x) = \operatorname{Tr} x$ is called the *generic trace* of x, and $\sigma_n(x) = N(x)$ is called the *generic norm* (or *generic determinant*) of x.

For any regular element $x \in V$, the generic norm and the minimum polynomial are related by the following determinant representation:
$$N(te - x) = m_x(t).$$

The determinant on an associative linear algebra A is a multiplicative map. The generic norm N of a Jordan algebra V is nonmultiplicative in general, but it is multiplicative on any ring $\mathbb{F}[x] \subset V$ generated by a single element $x \in V$, i.e.,
$$N(u \bullet v) = N(u) N(v), \qquad \forall u, v \in \mathbb{F}[x];$$
see, for instance, Proposition II.2.2 in [85].

An element $x \in V$ is said to be *invertible* if there exists $y \in \mathbb{F}[x]$ such that $x \bullet y = e$. Since $\mathbb{F}[x]$ is associative, the element y is unique. It is called the *inverse* element of x, denoted by x^{-1}. It is easily seen that if the multiplication operator $L(x)$ is invertible as an endomorphism of V, then x is invertible too and $x^{-1} = L(x)^{-1} e$.

REMARK 3.2.1. The following example [85, p. 30] shows that the equation $x \bullet y = e$ does not imply in general that y is the inverse of x. Let V be the Jordan algebra $\operatorname{Sym}_2(\mathbb{R}, \bullet)$ and let
$$x = \begin{pmatrix} 1 & 0 \\ 0 & -1 \end{pmatrix}, \quad y = \begin{pmatrix} 1 & a \\ a & -1 \end{pmatrix}.$$

Then $x \bullet y = \frac{1}{2}(xy + yx) = e$, the unit matrix, but $y \notin \mathbb{R}[x]$ if $a \neq 0$.

In general, one has the following criterion [**85**, p. 31], which is reminiscent of the corresponding matrix invertibility criterion: An element $x \in V$ is invertible if and only if $N(x) \neq 0$, and the inverse element, if it exists, is given by

$$(3.2.7) \quad x^{-1} = (-1)^{r-1} N(x)^{-1} \left(x^{r-1} - \sigma_1(x) x^{r-2} + \cdots + (-1)^{r-1} \sigma_{r-1}(x) e \right).$$

The isomorphism of two algebras is defined as usual. We shall exploit the following simple criterion of isomorphy.

PROPOSITION 3.2.1. $\phi : (J, \bullet) \to (\widetilde{J}, \circ)$ *is an isomorphism of the Jordan algebras if and only if ϕ is an invertible linear map satisfying*

$$(3.2.8) \qquad\qquad\qquad \phi(x^{\bullet 2}) = \phi(x)^{\circ 2}.$$

PROOF. It suffices to prove only the "if" part. To this end we suppose that ϕ is invertible and satisfies (3.2.8). Then polarizing this identity, one finds

$$\begin{aligned} 2\phi(x \bullet y) &= \phi((x+y)^{\bullet 2}) - \phi(x^{\bullet 2}) - \phi(y^{\bullet 2}) \\ &= \phi(x+y)^{\circ 2} - \phi(x)^{\circ 2} - \phi(y)^{\circ 2} \\ &= 2\phi(x) \circ \phi(y), \end{aligned}$$

which yields the desired conclusion. □

3.2.3. Examples of Jordan algebras.

EXAMPLE 3.2.1 (Special Jordan algebras). Let us consider an associative (not necessarily commutative) algebra A over the ground field $\mathbb{F}$ with the product $xy : A \times A \to A$. We define a new algebra A^+ whose vector space is the same as that of A and whose multiplicative structure is defined by the product

$$(3.2.9) \qquad\qquad\qquad x \bullet y = \frac{1}{2}(xy + yx).$$

It is easily verified that the (commutative) bullet product satisfies the Jordan identity. Indeed, $x^{\bullet 2} = x^2$ and by using associativity in A, one obtains

$$4L(x)L(x^{\bullet 2})y = x^3 y + x^2 y x + x y x^2 + y x^3 = 4L(x^{\bullet 2})L(x)y.$$

The Jordan algebra obtained from A by forgetting the associative structure is denoted by $A^+ = (A, \bullet)$.

DEFINITION 3.2.3. A Jordan algebra V is called special if it can be realized as a Jordan subalgebra of some A^+; otherwise it is exceptional.

EXAMPLE 3.2.2 (Jordan matrix algebras). The simplest special Jordan algebra is the vector space of $n \times n$ matrices with entries in $\mathbb{F}$, denoted by $M_n(\mathbb{F})$, with the Jordan product defined by (3.2.9). The minimum polynomial of $x \in M_n(\mathbb{F})$ is the characteristic polynomial of the matrix x. In particular, $M_n(\mathbb{F})$ has rank $r = n$ and the trace and determinant of x are given by $\mathrm{Tr}(x) = \mathrm{tr}\, x$ and $N(x) = \det x$, respectively. The inverse element of x coincides with the inverse matrix x^{-1}.

EXAMPLE 3.2.3 (Hermitian Jordan algebras). The most important examples of special Jordan subalgebras are the algebras of Hermitian or skew elements of an associative algebra with involution. Let A be an associative algebra over $\mathbb{F}$ with an involution $* : A \to A$, and let $\mathcal{H}(A) = \{x \in A : x^* = x\}$ denote the subspace of the Hermitian (selfadjoint) elements equipped with the Jordan product (3.2.9). It is readily verified that $\mathcal{H}(A)$ is a Jordan algebra. Observe that $\mathrm{Herm}_n(\mathbb{F}_d) = \mathcal{H}(M_n(\mathbb{F}_d))$, $d = 1, 2, 4$.

EXAMPLE 3.2.4 (The spin factor of a quadratic form). Let (V, Q) be a quadratic vector space over $\mathbb{F}$, and let
$$Q(x;y) = \frac{1}{2}(Q(x+y) - Q(x) - Q(y))$$
be a polarization of Q. Suppose that Q possesses a basepoint $c \in V$, i.e., $Q(c) = 1$. Then one can turn (V, Q) into a Jordan algebra by defining the product on V as
$$(3.2.10) \qquad x \bullet y = Q(x;c)y + Q(y;c)x - Q(x;y)c.$$

One has $x \bullet c = Q(c;c)x = x$; hence c is the unit element. It is straightforward to verify that the bullet product does satisfy the Jordan identity. The Jordan algebra obtained in this way is denoted by $\mathcal{S}(V, Q, c)$ and is called a *spin factor* (or quadratic factor).

It follows from the definition that
$$(3.2.11) \qquad x^{\bullet 2} = 2Q(x;c)x - Q(x)c;$$
thus if $\dim_\mathbb{F} V \geq 2$, then $\mathcal{S}(V, Q, c)$ is a rank 2 Jordan algebra. Moreover, (3.2.11) yields the explicit expressions for the generic trace and the generic norm:
$$\mathrm{Tr}(x) = 2Q(x;c) \quad \text{and} \quad N(x) = Q(x).$$
It also follows from (3.2.7) that the inverse element is given by
$$x^{-1} = \frac{2Q(x;c)e - x}{Q(x)}.$$

EXAMPLE 3.2.5 (The Minkowski spin factor). An important example of the spin factor is that associated with the Minkowski space $V = \mathbb{R}^n_1 = \mathbb{R}^{n-1} \oplus \mathbb{R}$ with the quadratic form
$$(3.2.12) \qquad M(x) = x_n^2 - x_1^2 - \cdots - x_{n-1}^2 \equiv x_n^2 - |\bar{x}|^2, \quad \bar{x} = (x_1, \ldots, x_{n-1}),$$
and obvious basepoint $e = (\bar{0}, 1)$. Then Example 3.2.4 yields the rank 2 *formally real* Jordan algebra $\mathcal{S}(\mathbb{R}^n_1, M, e)$. Indeed, by (3.2.10) the Jordan structure on this algebra takes the form
$$x \bullet y = (x_0 \bar{y} + y_0 \bar{x}, \ x_0 y_0 + \langle \bar{x}, \bar{y} \rangle), \quad x = (\bar{x}, x_n), \ y = (\bar{y}, y_n),$$
where $\langle \bar{x}, \bar{y} \rangle = x_1 y_1 + \cdots + x_{n-1} y_{n-1}$ is the standard inner product in $\mathbb{R}^{n-1}$, which shows that the bilinear trace form
$$\mathrm{Tr}(x \bullet y; e) = x_0 y_0 + \langle \bar{x}, \bar{y} \rangle = \sum_{i=0}^{n} x_i y_i$$
is positively definite. One also has for the generic trace and the generic norm on $\mathcal{S}(\mathbb{R}^n_1, M, e)$
$$(3.2.13) \qquad \mathrm{Tr}(x) = 2x_0 \quad \text{and} \quad N(x) = x_0^2 - \langle \bar{x}, \bar{x} \rangle.$$

EXAMPLE 3.2.6 (The exceptional Albert algebra). In [5] A. A. Albert published his study of the Jordan algebra $\mathbb{A} = \mathrm{Herm}_n(\mathbb{F}_8)$ consisting of the Hermitian 3×3 matrices with entries in the real algebra of octonions. Albert established that this 27-dimensional rank 3 Jordan algebra cannot be imbedded in any associative algebra; hence it is exceptional. Being too small to produce a satisfactory space for quantum mechanics, the Albert exceptional algebra $\mathbb{A}$ turned out to have many surprising and important connections with diverse branches of mathematics and physics from the Jordan-Freudenthal construction of the projective octonionic plane, exceptional Lie algebras, and the Freudenthal-Tits magic square to supersymmetry and supergravity theories.

3.2.4. The quadratic representation and semisimple Jordan algebras.
The map $P: V \to V$ defined by

$$P(x) := 2L^2(x) - L(x^{\bullet 2}), \qquad x \in V,$$

is called the *quadratic representation* of V. If V is the special algebra A^+ of an associative algebra A, the quadratic representation P is written in the associative product by $P(x)y = xyx$. For a general Jordan algebra V and any $x, y \in V$ the following fundamental formula holds:

$$P(P(x)y) = P(x)P(y)P(x).$$

Furthermore, $x \in V$ is invertible if and only if $P(x)$ is invertible, and in this case

$$P(x^{-1}) = P(x)^{-1} \quad \text{and} \quad P(x)x^{-1} = x.$$

Any Jordan algebra possesses a natural inner product defined by virtue of the trace form as follows:

(3.2.14) $$\tau(x; y) := \operatorname{tr} L(x \bullet y),$$

where tr is the usual trace of L as an endomorphism of V. The inner product is symmetric, $\tau(x; y) = \tau(y; x)$, and polarizing the Jordan identity one can also prove that τ satisfies a remarkable associative property, i.e.,

$$\tau(x \bullet y; z) = \tau(x; y \bullet z).$$

A Jordan algebra V is called *semisimple* if τ is nonsingular, i.e., $\tau(x; y) = 0$ for some $x \in V$ and all $y \in V$ implies $x = 0$. A Jordan algebra V is called *simple* if V is semisimple and there is no direct sum decomposition of V with two nontrivial subalgebras.

One has the following fundamental result; see for instance [130, p. 65].

THEOREM 3.2.1. *Every semisimple Jordan algebra V is a direct sum of simple Jordan algebras, and two decompositions of a semisimple Jordan algebra into direct sums of simple Jordan algebras are equal up to a permutation of the summands.*

In a simple Jordan algebra, the quadratic representation P, the generic norm N, and the trace form τ of (3.2.14) are related by virtue of the following relations:

$$\det P(x) = N(x)^{2n/r},$$

$$\tau(x; y) = \frac{n}{r} \operatorname{Tr}(x \bullet y),$$

where $n = \dim V$, r is the rank of V, and det stands for the usual determinant of $P(x)$ as a linear endomorphism of V; see for instance [85, p. 52].

3.2.5. The Euclidean Jordan algebras.
A Jordan algebra V over $\mathbb{F} = \mathbb{R}$ is said to be *Euclidean* if there exists a positive definite symmetric bilinear form on V which is associative. In other words, there exists an inner product b on V such that the multiplication operator $L(x)$ is selfadjoint with respect to b: $b(L(x)y; z) = b(y; L(x)z)$.

It can be shown that for a Jordan algebra V over $\mathbb{R}$ the following conditions are equivalent (see, for instance, pp. 46 and 153 in [**85**]):

- V is Euclidean.
- V is formally real, i.e., $x^{\bullet 2} + y^{\bullet 2} = 0$ implies $x = y = 0$.
- The bilinear form $\mathrm{Tr}(x \bullet y)$ is positive definite.
- The bilinear form $\mathrm{tr}\, L(x \bullet y)$ is positive definite.

The Hermitian matrix algebras $\mathrm{Herm}_k(\mathbb{F}_d)$ are an obvious example of Euclidean Jordan algebras for all admissible pairs (k,d). Another example is the spin factor $S(V, Q, c)$ with $Q(x)$ such that $Q^*(x; y) := 2Q(x; c)Q(y; c) - Q(x; y)$ is positively definite. Indeed, we find by (3.2.11) and (3.2.13) that

$$\mathrm{Tr}(x \bullet y) = Q(x; c)\mathrm{Tr}\, y + Q(y; c)\mathrm{Tr}\, x - 2Q(x; y) = 4Q(x; c)Q(y; c) - 2Q(x; y).$$

There is a remarkable relationship between formally real Jordan algebras, self-dual homogeneous cones, and symmetric upper half-planes in finite dimensions. Recall that an open cone Ω in a vector space V with the inner product $b(x; y)$ is called symmetric if the group of automorphisms of Ω acts on it transitively and Ω is self-dual, i.e., $\Omega = \Omega^* := \{y \in V : b(x; y) > 0, \forall x \in \overline{\Omega} \setminus 0\}$. Then the fundamental result by M. Koecher [**130**] and E. Vinberg [**262**], [**264**] states that the interior of the cone of squares in a Euclidean Jordan algebra is a symmetric cone, i.e., a homogeneous and self-dual open cone, and every symmetric cone is obtained in this way.

3.2.6. The Peirce decomposition. An element $c \in V$ is called an *idempotent* if $c^{\bullet 2} = c$. Two idempotents c_1 and c_2 are called orthogonal if $c_1 \bullet c_2 = 0$.

An idempotent c is called *primitive* if it is nonzero and cannot be written as the sum of two nonzero idempotents. A system of idempotents $\{c_1, \ldots, c_m\}$ is called a *Jordan frame* if each c_i is a primitive idempotent, $c_i \bullet c_j = 0$ if $i \neq j$, and $\sum_{i=1}^m c_i = e$.

If $c \in V$ is an idempotent, then $L(c^{\bullet k}) = L(c)$; hence by (3.2.5), $L(c) = 3L(c)^2 - 2L(c)^3$, which yields

(3.2.15) $$L(c)(L(c) - \mathbf{1})(2L(c) - \mathbf{1}) = 0,$$

where $\mathbf{1} = L(e)$ is the identity map. This implies the so-called *Peirce decomposition* of V with respect to the idempotent c:

$$V = V_c(0) \oplus V_c(1/2) \oplus V_c(1),$$

where $V_c(\alpha)$ is the α-eigenspace of the operator $L(c)$, $\alpha \in \{0, 1/2, 1\}$. One can deduce from (3.2.15) that $V_c(\alpha) \bullet V_c(\alpha) \subset V_c(\alpha)$ for $\alpha = 0$ and $\alpha = 1$; hence $V_c(0)$ and $V_c(1)$ are Jordan algebras. Furthermore, one has

(3.2.16) $$\begin{aligned} V_c(1) \bullet V_c(0) &= \{0\}, \\ V_c(1/2) \bullet V_c(1/2) &\subset V_c(0) + V_c(1), \\ (V_c(0) + V_c(1)) \bullet V_c(1/2) &\subset V_c(1/2). \end{aligned}$$

More generally, if $e = \sum_{i=1}^m c_m$ is a Jordan frame, then one has the Peirce decomposition

$$V = \bigoplus_{i \leq j} V_{ij}$$

where $L(c_i)$ acts on V_{jk} as multiplication by 1 if $i = j = k$, multiplication by 0 if all i, j, k are pairwise distinct, and multiplication by $\frac{1}{2}$ otherwise. If additionally each c_i is primitive, then $\dim V_{ii} = 1$; see [**85**, p. 65].

3.2.7. Cubic Jordan algebras and the Springer construction. Any simple Jordan algebra of rank $r = 2$ is known to be isomorphic to a spin factor. In what follows, we shall mainly be interested in the class of Jordan algebras of rank $r = 3$, which are commonly called *cubic Jordan algebras*. An example of such an algebra is the Jordan algebra of symmetric 3×3 matrices over $\mathbb{R}$ with respect to the bullet product. Cubic Jordan algebras, and especially the 27-dimensional exceptional simple Jordan algebras known as Albert algebras, are very distinguished in many aspects. Cubic Jordan algebras have an important property; namely, their bullet product can be constructed starting with an appropriate cubic form which then appears as the generic norm $N(x)$ of the generated Jordan algebra. This construction of a Jordan algebra from a cubic form on a vector space is also called the *Springer construction*. Observe that only certain very special cubic forms are admissible for the Springer construction, and we shall see in the next sections that the admissible cubic forms N are exactly those arising as the cubic eiconals.

Let us consider the Springer construction in more detail, following [**158**, II.4.2]. Let V denote a finite-dimensional real vector space and let $u(x) \not\equiv 0$ be an arbitrary homogeneous cubic form on V. Since u is an odd function, the set $u^{-1}(1) = \{x \in V : u(c) = 1\}$ is obviously nonempty. Any $c \in u^{-1}(1)$ is called a basepoint.

Let us consider the decomposition

$$(3.2.17) \quad u(x + ty) = u(x) + 3tu(x; y) + 3t^2 u(y; x) + t^3 u(y), \qquad x, y \in V, \ t \in \mathbb{F},$$

where the first linearization

$$3u(x; y) = \partial_y u|_x$$

is the directional derivative of u in the direction y evaluated at x. Since $u(x; y)$ is a quadratic form with respect to the first variable, one can polarize it to obtain the full linearization of the cubic form u:

$$(3.2.18) \quad u(x; y; z) = \frac{1}{2} \left(u(x + y; z) - u(x; z) - u(y; z) \right).$$

Euler's homogeneous function theorem implies

$$(3.2.19) \quad u(x) = u(x; x) = u(x; x; x).$$

Now we fix some cubic form $N(x) \not\equiv 0$ on V and its basepoint $c \in N^{-1}(1)$. Linearizing N as above, one defines the *linear trace* form and the *quadratic spur* form to be

$$(3.2.20) \quad \mathrm{Tr}(x) = 3N(c; x) = 3N(c; c; x)$$

and

$$(3.2.21) \quad S(x) = 3N(x; c) = 3N(x; x; c),$$

respectively. Then

$$(3.2.22) \quad \mathrm{Tr}(c) = S(c) = 3N(c) = 3.$$

Polarizing the quadratic spur form

$$S(x; y) = 6N(x; y; c) \equiv S(x + y) - S(x) - S(y),$$

we define the bilinear *trace form*[1] by

(3.2.23) $$T(x;y) = \frac{1}{3}(\operatorname{Tr}(x)\operatorname{Tr}(y) - S(x;y)) \equiv -\partial_x \partial_y \left(\frac{1}{3}\log N\right)|_c.$$

DEFINITION 3.2.4. Let V be a finite-dimensional vector space over $\mathbb{F}$ and let $N: V \to \mathbb{R}$ be a cubic form with basepoint c. N is said to be a Jordan cubic form if the following are satisfied:
  (i) The trace bilinear form $T(x;y)$ is a nonsingular bilinear form.
  (ii) The quadratic sharp map $\#: V \to V$, defined uniquely by

(3.2.24) $$T(x^\#; y) = N(x; y),$$

satisfies the adjoint identity

(3.2.25) $$(x^\#)^\# = N(x)x.$$

EXAMPLE 3.2.7. Consider the Jordan algebra of symmetric 3×3 matrices $V = \operatorname{Sym}_3(\mathbb{R})$ with basepoint $c = \mathbf{1}$ being the unit matrix. Then the generic norm $N(x) = \det x$ and the generic trace $\operatorname{Tr}(x) = \operatorname{tr} x$. By virtue of the Newton identities, one has for the spur form

$$S(x) = \partial_c N|_x = (\det(x + t\mathbf{1}))'_{t=0} = \frac{1}{2}((\operatorname{tr} x)^2 - \operatorname{tr} x^2),$$

which yields $S(x;y) = \operatorname{tr} x \operatorname{tr} y - \operatorname{tr} x \bullet y$. Thus,

$$3T(x;y) = \operatorname{tr} x \bullet y \equiv \operatorname{tr} xy.$$

Observe that $T(x;y) = \frac{1}{3}\operatorname{tr} xy$ is a nonsingular bilinear form which yields (i) in Definition 3.2.4. Besides, if x is an invertible matrix, then

$$3N(x;y) = \partial N_y|_x = (\det(x+ty))'_{t=0} = (\det x \, \det(\mathbf{1} + tyx^{-1}))'_{t=0} = \det x \, \operatorname{tr}(yx^{-1}),$$

which by (3.2.24) yields that the adjoint map $x^\# = x^{-1}\det x$ is the adjoint matrix in the usual sense of linear algebra. Then $x^{\#\#} = x \det x$, which proves the adjoint identity (3.2.25) and thereby establishes (ii) in Definition 3.2.4.

The multiplicative Jordan structure on a vector space with a cubic Jordan form is given explicitly by the following construction.

PROPOSITION 3.2.2 (The Springer construction, [**158**, p. 77]). *To any admissible cubic $N: V \to \mathbb{F}$ with basepoint, one can associate a Jordan algebra $\mathcal{J}(V, N, c)$ of rank 3 with the unit c and the product determined by the formula*

(3.2.26) $$x \bullet y = \frac{1}{2}(x\#y + \operatorname{Tr}(x)y + \operatorname{Tr}(y)x - S(x;y)c),$$

where the Freudenthal product $\#$ is defined by $x\#y = (x+y)^\# - x^\# - y^\#$. All elements of $\mathcal{J}(V, N, c)$ satisfy the cubic characteristic identity

(3.2.27) $$x^{\bullet 3} - \operatorname{Tr}(x)x^{\bullet 2} + S(x)x - N(x)c = 0$$

and the sharp identity

(3.2.28) $$x^\# = x^{\bullet 2} - \operatorname{Tr}(x)x + S(x)c.$$

An element x is invertible iff $N(x) \neq 0$, and in this case $x^{-1} = N(x)^{-1}x$.

---

[1] We use the normalization factor $\frac{1}{3}$ in the bilinear form $T(x;y)$ to ensure that $T(c;c) = 1$.

We mention some useful consequences of the above construction. By (3.2.24) one obtains $T(x^\#; c) = N(x; c) = \frac{1}{3}S(x)$. Polarizing this identity one has

$$T(x\#y; c) = \frac{1}{3}S(x; y).$$

Furthermore, (3.2.23) and (3.2.22) yield

$$T(x; c) = \mathrm{Tr}(x) - \frac{1}{3}S(x; c) = \mathrm{Tr}(x) - 2N(x; c; c) = \frac{1}{3}\mathrm{Tr}(x);$$

hence

(3.2.29) $$\mathrm{Tr}(x\#y) = 3T(x\#y; c) = S(x; y).$$

Applying the trace to (3.2.28) and using the last identity one readily finds that

(3.2.30) $$S(x; y) = \mathrm{Tr}(x)\mathrm{Tr}(y) - \mathrm{Tr}(x \bullet y),$$

which in its turn yields

(3.2.31) $$\mathrm{Tr}(x \bullet y) = 3T(x; y).$$

EXAMPLE 3.2.8 (The reduced cubic factor). One can construct the so-called *reduced cubic factor* by adding the ground field $\mathbb{F}$ to a quadratic spin factor over $\mathbb{F}$. This construction plays an important role in many applications and we recall it in more detail here. Let $J_1 = \mathbb{F}$ be the Jordan algebra of real or complex numbers with the standard multiplication, and let $J_2 = \mathcal{S}(V, Q, e)$ be the spin factor defined as in Example 3.2.4 above. Let us define a Jordan algebra

$$J = J_1 \oplus J_2$$

with the componentwise operations and the obvious unit $\mathbf{e} = 1 \oplus e$. Then by (3.2.11) for any $x \in J_2$

$$x^{\bullet 2} = 2Q(x; e)x - Q(x)e$$

and

$$x^{\bullet 3} = 2Q(x, c)x^{\bullet 2} - Q(x)x = (4Q^2(x; c) - Q(x))x - 2Q(x, c)Q(x)e.$$

Since $\mathbf{x}^{\bullet k} = x_0^k \oplus x^{\bullet k}$, it is easily verified that any $\mathbf{x} = (x_0, x) \in J$ satisfies the cubic equation

$$\mathbf{x}^{\bullet 3} - \mathrm{Tr}(\mathbf{x})\mathbf{x}^{\bullet 2} + S(\mathbf{x})\mathbf{x} - N(\mathbf{x})\mathbf{e} = \mathbf{0},$$

with

$$\mathrm{Tr}(\mathbf{x}) = x_0 + 2Q(x; e), \quad S(\mathbf{x}) = Q(x) + 2x_0 Q(x; e), \quad N(\mathbf{x}) = x_0 Q(x).$$

Observe that the generic norm is reducible, $N(\mathbf{x}) = N_1(x_0)N_2(x)$, where $N_1(x_0) = x_0$ is the generic norm on $J_1 = \mathbb{F}$ and $N_2(x) = Q(x)$ is the generic norm on the spin factor J_2. From (3.2.23) we also find the bilinear form,

(3.2.32) $$T(\mathbf{x}; \mathbf{y}) = T(\mathbf{x} \bullet \mathbf{y}; \mathbf{e}) = \frac{1}{3}\mathrm{Tr}(\mathbf{x} \bullet \mathbf{y}) = \frac{1}{3}(x_0 y_0 + 4Q(x; e)Q(y; e) - 2Q(x; y)).$$

Alternatively, one can obtain the reduced cubic factor from the Jordan cubic norm $N(x) = x_0 Q(x)$ by virtue of the Springer construction:

$$\mathbb{F} \oplus \mathcal{S}(V, Q, e) = \mathcal{J}(V, N, 1 \oplus e).$$

3.3. The eiconal equation and cubic Jordan algebras

In this section, we prove Theorem 3.1.1. We split the proof of this result into several propositions, each one given with minimal hypothesis, and start by establishing an eiconal type equation for the generic determinant of an arbitrary cubic Jordan algebra.

3.3.1. From a cubic Jordan algebra to a cubic eiconal.

Let J denote the cubic Jordan algebra $\mathcal{J}(V, N, e)$ associated with a Jordan form $N : V \to \mathbb{F}$ with basepoint e (which becomes automatically the unit in J). By the Springer construction, the bilinear trace form $T(x; y)$ is nonsingular; hence it induces a natural inner product on V. Observe that the basepoint e becomes a unit vector in the induced inner structure. Let us denote by $\overline{\nabla}$ the covariant derivative on (V, T). Then the sharp map definition (3.2.24) yields

$$T(x^\#; y) = N(x; y) = \frac{1}{3}\partial_y N|_x = \frac{1}{3}T(\overline{\nabla}N(x); y),$$

and by the nonsingularity of T,

(3.3.1) $$\overline{\nabla}N(x) = 3x^\#.$$

Let us denote

$$t = T(x; e) = \frac{1}{3}\operatorname{Tr}(x).$$

This coordinate is distinguished in many relations and can be thought of as a "time" on V. Notice that the orthogonal complement to e coincides with the subspace of trace free elements of J,

$$e^\perp := \{x \in J : T(x; e) = 0\} \equiv \{x \in J : \operatorname{Tr}(x) = 0\},$$

which can be thought of as the spatial subspace. We have the following orthogonal decomposition:

$$V = \mathbb{F}e \oplus e^\perp.$$

Let ∇ be the covariant derivative induced by $\overline{\nabla}$ on $e^\perp$. Then $\overline{\nabla}$ is decomposed into the time component and the spatial component as follows:

(3.3.2) $$\overline{\nabla} = \partial_t \oplus \nabla.$$

Given a smooth function $u : e^\perp \to \mathbb{F}$, the gradient $\nabla u(x)$ coincides with the orthogonal projection of $\overline{\nabla}\widetilde{u}(x)$ onto $e^\perp$, where $\widetilde{u}$ is an arbitrary smooth extension of u onto a neighborhood of x in J. It is well known that the choice of the extension does not affect the resulting projection.

We have the following time-evolution equation for the generic norm.

PROPOSITION 3.3.1. *It is true that*

(3.3.3) $$T(\nabla N(x)) - 2N_t'^2(x) = -18tN(x).$$

PROOF. By (3.2.31),

(3.3.4) $$N_t'(x) \equiv T(\overline{\nabla}N(x); e) = 3T(x^\#; e) = \operatorname{Tr}(x^\#) = S(x),$$

and using (3.2.30),

(3.3.5) $$N_t'(x) = S(x) = \frac{1}{2}S(x;x) = \frac{3(3t^2 - T(x))}{2}.$$

Rewriting (3.2.28) by virtue of (3.3.1) as $\overline{\nabla}N(x) + 9tx = 3x^{\bullet 2} + 3N'_t c$ and applying the quadratic form T to the obtained identity, we find

(3.3.6) $\quad T(\overline{\nabla}N(x)) + 54tN(x) + 81t^2\, T(x) = 9T(x^{\bullet 2}) + 18N'_t T(x) + 9N'^{2}_t(x).$

Here we have made use of the homogeneity of N: $T(\overline{\nabla}N(x); x) = \partial_x N|_x = 3N(x)$.

On the other hand, rewriting (3.2.27) by virtue of (3.3.4) and then multiplying the obtained identity by x yields

$$x^{\bullet 4} = 3t\, x^{\bullet 3} - N'_t(x) x^{\bullet 2} + N(x) x.$$

Applying the trace form to both sides of the latter relation, we find

$$\begin{aligned}
9T(x^{\bullet 2}) &= 9T(x^{\bullet 2}; x^{\bullet 2}) = 9T(x^{\bullet 4}; c) \\
&= 27tT(x^{\bullet 3}; c) - 9N'_t(x) T(x^{\bullet 2}; c) + 9N(x) T(x; c) \\
&= 27t(3tT(x) - tN'_t(x) + N(x)) - 9N'_t(x) T(x) + 9tN(x) \\
&= 81t^2 T(x) + 36tN(x) - (27t^2 + 9T(x)) N'_t(x).
\end{aligned}$$

Substituting the resulting relations into (3.3.6) yields by (3.2.30)

$$T(\overline{\nabla}N(x)) = -18tN(x) - 9(3t^2 - T(x)) N'_t(x) + 9N'^{2}_t(x).$$

Rewriting (3.3.5) by virtue of (3.3.4) as $3t^2 - T(x) = \frac{2}{3} N'_t(x)$ and substituting this into the last identity yields

(3.3.7) $\quad\quad\quad\quad T(\overline{\nabla}N(x)) - 3N'^{2}_t(x) = -18tN(x).$

By (3.3.2), $\overline{\nabla}N(x) = \nabla N(x) + N'_t(x)\, e$; thus $T(\overline{\nabla}N(x)) = T(\nabla N(x)) + N'^{2}_t(x)$, which yields by virtue of (3.3.7) the required formula (3.3.3). $\square$

Now we are ready to associate a cubic eiconal to a given cubic Jordan algebra $J = \mathcal{J}(V, N, e)$. Rewrite (3.3.3) by virtue of (3.3.5) as

(3.3.8) $\quad\quad\quad\quad T(\nabla N(x)) = \dfrac{9(3t^2 - T(x))^2}{2} - 18tN(x)$

and define the cubic form

(3.3.9) $\quad\quad\quad\quad u_J(x) := \sqrt{2}\, N(x)|_{e^\perp}, \quad\quad x \in e^\perp,$

to be the restriction of the normalized generic norm on the trace free subspace. Then

$$\nabla u_J(x) = \sqrt{2}\, (\overline{\nabla} N(x))^{e^\perp} = \sqrt{2}\, \nabla N(x) \quad\text{for any } x \in e^\perp;$$

hence setting $t = 0$ in (3.3.8) yields

(3.3.10) $\quad\quad\quad\quad T(u_J(x)) = 9T^2(x).$

Thus, we have obtained the following result.

PROPOSITION 3.3.2. *Let $J = \mathcal{J}(V, N, e)$ be a cubic Jordan algebra with the bilinear trace form T and let $e^\perp$ be the vector subspace of trace free elements equipped with the inner product $T|_{e^\perp}$. Then $u_J(x)$ is a cubic eiconal in $(e^\perp, T|_{e^\perp})$.*

3.3.2. From a cubic eiconal to a cubic Jordan algebra.

Now we shall establish the converse correspondence. First we introduce some notation. We assume that (W, Q) is a quadratic vector space and that $u : W \to \mathbb{F}$ is an arbitrary cubic form, not identically zero. Then the covariant gradient of u is a quadratic endomorphism of W and its polarization

$$(3.3.11) \qquad \operatorname{hess} u(x; y) = \frac{1}{2}(\nabla u(x + y) - \nabla u(x) - \nabla u(y))$$

is obviously a symmetric $\mathbb{F}$-bilinear map. The latter map is said to be the mixed Hessian of u. It readily follows from (3.2.17) that the mixed Hessian and the usual one, $\operatorname{Hess}_x u \in \operatorname{End}(V)$, are connected to each other by

$$2 \operatorname{hess} u(x; y) = \operatorname{Hess}_x(u)y + \operatorname{Hess}_y(u)x.$$

From (3.3.11) and the homogeneity of $\nabla u(x)$ it follows that

$$(3.3.12) \qquad \operatorname{hess} u(x; x) = \nabla u(x)$$

and

$$Q(x; \operatorname{hess} u(y; z)) = 3u(x; y; z), \qquad \forall x, y, z \in V,$$

where $u(x; y; z)$ is the full linearization of u defined by (3.2.18). In particular,

$$(3.3.13) \qquad Q(x; \operatorname{hess} u(y; z)) = Q(\operatorname{hess} u(x; y)\, z).$$

PROPOSITION 3.3.3. *Let $u(x) \in \mathbb{F}[x]$ be a cubic eiconal on a (nonsingular) quadratic space (W, Q). Consider the new vector space $V = \mathbb{F} \times W$ and define*

$$(3.3.14) \qquad N(\mathbf{x}) \equiv N_u(\mathbf{x}) := x_0^3 - \frac{3x_0 Q(x)}{2} + \frac{u(x)}{\sqrt{2}}, \qquad \mathbf{x} = (x_0, x).$$

Then $N(\mathbf{x})$ is a Jordan cubic form on V with respect to the basepoint $e = (1, 0)$. The bilinear trace form T associated with N is given by

$$(3.3.15) \qquad T(\mathbf{x}; \mathbf{y}) = x_0 y_0 + Q(x; y);$$

hence $Q = T|_{e^\perp}$.

PROOF. Let us denote for short $h(x; y) := \operatorname{hess} u(x; y)$ and (cf. with (3.3.12))

$$(3.3.16) \qquad h(x) := h(x; x) = \nabla u(x).$$

Then (3.1.8) becomes

$$(3.3.17) \qquad Q(h(x)) = 9Q^2(x).$$

Polarizing in the latter equation $Q(h(x)) = Q(h(x); h(x))$ and applying the directional derivative with respect to y yields

$$(3.3.18) \qquad Q(h(x); h(x; y)) = 9Q(x)\, Q(y; x) \equiv 9Q(y; Q(x)x).$$

Rewriting (3.3.13) as $Q(h(x); h(x; y)) = Q(y; h(x; h(x)))$, we obtain from (3.3.18) by the nondegeneracy of Q that

$$(3.3.19) \qquad h(x; h(x)) = 9Q(x)x.$$

Polarizing the latter identity, we obtain

$$(3.3.20) \qquad h(y; h(x)) + 2h(x; h(x; y)) = 18Q(x; y)x + 9Q(x)y.$$

Since $h(h(x)) = h(h(x); h(x))$, we find from (3.3.19)

$$h(x; h(x; h(x))) = 9Q(x)h(x; x) = 9Q(x)h(x).$$

We also find by (3.3.19) and (3.2.19) that

(3.3.21) $$Q(x; h(x)) \equiv Q(x; h(x,x)) = \frac{1}{2}u(x;x;x) = 3u(x).$$

Hence, setting $y = h(x)$ in (3.3.20) yields

(3.3.22) $$h(h(x)) = 18Q(x; h(x))x - 9Q(x)h(x) = 54u(x)x - 9Q(x)h(x).$$

Now we prove that the cubic form (3.3.14) is Jordan. Observe that $N(e) = 1$ by definition; hence e is a basepoint of N. Let $\mathbf{x} = (x_0, x)$ and $\mathbf{y} = (y_0, y)$ be two arbitrary vectors in $V = \mathbb{F} \times W$. Since by (3.3.13)

$$Q(x; h(x; y)) = Q(y; h(x; x)) = Q(y; h(x)),$$

we find

(3.3.23) $$\partial_{\mathbf{y}} N|_{\mathbf{x}} = 3\left(x_0^2 - \frac{Q(x)}{2}\right)y_0 - 3x_0 Q(x; y) + \frac{Q(y; h(x))}{\sqrt{2}}.$$

Setting $\mathbf{x} = e$ in the last identity we find by (3.2.20) that $\operatorname{Tr}(\mathbf{y}) = \partial_{\mathbf{y}} N|_e = 3y_0$, and setting $\mathbf{y} = e$ in (3.3.23), we find by (3.2.21) that $S(\mathbf{x}) = \partial_e N|_{\mathbf{x}} = 3(x_0^2 - \frac{Q(x)}{2})$, which yields the corresponding bilinear spur form

$$S(\mathbf{x}; \mathbf{y}) = S(\mathbf{x} + \mathbf{y}) - S(\mathbf{x}) - S(\mathbf{y}) = 6x_0 y_0 - 3Q(x; y),$$

which proves (3.3.15) by virtue of (3.2.23):

$$T(\mathbf{x}; \mathbf{y}) = \frac{1}{3}(\operatorname{Tr}(\mathbf{x}) \operatorname{Tr}(\mathbf{y}) - S(\mathbf{x}; \mathbf{y})) = x_0 y_0 + Q(x; y).$$

On the other hand, $\operatorname{Tr}(\mathbf{x}) = 3x_0$ yields $e^{\perp} = \{0\} \times W$; hence $T|_{e^{\perp}} = Q$.

It remains to establish the adjoint identity (3.2.25). To this end, note that by our assumptions Q is nonsingular; hence T is nonsingular too. Thus $\mathbf{x}^{\#}$ is well-defined. To find it explicitly we rewrite (3.3.23) as

(3.3.24) $$N(\mathbf{x}; \mathbf{y}) = \frac{1}{3}\partial_{\mathbf{y}} N|_{\mathbf{x}} = \left(x_0^2 - \frac{Q(x)}{2}\right)y_0 + Q\left(\frac{h(x)}{3\sqrt{2}} - x_0 x; y\right),$$

which yields by (3.2.24) and (3.3.15) that

(3.3.25) $$\mathbf{x}^{\#} \equiv (x_0^{\#}, x^{\#}) = \left(x_0^2 - \frac{Q(x)}{2}, \frac{h(x)}{3\sqrt{2}} - x_0 x\right).$$

Let us denote $\mathbf{x}^{\#\#} \equiv (x_0^{\#\#}, x^{\#\#})$. Then by (3.3.25)

(3.3.26) $$x_0^{\#\#} = (x_0^{\#})^2 - \frac{Q(x^{\#})}{2} = \left(x_0^2 - \frac{Q(x)}{2}\right)^2 - \frac{Q(x^{\#})}{2}$$

where the latter term is found using (3.3.21) and (3.3.17) to be

$$Q(x^{\#}) = Q\left(\frac{h(x)}{3\sqrt{2}} - x_0 x\right)$$
$$= x_0^2 Q(x) - \frac{\sqrt{2}}{3}x_0 Q(x; h(x)) + \frac{Q(h(x))}{18}$$
$$= x_0^2 Q(x) - \sqrt{2}x_0 u(x) + \frac{Q(x)^2}{2},$$

and substituting this into (3.3.26) yields

(3.3.27) $$x_0^{\#\#} = N(\mathbf{x})x_0.$$

Arguing as above we also obtain

$$x^{\#\#} = -x_0^\# x^\# + \frac{1}{3\sqrt{2}} h(x^\#)$$

(3.3.28)
$$= \left(x_0^2 - \frac{Q(x)}{2}\right)\left(x_0 x - \frac{h(x)}{3\sqrt{2}}\right) + \frac{1}{3\sqrt{2}} h\left(\frac{h(x)}{3\sqrt{2}} - x_0 x\right).$$

Using (3.3.19) and (3.3.22), we have

$$h\left(\frac{h(x)}{3\sqrt{2}} - x_0 x\right) = \frac{h(h(x))}{18} - \frac{\sqrt{2}\,x_0}{3} h(x; h(x)) + x_0^2 h(x)$$

$$= \frac{6u(x)x - Q(x)h(x)}{2} - 3\sqrt{2}\,x_0 Q(x)x + x_0^2 h(x)$$

$$= 3(u(x) - \sqrt{2}\,x_0 Q(x))x + \left(x_0^2 - \frac{Q(x)}{2}\right)h(x),$$

which yields by (3.3.28) that $x^{\#\#} = N(\mathbf{x})x$. Combining this with (3.3.27) yields the required adjoint identity $\mathbf{x}^{\#\#} = N(\mathbf{x})\mathbf{x}$, and the proposition follows. $\square$

Now we exhibit the multiplication structure associated with the norm described in Proposition 3.3.3 explicitly. For short let us denote by

$$J(u) = \mathcal{J}(V, N_u, c)$$

the Jordan algebra obtained from N_u by the Springer construction.

PROPOSITION 3.3.4. *The Jordan product on $J_\alpha(u)$ is found by*

(3.3.29) $\quad (x_0, x) \bullet (y_0, y) = \left(x_0 y_0 + Q(x, y),\ x_0 y + y_0 x + \frac{1}{3\sqrt{2}} \operatorname{hess} u(x; y)\right).$

The cubic eiconal u is recovered from the norm N_u by forgetting the x_0-coordinate in the determinant-like representation

(3.3.30) $\qquad\qquad u(x) = \sqrt{2}\,N_u(i(x)),$

where $i(x) = (0, x) : W \to V$ is the standard embedding. Furthermore, if u_1 and u_2 are two congruent cubic eiconals, then the corresponding Jordan algebras $J(u_1)$ and $J(u_2)$ are isomorphic.

PROOF. We find by (3.3.25) that

$$\mathbf{x}\#\mathbf{y} = (\mathbf{x}+\mathbf{y})^\# - \mathbf{x}^\# - \mathbf{y}^\# = \left(2x_0 y_0 - Q(x; y),\ \frac{\sqrt{2}}{3} \operatorname{hess} u(x; y) - x_0 y - y_0 x\right),$$

which yields (3.3.29) by virtue of (3.2.26). Setting $x_0 = 0$ in (3.3.14) yields (3.3.30).

Now suppose that two cubic forms $u(x)$ on (W, Q) and $\widetilde{u}(\widetilde{x})$ on $(\widetilde{W}, \widetilde{Q})$ are congruent, i.e., there exists a linear map $O : (W, Q) \to (\widetilde{W}, \widetilde{Q})$ such that O is invertible, $\widetilde{Q}(Ox) = Q(x)$ for all $x \in W$, and $u(x) = \widetilde{u}(Ox)$. Denote by $\bullet$ and $\circ$ the products on $J(u)$ and $J(\widetilde{u})$, respectively. The map $\psi : J(u) \to J(\widetilde{u})$ by $\psi(x_0, x) = (x_0, O\widetilde{x})$ is obviously an isomorphism of W and $\widetilde{W}$ on the level of vector spaces. Furthermore, applying (3.3.29) and (3.3.16) we obtain

$$\psi(\mathbf{x})^{\circ 2} = \left(x_0^2 + \widetilde{Q}(Ox; Ox), 2x_0\, Ox + \frac{1}{3\sqrt{2}}\operatorname{hess}\widetilde{u}(Ox; Ox)\right)$$

(3.3.31)
$$= \left(x_0^2 + \widetilde{Q}(Ox), 2x_0\, Ox + \frac{1}{3\sqrt{2}}\nabla\widetilde{u}(Ox)\right).$$

On the other hand,

$$\partial_y u|_x \equiv \lim_{t\to 0}\frac{u(x+ty) - u(x)}{t} = \lim_{t\to 0}\frac{\widetilde{u}(Ox + tOy) - \widetilde{u}(Ox)}{t} \equiv \partial_{Oy}\widetilde{u}|_{Ox},$$

which yields by the definition of the covariant gradient

$$Q(\nabla u(x); y) = \partial_y u|_x = \partial_{Oy}\widetilde{u}|_{Ox} = \widetilde{Q}(\nabla\widetilde{u}(Ox); Oy) = Q(O^{-1}\nabla\widetilde{u}(Ox); y);$$

hence by nonsingularity of Q, $\nabla u(x) = O^{-1}\nabla\widetilde{u}(Ox)$. Hence (3.3.31) yields

$$\psi(\mathbf{x})^{\circ 2} = \left(x_0 + Q(x), O\left(2x_0 x + \frac{1}{3\sqrt{2}}\nabla u(x)\right)\right) = \psi(\mathbf{x}^{\bullet 2}),$$

which proves by virtue of Proposition 3.2.1 that ψ is actually an isomorphism of the Jordan algebras, as desired. $\square$

3.3.3. The correspondence. To finish the proof of Theorem 3.1.1 one uses

PROPOSITION 3.3.5. (a) *If J is a cubic Jordan algebra, then the Jordan algebras $J(u_J)$ and J are isomorphic.*

(b) *If J_1 and J_2 are isomorphic cubic Jordan algebras, then u_{J_1} and u_{J_2} are congruent cubic eiconals.*

PROOF. (a) Let $J = \mathcal{J}(V, N, e)$ and let $u_J(x)$ be the cubic eiconal associated to J by (3.3.9). By (3.3.10), $u_J(x)$ is a cubic eiconal on $(e^\perp, T|_{e^\perp})$. Now let $J(u_J) = \mathcal{J}(\mathbb{F}\times e^\perp, \widehat{N}, c)$ be the Jordan algebra associated to $u_J(x)$ by virtue of Proposition 3.3.4, where $\widehat{N}$ is defined by (3.3.14) as

$$\widehat{N}(\mathbf{x}) = x_0^3 - \frac{3}{2}x_0 T(x) + \frac{1}{\sqrt{2}}u_J(x), \qquad \mathbf{x} = (x_0, x) \in \mathbb{F}\times e^\perp.$$

Then the mapping $\phi(\mathbf{x}) = x_0 e + x : J(u_J) \to J$ establishes an isomorphism between $J(u_J)$ and J on the level of vector spaces. In order to prove that ϕ is actually an isomorphism of the Jordan algebras it is sufficient by Proposition 3.2.1 to show[2] that $\phi(\mathbf{x}^{\bullet 2}) = \phi(\mathbf{x})^{\bullet 2}$. To this end, we find from (3.3.29) using (3.3.16) and (3.3.12) that

$$\mathbf{x}^{\bullet 2} = \left(x_0^2 + Q(x), 2x_0 x + \frac{1}{3\sqrt{2}}\operatorname{hess} u_J(x; x)\right)$$
$$= \left(x_0^2 + Q(x), 2x_0 x + \frac{1}{3\sqrt{2}}\nabla u_J(x)\right).$$

On the other hand, by the definition of ϕ,

$$\phi(\mathbf{x})^{\bullet 2} = x_0^2 e + 2x_0 x + x^{\bullet 2};$$

hence

(3.3.32) $$\phi(\mathbf{x}^{\bullet 2}) - \phi(\mathbf{x})^{\bullet 2} = \frac{1}{3\sqrt{2}}\nabla u_J(x) - x^{\bullet 2} + T(x)e.$$

---

[2]We somewhat abuse notation here by using here and below the same bullet notation for the different products in the Jordan algebras.

By definition of $u_J(x)$ and (3.3.1),
$$\nabla u_J(x) = \sqrt{2}\nabla N(x)|_{e^\perp} = \sqrt{2}(\overline{\nabla} N(x))^{e^\perp} = 3\sqrt{2}(x^\#)^{e^\perp},$$
and using $\mathrm{Tr}(x) = 0$ and (3.2.28),
$$\begin{aligned}(x^\#)^{e^\perp} &= (x^{\bullet 2} - \mathrm{Tr}(x)x + S(x)e)^{e^\perp} = (x^{\bullet 2})^{e^\perp} \\ &= x^{\bullet 2} - T(x^{\bullet 2}; e)e = x^{\bullet 2} - T(x; x)e \\ &= x^{\bullet 2} - T(x)e.\end{aligned}$$
Thus $\frac{1}{3\sqrt{2}}\nabla u_J(x) = x^{\bullet 2} - T(x)e$; hence (3.3.32) yields $\phi(\mathbf{x}^{\bullet 2}) - \phi(\mathbf{x})^{\bullet 2} = 0$, which implies the desired isomorphy of ϕ.

In order to verify (b) we suppose that $\phi : J_1 \to J_2$ is an isomorphism of Jordan algebras. Since J_1 is a cubic Jordan algebra, any element $\xi \in J_2$ satisfies the characteristic relation
$$\xi^{\bullet 3} - \mathrm{Tr}_2(\xi)\xi^{\bullet 2} + S_2(\xi)\xi - N_2(\xi)e_2 = 0,$$
where Tr_k, S_k, N_k, and e_k denote the generic trace, the generic spur, the generic norm, and the unit element in the algebra J_k, respectively. Applying ϕ^{-1} to the latter relation we find by isomorphy of ϕ that
$$(\phi^{-1}(\xi))^{\bullet 3} - \mathrm{Tr}_2(\xi)(\phi^{-1}(\xi))^{\bullet 2} + S_2(\xi)\phi^{-1}(\xi) - N_2(\xi)e_1 = 0.$$
Since ϕ is an isomorphism, any element $x \in J_1$ satisfies the cubic relation
$$x^{\bullet 3} - \mathrm{Tr}_2(\phi(x))x^{\bullet 2} + S_2(\phi(x))x - N_2(\phi(x))e_1 = 0,$$
and we deduce from the uniqueness of the minimum polynomial [**120**] that $\mathrm{Tr}_1(x) = \mathrm{Tr}_2(\phi(x))$, $S_1(x) = S_2(\phi(x))$, and

(3.3.33) $$N_1(x) = N_2(\phi(x))$$

for any $x \in J_1$. The first two identities imply by virtue of (3.2.23) for the bilinear trace forms that $T_1(x) = T_2(\phi(x))$ and polarizing this identity yields
$$T_1(x; y) = T_1(\phi(x); \phi(y)),$$
i.e., ϕ is actually an isometry of the corresponding (quadratic) vector spaces. Since $\phi(e_1) = e_2$, one has $\phi(e_1^\perp) = e_2^\perp$; hence $\phi : e_1^\perp \to e_2^\perp$ is also an isometry. Thus, (3.3.33) implies
$$u_{J_1}(x) = N_1|_{e_1^\perp}(x) = N_2|_{e_2^\perp}(\phi(x)) \equiv u_{J_2}(\phi(x));$$
hence $u_{J_1}(x)$ is congruent to $u_{J_2}(\phi(x))$, which proves (b). $\square$

3.3.4. The Euclidean eiconal cubics revisited. We demonstrate the above constructions by giving an alternative proof of the recent classification result established in [**250**].

COROLLARY 3.3.1. *Any cubic form $u(x) : \mathbb{R}^n \to \mathbb{R}$ satisfying the eiconal equation*

(3.3.34) $$|\nabla u(x)|^2 = 9|x|^4, \quad x \in \mathbb{R}^n,$$

is congruent to either $C_d(x)$ for $d = 1, 2, 4, 8$ or to a reduced form r_n; see (3.1.1) *and* (3.1.6).

PROOF. Suppose $u(x)$ is a solution of (3.3.34). Then by Proposition 3.3.3 there exists a cubic Jordan algebra structure on $J := \mathbb{R} \times \mathbb{R}^n$ such that $u(x) = N(x)|_{e^\perp}$, where $N(x)$ is the generic norm in J, $e^\perp = \{x \in J : \mathrm{Tr}(x) = 0\}$, and the trace form is given by $T(x; y) = x_0 y_0 + Q(x; y) \equiv \sum_{i=0}^{n} x_i y_i$. Observe that T is positive definite; hence J is actually a Euclidean Jordan algebra, and thus it is formally real (see Subsection 3.2.5). By the Jordan-von Neumann-Wigner classification [**124**], $J = \bigoplus_\alpha J_\alpha$, where any J_α is either a Jordan algebra of Hermitian matrices $\mathrm{Herm}_k(\mathbb{F}_d)$ of rank $k \leq 3$ over $\mathbb{F}_d$, where $d = 1, 2, 4, 8$, or it is a Euclidean spin factor $\mathcal{S}(\mathbb{R}_1^n, M, e)$ considered in Example 3.2.5. Since J has rank 3, the only possibilities are the following:

(i) $J = \mathrm{Herm}_3(\mathbb{F}_d)$, $d = 1, 2, 4, 8$.
(ii) J is the reduced cubic factor $\mathbb{R}\mathcal{S}(\mathbb{R}_1^n, M, e) = \mathbb{R} \oplus \mathcal{S}(\mathbb{R}_1^n, M, e)$ in the notation of Example 3.2.8.
(iii) J is the trivial reduced cubic factor $\mathbb{F}_1^3 = \mathbb{R} \oplus \mathbb{R} \oplus \mathbb{R}$ with the coordinatewise multiplication.

In the first case, $u(x)$ is the restriction of the generic norm in $\mathrm{Herm}_3(\mathbb{F}_d)$ on the trace free subspace, i.e., a Cartan isoparametric cubic. Moreover, the ambient dimension can only take the following values: $n = 3d + 2$, $d = 1, 2, 4, 8$.

Case (iii) is trivially verified to be a subcase of (ii) for $n = 2$. Thus it suffices to examine case (ii). Applying the constructions of Example 3.2.8, we have for the generic norm

$$N(\mathbf{x}) = x_0 M(x) = x_0(x_n^2 - x_1^2 - \cdots - x_{n-1}^2), \quad \mathbf{x} = x_0 \oplus x \in \mathbb{R} \oplus \mathbb{R}_1^n,$$

and M is found by (3.2.12). The basepoint of J is $\mathbf{e} = (1, e)$, where $e = (0, \ldots, 0, 1)$ is the basepoint of the quadratic spin factor $\mathcal{S}(\mathbb{R}_1^n, M, e)$. Then using (3.2.32)

$$T(\mathbf{x}; \mathbf{y}) = \frac{1}{3}(x_0 y_0 + 4M(x; e)M(y; e) - 2M(x; y)) = \frac{1}{3}(x_0 y_0 + 2x_1 y_1 + \cdots + 2x_n y_n).$$

This yields $\mathbf{e}^\perp = \{\mathbf{x} \in J : x_0 = -2x_n\}$. Then

$$\xi = \frac{1}{\sqrt{2}} \cdot (2\xi_n, \sqrt{3}\,\xi_1, \ldots, \sqrt{3}\,\xi_{n-1}, -\xi_n) : \mathbb{R}^n \to \mathbf{e}^\perp$$

is an isometry of the quadratic spaces because $T(\xi) = \xi_1^2 + \cdots + \xi_n^2$; hence the desired cubic eiconal is found by

$$u(\xi) = \sqrt{2} N(\xi) = \xi_n^3 - 3\xi_n(\xi_1^2 + \cdots + \xi_{n-1}^2),$$

which yields (3.1.6) and proves the corollary. □

CHAPTER 4

Solutions from Trialities

In this chapter we consider the functions
$$w_{j,\delta}(x) = \frac{P_j(x)}{|x|^\delta}$$
for the triality cubic polynomials $P_j(x), x \in \mathbb{R}^j$, $j = 3, 6, 12, 24$, and $\delta \in [1, 2[$. We will show that $w_{j,\delta}(x)$ for $j = 12$ and $j = 24$ is a solution of a fully nonlinear uniformly elliptic equation. For $j = 3$ the function $w_{3,\delta}$ is not a solution of an elliptic equation, and for $j = 6$ one can show that $w_{6,1}$ is a solution of a strictly, but not uniformly, elliptic equation which is not very interesting. In Section 4.1 we prove that $w_{12}(x) = w_{12,1}(x)$ is a solution to a uniformly elliptic equation, thus giving a first example of a nonclassical viscosity solution. In that section the arguments are fairly elementary and direct and use neither advanced theory nor complicated calculations. In Section 4.2 we show that $w_{24,\delta}(x)$ is a solution of a uniformly elliptic Hessian equation, which needs some more advanced algebraic theory, essentially that of Chapter 2, to calculate the corresponding eigenvalues; another possibility is to use more heavy (computer) computations. Section 4.3 is devoted to the function $w_{12,\delta}(x)$ which is a solution of a fully nonlinear uniformly elliptic Hessian equation as well; however, to prove this we need rather elaborate computer calculations in MAPLE along with some laborious estimations. In Section 4.4 we prove that the equations in Sections 4.1–4.3 can be written as Isaacs equations. All equations in this chapter are of the simplest form $F(D^2 w) = 0$, i.e., with "constant coefficients".

4.1. The simplest nonclassical construction

We begin with w_{12}.

THEOREM 4.1.1. *The function*
$$w_{12}(x) := \frac{\operatorname{Re}(q_1 q_2 q_3)}{|x|} = \frac{P_{12}(x)}{|x|},$$
where $q_i \in \mathbb{H}$, $i = 1, 2, 3$, are Hamiltonian quaternions, and $x = (q_1, q_2, q_3) \in \mathbb{H}^3 = \mathbb{R}^{12}$ is a viscosity solution in $\mathbb{R}^{12}$ of a uniformly elliptic equation $F(D^2 w) = 0$ with a smooth F.

The elliptic operator F is defined in a constructive way in Section 4.1.2, and its ellipticity constant $\Lambda < 10^8$.

As an immediate consequence of the theorem we have

COROLLARY 4.1.1. *Let $\Omega = B_1 \subset \mathbb{R}^{12}$ be the unit ball and let $\varphi = w_{12}$ on $\partial \Omega = \mathbb{S}_1^{11}$. Then there exists a smooth uniformly elliptic F such that the Dirichlet problem (1.1.12) has no classical solution.*

4. SOLUTIONS FROM TRIALITIES

4.1.1. Ellipticity criterion. Let w be a homogeneous function of order 2, defined on $\mathbb{R}^n$ and smooth in $\mathbb{R}^n \setminus \{0\}$. Then the Hessian of w is homogeneous of order 0 and defines a map

$$H : \mathbb{S}^{n-1} \to Q(\mathbb{R}^n), \quad H(a) := D^2 w(a)$$

where $Q(\mathbb{R}^n)$ denotes the space of quadratic forms on $\mathbb{R}^n$, which we identify with the space of symmetric $n \times n$ matrices: $Q(\mathbb{R}^n) \simeq \mathrm{Sym}_n(\mathbb{R})$. The inner product of $a, b \in Q(\mathbb{R}^n)$ is given by $a \cdot b = \mathrm{tr}(ab)$.

Hessian Problem (H). We say that w *satisfies property* (H), or that w is a solution of the *Hessian Problem*, if the following hold:
 A) The map $H : \mathbb{S}^{n-1} \to Q(\mathbb{R}^n)$ is a smooth embedding.
 B) There exists a constant $M \geq 1$ such that for any two points $a, b \in H(\mathbb{S}^{n-1}), a \neq b$, if $\mu_1 \geq \cdots \geq \mu_n$ denote the eigenvalues of the quadratic form $a - b$, then

$$1/M < -\mu_1/\mu_n < M.$$

LEMMA 4.1.1. *If function w satisfies hypotheses* (H), *then w is a viscosity solution in $\mathbb{R}^n$ of a uniformly elliptic equation.*

PROOF. Let us choose in the space $Q(\mathbb{R}^n)$ an orthonormal coordinate system $z_1, \ldots, z_k, s$, with $k = \frac{n(n+1)}{2} - 1$ such that $\sqrt{n}s$ is the trace. Let $\pi : Q(\mathbb{R}^n) \to Z$ be the orthogonal projection of $Q(\mathbb{R}^n)$ onto the z-space. For $\lambda \geq 1$, we denote by K_λ the cone

$$K_\lambda = \{a \in Q(\mathbb{R}^n) \text{ such that } \mathrm{Spec}(a) \subset [C/\lambda, C\lambda] \text{ for some } C > 0 \,\}.$$

Since on $Q(\mathbb{R}^n)$ the maximal eigenvalue of a quadratic form is a convex function and the minimal eigenvalue is a concave function, we see that K_λ is a convex cone.

Let K_λ^* denote the adjoint cone of K_λ; that is,

$$K_\lambda^* = \{b \in Q(\mathbb{R}^n) : \langle b, c \rangle \geq 0 \text{ for all } c \in K_\lambda\}.$$

As an adjoint to a convex cone, the cone K_λ^* is a convex cone itself. Let us define

$$L_\lambda = Q(\mathbb{R}^n) \setminus (K_\lambda^* \cup -K_\lambda^*).$$

Notice that $a \in L_\lambda$ is equivalent to $\langle a, b \rangle = 0$ for some $b \in K_\lambda$, i.e., L_λ is a union of all hyperplanes in $Q(\mathbb{R}^n)$ with normals in K_λ. Since the quadratic forms of K_λ are positively defined, it follows that the vector $I \in K_\lambda^*$. Let $K \subset Q(\mathbb{R}^n)$ be a cone with a smooth strictly convex base such that $K_{2\lambda} \subset K \subset K_\lambda$. Let $e_1, \ldots, e_k, I$ be an orthonormal basis of $Q(\mathbb{R}^n)$ corresponding to the coordinates $z_1, \ldots, z_k, s$. Then any $b \in Q(\mathbb{R}^n) \simeq \mathrm{Sym}_n(\mathbb{R})$ can be written as

$$b = sI + \sum_{j=1}^{k} z_j e_j.$$

Now define

$$x(z) := \inf\{c : (a + cI) \in K^*\}$$

for $z := \sum_{j=1}^{k} z_j e_j$. The graph of the function $s = x(z)$ represents the boundary of the cone K. Clearly $x(\cdot)$ is Lipschitz, convex, homogenous, smooth outside the origin and $x(0) = 0$. By a simple computation we get that $|\nabla x| < \sqrt{n}$.

Let $G \subset Q(\mathbb{R}^n)$. We say that G satisfies the λ-*cone condition* if for any two points $a, b \in G$, the matrix $a - b \in L_\lambda$.

LEMMA 4.1.2. *Let $\Sigma \subset Q(\mathbb{R}^n)$ be a smooth compact $(n-1)$-dimensional manifold. Assume that Σ satisfies the λ-cone condition. Then there exists a smooth function F on $Q(\mathbb{R}^n)$ satisfying the ellipticity condition with the ellipticity constant $\Lambda < 4\lambda^2 \sqrt{n}$ such that $F(\Sigma) = 0$.*

PROOF. Set $\sigma = \pi(\Sigma)$. We prove that Σ is a graph of a Lipschitz continuous function,
$$\Sigma = \{z \in \sigma : s = g(z)\}.$$
Let $z, z' \in \Sigma$, where
$$z = sI + \sum_{j=1}^k z_j e_j, \quad z' = \hat{s}I + \sum_{j=1}^k z'_j e_j.$$
Since $z - z' \in L_\lambda$, we have
$$-x(z - z') \leq \hat{s} - s \leq x(z - z').$$
Since $x(0) = 0$, $g(z) := s$ is single-valued. Also
$$|g(z) - g(z')| = |s - \hat{s}| \leq |x(z - z')| \leq C|z - z'|.$$
Since Σ is a smooth surface, g is a smooth function and hence σ is a smooth surface as well.

Let G_k^m be the Grassmannian manifold of m-dimensional subspaces of the k-dimensional subspace z of $Q(\mathbb{R}^n)$. Let $l \in G_k^{n-1}$ and let $t : l \to s$ be a linear function on l such that the graph of t satisfies the λ-cone condition. We denote by τ the set of all such linear functions t defined for all $l \in G_k^{n-1}$. Let $t \in \tau$ be defined on $l \in G_k^{n-1}$ with $\nabla t \neq 0$. Then there exist a constant $c' > 1$ with $c't \leq x$ on l and a point $z' \in l, |z'| = 1$ with $c't(z') = x(z')$. Since K^* is a strictly convex cone, the vector z' is unique. Denote
$$\eta(t) = \{z \in l, t(z) = 0\}.$$
Then $\eta(t) \perp \nabla t$. Since $\eta(t) \subset Z$ and $\eta(t)$ is tangent to the cone K^* at z', it follows that $\eta(t) \perp \nabla x(z')$.

Let θ be a smooth function defined on $[1, \infty)$ with $0 \leq \theta \leq 1, \theta = 1$ on $[1, A]$, $\theta = 0$ on $[2A, \infty)$, where a sufficiently large constant A will be chosen later. Set
$$\nu(t) = \theta(c')\nabla x(z') + (1 - \theta(c'))\nabla t.$$

For $z \in \sigma$ we denote by $l(z) \in G_k^{(n-1)}$ the tangent subspace to σ at z. Let $x \in l(z)$ and let $t_z(x)$ be the differential of g at z.

Let $z \in \sigma$ and $\nabla t_z \neq 0$. Denote by $\Psi(z)$ the $(n-1)$-dimensional subspace spanned by $\eta(t_z)$ and $\nu(t_z)$. If $\nabla t_z = 0$, we set $\Psi(z) = l$. We get a smooth map
$$\Psi : \sigma \to G_k^{n-1}.$$

Let us then denote by $\gamma \subset Z$ a closed neighborhood of σ diffeomorphic to $\sigma \times B$, where B is the $(k-n+1)$-dimensional disk. We define a projection $\gamma \to \sigma$ such that the fiber $p^{-1}(z)$ is orthogonal to $\Psi(z)$ at $z \in \sigma$. Since $(\nabla x(z'), z') > 0$, the fiber $p^{-1}(z)$ is transversal to σ at z. Extend then the function g to γ by $g(y) = g(p(y))$. Let Γ be the graph of g over γ, let $z \in \sigma$, and let $dg(z)$ be the differential of g over γ. If A is a sufficiently large constant, the following alternative holds: either

$|\nabla g(z)|$ is sufficiently small or the graph of $c'dg(z)$ is tangent to the cone K^*. In both cases the graph of $dg(z)$ satisfies the 2λ-cone condition. Since $g \in C^1(Z)$ and the function g is constant along the fibers, we may assume the neighborhood γ to be sufficiently small so that Γ satisfies the λ-cone condition.

Since $K_{2\lambda} \subset K$ and $g \in C^1(\sigma)$, the function g has an extension $\widetilde{g}$ from the set γ to $\mathbb{R}^k$ such that $\widetilde{g}$ is a Lipschitz function and the graph of $\widetilde{g}$ satisfies the 2λ-cone condition. One can define such an extension $\widetilde{g}$ simply by

$$\widetilde{g}(z) := \inf_{w \in \gamma} \{g(w) + x(z-w)\}.$$

To demonstrate that this formula works let two points $(z, \widetilde{g}(z)), (z', \widetilde{g}(z'))$ lie on the graph $\widetilde{g}$. We must show

$$-x(z-z') \leq \widetilde{g}(z) - \widetilde{g}(z') \leq x(z-z').$$

Notice that
$$\widetilde{g}(z') = g(w) + x(z'-w)$$
for some $w \in \gamma$. Thus one deduces

$$\widetilde{g}(z) - \widetilde{g}(z') \leq g(w(z')) + x(z-w(z')) - (g(w(z')) + x(z'-w(z'))) \leq x(z-z')$$

since $x(a+b) \leq x(a) + x(b)$, as $x(\cdot)$ is convex and homogenous. Similarly

$$\widetilde{g}(z) - \widetilde{g}(z') \geq -x(z-z').$$

Let $D_1, D_2 \subset Z$ be bounded domains such that $\sigma \subset D_1 \subset D_2 \Subset \gamma$. Next, let $v(z) \in C^\infty(\mathbb{R}^k)$ be supported on the unit ball $B_1 \subset \mathbb{R}^k$, $\int_{\mathbb{R}^k} v(z)\,dz = 1$, and set

$$v_\delta(z) = \frac{1}{\delta^k} v\left(\frac{z}{\delta}\right).$$

Let $h \in C^\infty(Z)$, $0 \leq h \leq 1$ on Z with $h = 1$ on D_1, while $h = 0$ on $Z \setminus D_2$.

Define the mollifiers for $\varepsilon > 0$ by

$$g_\varepsilon = \widetilde{g} * v_\varepsilon, \quad G_\varepsilon = hg + (1-h)g_\varepsilon.$$

Since the graph of the function g satisfies the λ-cone condition, it follows that the upper normals to the graph belong to the cone K. Since K^* is a convex cone, the upper normals to the graphs of the functions g_ε satisfy the 2λ-cone condition for all small $\varepsilon > 0$, and hence the graphs of the linear function $d_z g_\varepsilon$ are in $L_{2\lambda}$ for all z where d_z is the differential at z. Since the functions g_ε are defined on the whole space $\mathbb{R}^k$, it follows that the graphs of the functions g_ε satisfy the λ-cone condition. Indeed, let a, b, $a \neq b$, be on the graph of g_ε. If $a - b \notin L_{2\lambda}$, then there is a point $\alpha \in [\pi(a), \pi(b)]$ such that $d_\alpha g_\varepsilon \notin L_{2\lambda}$. For any $k > 0$ the function $g_\varepsilon \to g$ in $C^k(D_2)$ as $\varepsilon \to 0$. Hence for a sufficiently small $\varepsilon_o > 0$ the graph of the function $G_{\varepsilon_0} := G$ will satisfy the 2λ-cone condition. Moreover G will be a smooth function on Z, $G = g$ on D_1, and $|\nabla G| < \sqrt{n}$ on Z.

Let us then set
$$F := s - G(z).$$

Denote
$$b := \nabla F = (-\nabla G, 1), \quad a := (\nabla G/|\nabla G|, |\nabla G|).$$

The vector a is tangent to the level surface of the function F, and $\mathrm{tr}(b) = \sqrt{n}$. Since the level surfaces of the function F satisfy the 2λ-cone condition and $a \cdot b = 0$, it follows that $a \in L_{2\lambda}$ and hence $b \in K_{2\lambda}$. Therefore the function F satisfies the ellipticity conditions with the ellipticity constant $\Lambda < 4\lambda^2 \sqrt{n}$. $\square$

We can now finish the proof of Lemma 4.1.1. Set $\lambda = (n-1)M$ and let ξ and η be the corresponding negative and nonnegative subspaces of the quadratic form $a - b$ in $\mathbb{R}^n$. Denote by $c \in Q(\mathbb{R}^n)$ the quadratic form $l|\xi|^2 + m|\eta|^2$, for $l, m > 0$ with $(a - b) \cdot c = 0$. Then $1/(n-1)M < l/m < (n-1)M$ by B) of (H) and hence the set $H(\mathbb{S}^{n-1})$ satisfies the λ-cone condition.

Let us then define the function F by Lemma 4.1.2 for $\Sigma = H(\mathbb{S}^{n-1})$. Then the function w satisfies the equation
$$F(D^2 w) = 0$$
on $\mathbb{R}^n \setminus \{0\}$.

Let us show then that w is a viscosity solution of $F(D^2 w) = 0$ on the whole space $\mathbb{R}^n$.

Let $p(x)$, $x \in \mathbb{R}^n$, be a quadratic form such that $p \leq w$ on $\mathbb{R}^n$. We choose any quadratic form $p'(x)$ such that $p \leq p' \leq w$ and there is a point $x' \neq 0$ at which $p'(x') = w(x')$. Then it follows that $F(p) \leq F(p') \leq 0$. Consequently for any quadratic form $p(x)$ the inequality $p \leq w$ (resp. $p \geq w$) implies that $F(p) \leq 0$ (resp. $F(p) \geq 0$). This proves that w is a viscosity solution of $F(D^2 w) = 0$ in $\mathbb{R}^n$ (see Proposition 2.4 in [**42**]). □

REMARK 4.1.1. For a real analytic manifold Σ one can obtain the existence of a real analytic function F after insignificant changes in the construction.

REMARK 4.1.2. The proof of the lemma holds if instead of the compactness of Σ we assume that Σ is a smooth closed manifold with a boundary.

4.1.2. Characteristic polynomial. In this subsection we calculate the characteristic polynomial of the Hessian of P_{12}. Let $V = (x, y, z) \in \mathbb{R}^{12}$ be a variable vector with x, y, and $z \in \mathbb{R}^4$. For any $t = (t_0, t_1, t_2, t_3) \in \mathbb{R}^4$ we denote
$$q_t = t_0 + t_1 i + t_2 j + t_3 k \in \mathbb{H}.$$

Recall that
$$\bar{q}_t q_t = q_t \bar{q}_t = |q_t|^2 = t_0^2 + t_1^2 + t_2^2 + t_3^2.$$

Let us define
$$\begin{aligned}P(x,y,z) = P_{12}(x,y,z) &= \mathrm{Re}(q_x\, q_y\, q_z) \\
&= x_0 y_0 z_0 - x_0 y_1 z_1 - x_0 y_2 z_2 - x_0 y_3 z_3 \\
&\quad - x_1 y_0 z_1 - x_1 y_1 z_0 - x_1 y_2 z_3 + x_1 y_3 z_2 \\
&\quad - x_2 y_0 z_2 + x_2 y_1 z_3 - x_2 y_2 z_0 - x_2 y_3 z_1 \\
&\quad - x_3 y_0 z_3 - x_3 y_1 z_2 + x_3 y_2 z_1 - x_3 y_3 z_0.\end{aligned}$$

Let $d = (a, b, c) \in \mathbb{R}^{12}$ be a vector with the norm $\sqrt{3}$, $|a|^2 + |b|^2 + |c|^2 = 3$. Define the quadratic form
$$Q_d = Q_{a,b,c} = Q_{a,b,c}(x, y, z)$$
by differentiating $P(x, y, z)$ in the direction d:
$$Q_{a,b,c}(x, y, z) = \sum_{i=0}^{4} a_i \frac{\partial P}{\partial x_i} + b_i \frac{\partial P}{\partial y_i} + c_i \frac{\partial P}{\partial z_i}.$$

A direct calculation shows that
$$Q_d(x, y, z) = x^t M_c y + x^t M_b^t z + y^t M_a z$$

where, in general, we define the matrix M_s for an arbitrary $s \in \mathbb{R}^4$ by

$$M_s = \begin{pmatrix} s_0 & -s_1 & -s_2 & -s_3 \\ -s_1 & -s_0 & -s_3 & s_2 \\ -s_2 & s_3 & -s_0 & -s_1 \\ -s_3 & -s_2 & s_1 & -s_0 \end{pmatrix}.$$

Direct (and easy) calculations show that M_s has the following properties:
1) $M_s \cdot M_s^t = M_s^t \cdot M_s = |s|^2 I_4$.

Thus, M_s is proportional to an orthogonal matrix. In particular, if $|s| = 1$, then M_s is orthogonal itself. In general, we write

$$M_s = |s| O_s \quad \text{where } O_s \in O(4).$$

2) $\det(M_s) = -|s|^4$, $\det(O_s) = -1$.
3) The characteristic polynomials $PM_s(\lambda)$ of M_s and $PO_s(\lambda)$ of O_s factor as

$$PM_s(\lambda) = (\lambda^2 - |s|^2)(\lambda^2 + 2s_0\lambda + |s|^2), \quad PO_s(\lambda) = (\lambda^2 - 1)(\lambda^2 + 2s_0^*\lambda + 1)$$

with $s_0 = \operatorname{Re}(q_s)$, $s_0^* = s_0/|q_s| = \operatorname{Re}(q_s/|q_s|)$.

4) Define the symmetric matrix $N_s = (O_s + O_s^t)$; then its characteristic polynomial and spectrum are

$$PN_s(\lambda) = (\lambda^2 - 4)(\lambda + 2s_0^*)^2, \qquad \operatorname{Spec}(N_s) = \{2, -2, -2s_0^*, -2s_0^*\}.$$

5) M_s is the matrix (with respect to the standard basis) of the endomorphism $\mathbb{H} \to \mathbb{H} : q \mapsto \bar{q} \cdot \bar{q}_s$.

Properties 3) and 5) applied to the product matrix $M_{rst} = M_r \cdot M_s \cdot M_t$ for $r, s, t \in \mathbb{R}^4 = \mathbb{H}$ give the following formula for the characteristic polynomial of M_{rst}:

$$PM_{rst}(\lambda) = (\lambda^2 - |q_r q_s q_t|^2)(\lambda^2 + 2P(r,s,t)\lambda + |q_r q_s q_t|^2)$$

with $P(r, s, t) = \operatorname{Re}(q_r q_s q_t)$ as above. Indeed, M_{rst} is conjugate to the matrix of the endomorphism $q \mapsto \bar{q} \cdot \bar{q}_r \cdot \bar{q}_s \cdot \bar{q}_t$.

For the corresponding orthogonal matrix O_{rst} we get the polynomial

$$PO_{rst}(\lambda) = (\lambda^2 - 1)(\lambda^2 + 2\widetilde{P}(r,s,t)\lambda + 1)$$

where $\widetilde{P}(r, s, t) = P(r, s, t)/|q_r q_s q_t|$ and for the corresponding symmetric matrix $N_{rst} = O_{rst} + O_{rst}^t$ the spectrum is

$$\operatorname{Spec}(N_{rst}) = \{2, -2, -2\widetilde{P}(r,s,t), -2\widetilde{P}(r,s,t)\}.$$

Let then $m = m(d) = m(a, b, c) = |q_a q_b q_c| = |a| \cdot |b| \cdot |c|$ and $n = n(d) = n(a, b, c) = P(a, b, c)$. By the inequality between the geometric and quadratic means one gets $|n(d)| \leq m(d) \leq 1$ since $|a|^2 + |b|^2 + |c|^2 = 3$.

PROPOSITION 4.1.1. *The characteristic polynomial $\chi_d(\lambda)$ of the quadratic form $2Q_d$ equals*

(4.1.1) $\qquad \chi_d(\lambda) = (\lambda^3 - 3\lambda + 2m)(\lambda^3 - 3\lambda - 2m)(\lambda^3 - 3\lambda + 2n)^2.$

PROOF. Conjugating H_0 by the orthogonal matrix

$$\begin{pmatrix} O_z^t & 0_4 & 0_4 \\ 0_4 & I_4 & 0_4 \\ 0_4 & 0_4 & O_x \end{pmatrix}$$

yields
$$\widetilde{H_0} = \begin{pmatrix} 0_4 & |z|I_4 & |y|O^t_{xyz} \\ |z|I_4 & 0_4 & |x|I_4 \\ |y|O_{xyz} & |z|I_4 & 0_4 \end{pmatrix}.$$

Let now $\lambda \in \mathrm{Spec}(\widetilde{H_0})$, $v_\lambda = (p_\lambda, q_\lambda, r_\lambda)$ being a corresponding eigenvector, normalized by the condition $|v_\lambda| = 1$. The condition $\widetilde{H}_0 \cdot v_\lambda = \lambda v_\lambda$ gives

$$\lambda p_\lambda = |z|q_\lambda + O^t_{xyz}|y|r_\lambda,$$

$$\lambda q_\lambda = |z|p_\lambda + |x|r_\lambda,$$

$$\lambda r_\lambda = |y|O_{xyz}p_\lambda + |x|q_\lambda.$$

Multiplying the second and the third equations by λ and inserting into the equations obtained, we get

$$(\lambda^2 - |z|^2)p_\lambda = (|x| \cdot |z| + O^t_{xyz}\lambda|y|)r_\lambda,$$
$$(\lambda^2 - |x|^2)r_\lambda = (|x| \cdot |z| + O_{xyz}\lambda|y|)p_\lambda,$$

which implies

$$(\lambda^2 - |x|^2)(\lambda^2 - |z|^2)p_\lambda = (|x| \cdot |z| + O^t_{xyz}\lambda|y|)(|x| \cdot |z| + O_{xyz}\lambda|y|)p_\lambda$$

and, after simplifying,

$$\lambda(\lambda^3 I_4 - \lambda I_4 - mN_{xyz})p_\lambda = 0$$

since $|x|^2 + |y|^2 + |z|^2 = 3$, $m = |x| \cdot |y| \cdot |z|$, $O_{xyz}O^t_{xyz} = I_4$, $N_{xyz} = O_{xyz} + O^t_{xyz}$. Hence, either $\lambda = 0$ or

$$(\lambda^3 - \lambda) \in m \cdot \mathrm{Spec}(N_{xyz}) = \{-2m, 2m, -2W, -2W\}.$$

This finishes the proof for $\lambda \neq 0$. If $\lambda = 0$, we get the conditions

$$0 = |z|q_\lambda + O^t_{xyz}|y|r_\lambda,$$
$$0 = |z|p_\lambda + |x|r_\lambda,$$
$$0 = |y|O_{xyz}p_\lambda + |x|q_\lambda,$$

implying immediately that $m = 0$ (since otherwise these equations give $p_\lambda = 0$) and the formula holds for this case as well. □

COROLLARY 4.1.2. *Let* $\alpha := \arccos(m)$, $\beta := \arccos(n)$. *Then*

$$\mathrm{Spec}(\widetilde{M_d}) = \left\{ 2\cos\left(\frac{\alpha + \pi k}{3}\right), 2\cos\left(\frac{\beta + \pi(2l+1)}{3}\right), 2\cos\left(\frac{\beta + \pi(2l+1)}{3}\right) \right\},$$

for $k = 0, 1, \ldots, 5$, $l = 0, 1, 2$.

PROOF. Indeed, if we put $\lambda = 2\cos\gamma$, the equations $\lambda^3 - 3\lambda + 2m = 0$, $\lambda^3 - 3\lambda - 2m = 0$, and $\lambda^3 - 3\lambda + 2n = 0$ become, respectively, $\cos(3\gamma) = -\cos\alpha$, $\cos(3\gamma) = \cos\alpha$, and $\cos(3\gamma) = -\cos\beta$, which implies the result. □

Let us now order the eigenvalues decreasingly:

$$\lambda_1 \geq \lambda_2 \geq \cdots \geq \lambda_{11} \geq \lambda_{12}.$$

COROLLARY 4.1.3. *One has the following*
1) $1 \le \lambda_4 \le \lambda_3 \le \lambda_2 \le \lambda_1 \le 2$.
2) $-2 \le \lambda_{12} \le \lambda_{11} \le \lambda_{10} \le \lambda_9 \le -1$.
3) $\lambda_{12} \le -\sqrt{3}$; $\lambda_1 \ge \sqrt{3}$.
4) *If* $\lambda_1/\lambda_3 = 2$ *(resp.* $\lambda_{12}/\lambda_{10} = 2$*), then the polynomial*
$$\chi_d(\lambda) = (\lambda+2)^3(\lambda-2)(\lambda+1)^2(\lambda-1)^6$$
in (4.1.1) *(resp.* $\chi_d(\lambda) = (\lambda-2)^3(\lambda+2)(\lambda-1)^2(\lambda+1)^6$*) and* $d = v_1$ *(resp.* $d = v_{12}$*), where* v_i *is a normalized eigenvector corresponding to* λ_i.

PROOF. All these conclusions except that concerning v_1 (resp. v_{12}) follow from Corollary 4.1.2 along with the following elementary lemma:

LEMMA 4.1.3. *Let* $F_m(x) = (x^3 - 3x - 2m)$ *with* $|m| \le 1$, *and let* $x_1 \ge x_2 \ge x_3$ *be its roots.*

1) *If* $0 \le m \le 1$, *then* $\sqrt{3} \le x_1 \le 2$, $x_3 \le -1$, *and each of the conditions* $x_1 = 2$, $x_3 = -1$ *implies* $m = 1$.
2) *If* $-1 \le m \le 0$, *then* $-2 \le x_3 \le -\sqrt{3}$, $1 \le x_1$, *and each of the conditions* $x_1 = 1$, $x_3 = -2$ *implies* $m = -1$.
3) *If* $|m| \le 0.75$, *then* $|x_1| > 1.38$, $|x_3| > 1.38$.

This lemma follows from the monotonicity of $\cos(x)$ on $[0, \pi]$ along with the inequalities
$$\cos\left(\frac{\arccos(0.75) + 2\pi}{3}\right) < -0.69, \qquad \cos\left(\frac{\arccos(0.75)}{3}\right) > 0.97.$$

To prove that $d = v_1$ one notes that $\lambda_1/\lambda_3 = 2$ implies $m(d) = n(d) = 1$ and hence the function P has an absolute maximum at d. Its derivative in the direction d equals 1, which gives $2Q_d(d) = 2$, i.e., $d = v_1$. The case of v_{12} is completely similar. □

COROLLARY 4.1.4. *Define*
$$\delta = \sup_{|d|=\sqrt{3}} \left(\max\left\{\frac{\lambda_+^\perp(d)}{\lambda_3(d)}, \frac{\lambda_-^\perp(d)}{\lambda_{10}(d)}\right\}\right)$$
where
$$\lambda_+^\perp(d) = 2 \sup_{v \perp d,\, |v|=\sqrt{3}} Q_d(v), \qquad \lambda_-^\perp(d) = 2 \inf_{v \perp d,\, |v|=\sqrt{3}} Q_d(v).$$
Then $\delta < \frac{3}{2}$.

PROOF. By 3) in Lemma 4.1.3 the claim is true for $|n| \le 0.75$ since $\frac{2}{1.38} < \frac{3}{2}$. Let then $n \ge 0.75$ (the case $n \le -0.75$ being symmetric). Suppose that $\frac{\lambda_+^\perp(d)}{\lambda_3(d)} \ge 3/2$, and hence $\lambda_+^\perp(d) \ge 1.5$ (since $\lambda_3(d) \ge 1$). We will show that the conditions $\lambda_+^\perp(d) \ge 1.5$ and $n \ge 0.75$ are incompatible. Indeed, define
$$T(x,y) = P(xd + y\sqrt{3}v_+^\perp(d)) = t_3 x^3 + t_2 x^2 y + t_1 xy^2 + t_0 y^3$$
where $v_+^\perp(d)$ is a vector with $|v_+^\perp(d)| = \sqrt{3}$ on which $Q_d(v)$ achieves its maximum. We get that $t_3 = T(1,0) = P(d) = n \ge 0.75$, $t_1 = T_x(0,1) = 3\lambda_+^\perp(d)/2 \ge 9/4$. For any (x,y) with $x^2 + y^2 = 1$ one has $|T(x,y)| = |P(xd + y\sqrt{3}v_+^\perp(d))| \le 1$. Let now

$x_0 = \frac{1}{\sqrt{2}}$, $y_0 = \pm \frac{1}{\sqrt{2}}$ where the sign of y_0 is chosen to ensure that $y_0(t_0 + t_2) \geq 0$. Then
$$|T(x_0, y_0)| \geq \frac{t_3 + t_1}{2\sqrt{2}} \geq \frac{3}{2\sqrt{2}} > 1,$$
which yields a contradiction. □

The following result will be used below to finish the proof.

COROLLARY 4.1.5. *Let $u \neq v \in \mathbb{R}^{12}$, $|u| = |v| = \sqrt{3}$. Then*
$$\frac{3\sqrt{3}}{4}\lambda_{10}(d)|u-v| \leq P(u) - P(v) \leq \frac{3\sqrt{3}}{4}\lambda_3(d)|u-v|,$$
where $d = \sqrt{3}(u-v)/|u-v|$.

PROOF. Denote $s = |u-v|/2$, $z = d^\perp \cap [u,v]$. Writing the Taylor expansion for the (cubic) function P, we get
$$P(u) - P(v) = 2s\left(\frac{Q_d(z)}{\sqrt{3}} + \frac{s^2 P_{ddd}}{18\sqrt{3}}\right).$$
Since
$$Q_d(z) \leq (1 - s^2/3) \sup_{x \perp d, |x| = \sqrt{3}} Q_d(x) \leq \tfrac{3}{2}(1 - \tfrac{1}{3}s^2)\lambda_+^\perp(d) \leq \tfrac{9}{4}(1 - \tfrac{1}{3}s^2)\lambda_3(d)$$
and
$$P_{ddd} \leq 2/\sqrt{3} \leq 2\lambda_3(d)/\sqrt{3},$$
we get
$$P(u) - P(v) \leq 2s\lambda_3(d)(9(1-s^2/3)/4\sqrt{3} + s^2/27) \leq 3\sqrt{3}\lambda_3(d)|u-v|/4.$$
The proof of the second inequality is completely similar. □

REMARK 4.1.3. Let us resume the spectral properties of $2Q_d$ when d varies over $\mathbb{S} = \mathbb{S}^{11}_{\sqrt{3}}$. We have a stratification
$$V_0 \subset \mathbb{S} \supset T \supset V = V_+ \cup V_-$$
where $T = \mathbb{S}^3_1 \times \mathbb{S}^3_1 \times \mathbb{S}^3_1$ is defined by the condition $m(d) = 1$, V_+ (resp. V_-) is defined by $P(d) = 1$ (resp. $P(d) = -1$), $V_0 = \{d : m(d) = 0\}$; each of V_+ and V_- is diffeomorphic to $\mathbb{S}^3_1 \times \mathbb{S}^3_1$. On V_+ (resp. V_-) the characteristic polynomial equals $(\lambda+2)^3(\lambda-2)(\lambda+1)^2(\lambda-1)^6$ (resp. $(\lambda-2)^3(\lambda+2)(\lambda-1)^2(\lambda+1)^6$); on $\mathbb{S} \setminus (T \cup V_0)$ we have
$$-\sqrt{3} < \lambda_{12}(d) < -2, \quad -\sqrt{3} < \lambda_9(d) < -1, \quad 1 < \lambda_4(d) < \sqrt{3}, \quad \lambda_1(d) < 2.$$
Finally, on V_0 the characteristic polynomial equals $\lambda^4(\lambda^2 - 3)^4$.

4.1.3. End of proof of Theorem 4.1.1. We finally show that the function
$$w(x) = \frac{P(x)}{|x|}$$
satisfies the desired properties, i.e., the map
$$H : \mathbb{S}^{11} \longrightarrow Q, \quad H(a) = D^2 w(a)$$
verifies the conditions of (H) in Subsection 4.1.1.

PROPOSITION 4.1.2. *Let* $a \neq b \in \mathbb{S}_1^{11}$. *Then there exist two vectors* $e, f \in \mathbb{S}_1^{11}$ *such that*
$$w_{ee}(a) - w_{ee}(b) \geq \frac{|a-b|}{4\sqrt{3}}, \quad w_{ff}(a) - w_{ff}(b) \leq -\frac{|a-b|}{4\sqrt{3}}.$$

PROOF. Let $d = \sqrt{3}(a-b)/|a-b|$. Recall that we denote by v_i, $i = 1, \ldots, 12$, the normalized eigenvectors of the form $2Q_d$ above (the eigenvalues $\lambda_i = \lambda_i(d)$ being written in decreasing order). Let V^+ be the 3-dimensional space generated by v_1, v_2, v_3 and let $e \in \mathbb{S}^{11} \cap V^+ \cap a^\perp \cap b^\perp$. It means in particular that $2Q_d(e) \geq \lambda_3(d)$. The conditions $b \perp e$, $a \perp e$ imply the equations $w_{ee}(a) = P_{ee}(a) - P(a)$, $w_{ee}(b) = P_{ee}(b) - P(b)$; hence
$$w_{ee}(a) - w_{ee}(b) = P_{ee}(a) - P_{ee}(b) - (P(a) - P(b)).$$
Since $P_{ee}(x)$ is a linear function, we get
$$P_{ee}(a) - P_{ee}(b) = \frac{|a-b|}{\sqrt{3}} P_{eed} = \frac{2|a-b|}{\sqrt{3}} Q_d(e) \geq \frac{|a-b|}{\sqrt{3}} \lambda_3(d).$$
Also we can write
$$P(a) - P(b) = \frac{P(a\sqrt{3}) - P(b\sqrt{3})}{3\sqrt{3}} \leq \frac{\sqrt{3}|a-b|\lambda_3(d)}{4}$$
by Corollary 4.1.5, and we get
$$w_{ee}(a) - w_{ee}(b) \geq \left(\frac{1}{\sqrt{3}} - \frac{\sqrt{3}}{4}\right)|a-b|\lambda_3(d) \geq \frac{|a-b|}{4\sqrt{3}}.$$
The second inequality is proven by replacing e by $f \in \mathbb{S}^{11} \cap V^- \cap a^\perp \cap b^\perp$ where V^- is generated by v_{10}, v_{11}, v_{12}. □

COROLLARY 4.1.6.
1) *The map* $H(a) = D^2 w(a) : \mathbb{S}^{11} \to Q$ *is a smooth embedding.*
2) *Let for* $a \neq b \in \mathbb{S}^{11}$
$$\mu_1 \geq \mu_2 \geq \cdots \geq \mu_{11} \geq \mu_{12}$$
be the eigenvalues of $H(a) - H(b)$. *Then*
$$M^{-1} = \frac{1}{1536\sqrt{3}} \leq -\frac{\mu_1}{\mu_{12}} \leq 1536\sqrt{3} = M.$$

PROOF. 1) This follows immediately from Proposition 4.1.2.
2) An easy calculation shows that $|w_{efg}(x)| \leq 2^5$ for any $e, f, g, x \in \mathbb{S}^{11}$. Hence
$$|w_{ef}(a) - w_{ef}(b)| \leq |w_{efd'}(d')| \cdot |a-b| \leq 2^5 |a-b|$$
for $d' = d/\sqrt{3}$.
Since all elements of the matrix $H(a) - H(b)$ are of absolute value $\leq 2^5 |a-b|$, all its eigenvalues are of absolute value $\leq 12 \cdot 2^5 |a-b|$. Using the inequalities of Proposition 4.1.2 we reach the conclusion. □

REMARK 4.1.4. We have the inequalities
$$2\lambda_3 \geq \lambda_1, \quad 2\lambda_{n-2} \leq \lambda_n, \quad n = 12 \text{ (or } 24\text{)},$$
which hold for the eigenvalues of P_{12} (and of P_{24} in the next section). They are

essential for the proofs in this section and are in fact the best possible. Indeed, one has the following result:

PROPOSITION 4.1.3. *Let P be a cubic form in $\mathbb{R}^n$. Then for any unit vector $d \in \mathbb{S}^{n-1} \subset \mathbb{R}^n$ the eigenvalues $\lambda_1 \geq \lambda_2 \geq \cdots \geq \lambda_n$ of the quadratic form $P_d := \sum_i d_i P_{x_i}$ satisfy*

$$\lambda_1 \geq 2\lambda_2, \quad 2\lambda_{n-1} \geq \lambda_n.$$

PROOF. Assume that at the point $a \in \mathbb{S}^{n-1}$ the cubic form $P \neq 0$ attains its supremum over $\mathbb{S}^{n-1}$. Since P is an odd function on $\mathbb{R}^n$, $P(a) > 0$. Choose $d = a$ and let $(x_1, \ldots, x_n)$ be orthonormal coordinates in $\mathbb{R}^n$ such that x_1 is directed along d. Since the form P attains at d its supremum over $\mathbb{S}^{n-1}$, it follows that in the coordinates x_i the cubic form P contains no monomials of the form $cx_1^2 x_i$, $i > 1$. Thus the quadratic form P_d contains no monomials of the form $cx_1 x_i$, $i > 1$, and hence the vector d is an eigenvector of the quadratic form P_d with the eigenvalue which we denote by λ. Let λ' be the maximal eigenvalue of P_d on the orthogonal complement of d attained on an eigenvector $b \in \mathbb{S}^{n-1}$. The proposition will follow if we prove that $\lambda \geq 2\lambda'$. We assume without loss of generality that $\lambda = 1$ and that x_2 is directed along b. Then the restriction of P_d on the plane $\{x_1, x_2\}$ can be written in the form

$$x_1^2 + \lambda' x_2^2$$

and thus the restriction of the cubic form P on this plane becomes

$$\frac{x_1^3}{3} + \lambda' x_1 x_2^2 + c x_2^3.$$

It is easy to see that if $\lambda' > 1/2$, then the supremum of the function P on the circle $x_1^2 + x_2^2 = 1$ is not at the point $(1, 0)$, which implies the result. $\square$

REMARK 4.1.5. One can "specialize" the polynomial P_{12} for purely imaginary q_i's. More precisely, let a variable vector $x = (X, Y, Z) \in \mathbb{R}^9$ be concatenated from X, Y, and $Z \in \mathbb{R}^3$. Let us then define $q'_X := X_1 \cdot i + X_2 \cdot j + X_3 \cdot k \in \operatorname{Im} \mathbb{H}$, and similarly for Y, Z. Set

$$\begin{aligned} P_9(X, Y, Z) &= \operatorname{Re}(q'_X \, q'_Y \, q'_Z) \\ &= X_1 Y_3 Z_2 - X_1 Y_2 Z_3 + X_2 Y_1 Z_3 - X_2 Y_3 Z_1 + X_3 Y_2 Z_1 - X_3 Y_1 Z_2 \\ &= -\det M_{XYZ}, \end{aligned}$$

where M_{XYZ} is the matrix with the lines X, Y, and Z. Then the function

$$w_9(x) := \frac{P_9(x)}{|x|}, \ x \in \mathbb{R}^9,$$

satisfies the above ellipticity criterion and thus is a nonclassical solution of a uniformly elliptic fully nonlinear equation in $\mathbb{R}^9$. However, this function does not verify the stronger necessary conditions for a *Hessian* equation and thus is not a solution of a Hessian uniformly elliptic fully nonlinear equation. We do not give a proof of that since the only proofs known to us are either too cumbersome or of a very calculatory nature (that of C. Smart, private communication). In any case, this result is superseded by those of Chapter 5.

4.2. The octonionic construction

In this section we show that $w_{24,\delta}$ provides singular solutions to Hessian uniformly elliptic equations in twenty-one (and more) dimensions. Moreover the following theorem holds:

THEOREM 4.2.1. *For any δ, $1 \leq \delta < 2$, and any coordinate plane $H' \subset \mathbb{R}^{24}$, $\dim H' = 21$, there exists a uniformly elliptic Hessian equation $F(D^2 w) = 0$ such that the restriction*

$$(w_{24,\delta})_{|H'} = \left(\frac{P_{24}(x)}{|x|^\delta}\right)_{|H'}$$

is a viscosity solution to a Hessian equation $F(D^2 w) = 0$ in the unit ball $B \subset \mathbb{R}^{21}$ with a Lipschitz functional F satisfying the uniform ellipticity condition for the triality polynomial

$$P_{24}(x) = \operatorname{Re}((o_1 \cdot o_2) \cdot o_3) = \operatorname{Re}(o_1 \cdot (o_2 \cdot o_3)),$$

where $o_i \in \mathbb{O}$, $i = 1, 2, 3$, $x = (o_1, o_2, o_3) \in \mathbb{O}^3 = \mathbb{R}^{24}$.

We call a plane $H' \subset \mathbb{R}^{24}$, $\dim H' = 21$, "coordinate" if it is defined by the equations

$$x_i = x_j = x_k = 0, \quad i \neq j \neq k \neq i.$$

In fact for the proof it is essential that the restriction of $P_{24}(x)$ to H' remains harmonic; it is true for infinitely many planes $H' \subset \mathbb{R}^{24}$, e.g., for those defined by three linear equations

$$l_1(x_1, \ldots, x_8) = l_2(x_9, \ldots, x_{16}) = l_3(x_{17}, \ldots, x_{24}) = 0$$

for any linear forms $l_1, l_2, l_3 \in (\mathbb{R}^8)^* \setminus \{0\}$. Therefore, the result remains true for such planes.

REMARK 4.2.1. For $\delta > 1$ the restriction of the solution on the ball B is essential: There are no homogeneous order $\alpha < 2$ solutions defined in the whole space $\mathbb{R}^n$ [**184**].

First we give a criterion for a solution of a Hessian equation. Let $w = w_\delta$ be a homogeneous function of order $3 - \delta$, $1 \leq \delta < 2$, defined on a unit ball $B = B_1 \subset \mathbb{R}^n$ and smooth in $B \setminus \{0\}$. Then the Hessian of w is homogeneous of order $(1 - \delta)$. Define the map

$$\Lambda : B \longrightarrow \mathbb{R}^n, \ x \mapsto \lambda(S) = \{\lambda_i : \lambda_1 \leq \cdots \leq \lambda_n\},$$

$\lambda(S)$ being the (ordered) set of eigenvalues of the matrix $S := D^2 w(x)$.

Let $K \subset \mathbb{R}^n$ be an open convex cone such that

$$\{x \in \mathbb{R}^n : x_i/|x| > c, \ i = 1, \ldots, n\} \subset K,$$

where $c > 0$ is a positive constant.

Set $L := \mathbb{R}^n \setminus (K \cup -K)$. We say that a set $E \subset \mathbb{R}^n$ satisfies the *K-cone condition* if $(a - b) \in L$ for any $a, b \in E$.

Let Σ_n be the group of permutations of $\{1, \ldots, n\}$. For any $\sigma \in \Sigma_n$, we denote by T_σ the linear transformation of $\mathbb{R}^n$ given by $x_i \mapsto x_{\sigma(i)}$, $i = 1, \ldots, n$.

4.2. THE OCTONIONIC CONSTRUCTION

LEMMA 4.2.1. *Assume that*

$$M := \bigcup_{\sigma \in \Sigma_n} T_\sigma \Lambda(B) \subset \mathbb{R}^n$$

satisfies the K-cone condition. If $\delta > 1$, we assume additionally that w changes sign in B. Then w is a viscosity solution in B of a uniformly elliptic Hessian equation.

PROOF. Let us choose in the space $\mathbb{R}^n$ an orthogonal coordinate system $z_1, \ldots, z_{n-1}, s$ such that $s = x_1 + \cdots + x_n$. Let $\pi : \mathbb{R}^n \to Z$ be the orthogonal projection of $\mathbb{R}^n$ onto the z-space. Let K^* denote the adjoint cone of K; that is, $K^* = \{b \in \mathbb{R}^n : b \cdot c \geq 0, \forall c \in K\}$. Notice that $a \in L$ if and only if $a \cdot b = 0$ for some $b \in K^*$. We represent the boundary of the cone K as the graph of a Lipschitz function $s = e(z)$, with $e(0) = 0$, the function e being convex, homogenous of order 1, and smooth outside the origin:

$$e(z) = \inf\{c : (z + cs) \in K\}.$$

Set $m = \pi(M)$. We prove that M is a graph of a Lipschitz function g defined on m,

$$M = \{s = g(z),\ z \in m\}.$$

Let $a, \hat{a} \in M, a = (z, s), \hat{a} = (z', \hat{s})$. Since $a - \hat{a} \in L$, we have

$$-e(z - z') \leq \hat{s} - s \leq e(z - z').$$

Since $e(0) = 0, g(z) := s$ is single-valued. Also

$$|g(z) - g(z')| = |s - \hat{s}| \leq |e(z - z')| \leq C|z - z'|.$$

The function g has an extension $\widetilde{g}$ from the set m to $\mathbb{R}^{n-1}$ such that $\widetilde{g}$ is a Lipschitz function and the graph of $\widetilde{g}$ satisfies the K-cone condition. One can define such an extension $\widetilde{g}$ simply by the formula

$$\widetilde{g}(z) := \inf_{w \in m} \{g(w) + e(z - w)\}.$$

Indeed, let $(z, \widetilde{g}(z)), (z', \widetilde{g}(z'))$ lie in the graph $\widetilde{g}$. We must show

$$-e(z - z') \leq \widetilde{g}(z) - \widetilde{g}(z') \leq e(z - z').$$

Now $\widetilde{g}(z') = g(w) + e(z' - w)$ for some $w \in m$. Thus

$$\widetilde{g}(z) - \widetilde{g}(z') \leq g(w) + e(z - w) - (g(w) + e(z' - w)) \leq e(z - z')$$

since $e(a + b) \leq e(a) + e(b)$, as $e(\cdot)$ is convex and homogenous of order 1. Similarly $\widetilde{g}(z) - \widetilde{g}(z') \geq -e(z - z')$.

Let us set

$$f' := s - \widetilde{g}(z).$$

Since the level surface of the function f' satisfies the K-cone condition, it follows that $\nabla f' \in K^*$ a.e., where K^* is the adjoint cone to K. Moreover, $f'_{|M} = 0$.

Set $f = \sum_{\sigma \in \Sigma_n} f'(\sigma(x))$. Then f is a Lipschitz function invariant under the action of the group Σ_n. Since K^* is a convex cone, it follows that $\nabla f \in K^*$ a.e., and since the set M is invariant under the action of the group Σ_n, it follows that $f_{|M} = 0$, and hence the function w satisfies the equation

$$f(\lambda(D^2 w)) = 0$$

on $B \setminus \{0\}$ (we recall that $\lambda(\cdot)$ is the spectrum of the quadratic form).

We show now that w is a viscosity solution of $F(D^2 w) = 0$ on the whole ball B.

Assume first that $\delta = 1$. Let $p(x)$, $x \in B$, be a quadratic form such that $p \leq w$ on B. We choose any quadratic form $p'(x)$ such that $p \leq p' \leq w$ and there is a point $x' \neq 0$ at which $p'(x') = w(x')$. Then it follows that $F(p) \leq F(p') \leq 0$. Consequently for any quadratic form $p(x)$ from the inequality $p \leq w$ ($p \geq w$) it follows that $F(p) \leq 0$ ($F(p) \geq 0$). This implies that w is a viscosity solution of $F(D^2 w) = 0$ in B (see Proposition 2.4 in [**42**]).

If $1 < \delta < 2$, then for any smooth function p in B the function $w - p$ changes sign in any neighborhood of 0. Hence, by the same proposition in [**42**], w is a viscosity solution in B. □

Next we need the following property of the eigenvalues $\lambda_1 \geq \lambda_2 \geq \cdots \geq \lambda_n$ of real symmetric matrices of order n, which is a classical result by H. Weyl [**272**]:

LEMMA 4.2.2. *Let A, B be two real symmetric matrices with the eigenvalues $\lambda_1 \geq \lambda_2 \geq \cdots \geq \lambda_n$ and $\lambda'_1 \geq \lambda'_2 \geq \cdots \geq \lambda'_n$, respectively. Then for the eigenvalues $\Lambda_1 \geq \Lambda_2 \geq \cdots \geq \Lambda_n$ of the matrix $A - B$ we have*

$$\Lambda_1 \geq \max_{i=1,\ldots,n} (\lambda_i - \lambda'_i), \quad \Lambda_n \leq \min_{i=1,\ldots,n} (\lambda_i - \lambda'_i).$$

4.2.1. Characteristic polynomial. First one notes that the triality polynomial $P(a) = P_{24}(a)$ is clearly harmonic and, moreover, its restriction onto a coordinate plane is also harmonic. Its principal property for us here is

PROPOSITION 4.2.1. *Let $a = (x, y, z) \in \mathbb{S}^{23}, |a| = 1, W := P(a), m := |x|\cdot|y|\cdot|z|$. Then the characteristic polynomial $G(\lambda)$ of $H(a) := D^2 P(a)$ is given by*

$$G(\lambda) = (\lambda^3 - \lambda + 2m)(\lambda^3 - \lambda - 2m)(\lambda^3 - \lambda + 2W)^6.$$

PROOF. In Section 2.3.2 we have seen that the automorphism group $Aut(t_8)$ of the triality t_8 coincides with $\mathrm{Spin}(8)$ and thus the triality polynomial $P(X, Y, Z)$ is $\mathrm{Spin}(8)$-invariant. Therefore the characteristic polynomial $G(\lambda)$ is invariant under the action of $\mathrm{Spin}(8)$, and we can suppose without loss of generality (applying the action) that the vectors $x \in \mathbb{R}$, $y \in \mathbb{R} + e_1 \mathbb{R}$, $z \in \mathbb{R} + e_1 \mathbb{R} + e_2 \mathbb{R} \subset \mathbb{O}$. Indeed the action of $\mathrm{Spin}(8)$ on

$$P : \mathbb{O} \times \mathbb{O} \times \mathbb{O} \longrightarrow \mathbb{R}$$

is given by three different projections of $\mathrm{Spin}(8)$ onto $SO(8)$ so that the tensor product of all three preserves the cubic form. One of these is the composition of the standard projection of $\mathrm{Spin}(8)$ onto $SO(8)$ with the standard action. The two others correspond to two other projections given by two other nontrivial elements of the center of $\mathrm{Spin}(8)$ isomorphic to $(\mathbf{Z}/2)^2$. To get the conclusion one moves x into $\mathbb{R}$ by the standard action of $SO(8)$, y going to y' by the second action and z going to z' by the third one. Next, by standard orthogonalization one gets $y' = y_0 + y_1 u_1$, $z' = z_0 + z_1 u_1 + z_2 u_2$, y_0, y_1, z_0, z_1, z_2 being real, and $u_1 \perp u_2$ being purely imaginary unit vectors. Then it is sufficient to note that the automorphism group G_2 of $\mathbb{O}$ acts transitively on the Stiefel manifold $V_{7,2}$ of orthonormal 2-frames in $\mathbb{R}^7$ (see Section 2.5) to bring u_1 to e_1, u_2 to e_2, thus obtaining the claim (note that the actions of $G_2 \subset \mathrm{Spin}(8)$ on $\mathbb{O}$ induced by the three above actions of $\mathrm{Spin}(8)$ coincide).

Therefore, $x, y, z \in \mathbb{H} \subset \mathbb{O}$ where $\mathbb{H}$ is generated by $\{1, e_1, e_2, e_4\}$. Brute force calculations give

$$H(a) = \begin{pmatrix} H_0 & 0 \\ 0 & H_1 \end{pmatrix}$$

4.2. THE OCTONIONIC CONSTRUCTION

for the Hessian $H(a)$ of P relative to the following ordering of coordinates in $\mathbb{R}^{24}$:

$$\{x_0, x_1, x_2, x_4, y_0, y_1, y_2, y_4, z_0, z_1, z_2, z_4, x_5, x_6, x_3, x_7, y_5, y_6, y_3, y_7, z_5, z_6, z_3, z_7\}$$

for the following matrices $H_0, H_1 \in M_{12}(\mathbb{R})$:

$$H_0 = \begin{pmatrix} 0_4 & M_z & M_y \\ M_z^t & 0_4 & M_x \\ M_y^t & M_x^t & 0_4 \end{pmatrix}, \qquad H_1 = \begin{pmatrix} 0_4 & L_z & L_y \\ L_z^t & 0_4 & L_x \\ L_y^t & L_x^t & 0_4 \end{pmatrix}$$

where

$$M_s = \begin{pmatrix} s_0 & -s_1 & -s_2 & -s_3 \\ -s_1 & -s_0 & -s_3 & s_2 \\ -s_2 & s_3 & -s_0 & -s_1 \\ -s_3 & -s_2 & s_1 & -s_0 \end{pmatrix}, \qquad L_s = \begin{pmatrix} -s_0 & -s_1 & s_2 & -s_3 \\ s_1 & -s_0 & -s_3 & -s_2 \\ -s_2 & s_3 & -s_0 & -s_1 \\ s_3 & s_2 & s_1 & -s_0 \end{pmatrix}$$

for an arbitrary $s = (s_0, s_1, s_2, s_3) \in \mathbb{R}^4$.

Direct easy calculations show that M_s, L_s have the following properties:

1) $M_s \cdot M_s^t = M_s^t \cdot M_s = L_s \cdot L_s^t = L_s^t \cdot L_s = |s|^2 I_4$.

Thus, M_s, L_s are proportional to orthogonal matrices. In particular, if $|s| = 1$, then M_s, L_s are orthogonal themselves. We write $M_s = |s| O_s$, $L_s = |s| O_s'$ with $O_s, O_s' \in O(4)$.

2) $\det(M_s) = -|s|^4$, $\det(O_s) = -1$, $\det(L_s) = |s|^4$, $\det(O_s') = 1$.

3) The characteristic polynomials $PM_s(\lambda), PL_s(\lambda)$ of M_s, L_s factor as

$$PM_s(\lambda) = (\lambda^2 - |s|^2)(\lambda^2 + 2s_0\lambda + |s|^2), \qquad PL_s(\lambda) = (\lambda^2 + 2s_0\lambda + |s|^2)^2$$

and those of O_s, O_s' as

$$PO_s(\lambda) = (\lambda^2 - 1)(\lambda^2 + 2s_0^*\lambda + 1), \qquad PO_s'(\lambda) = (\lambda^2 + 2s_0^*\lambda + 1)^2$$

with $s_0^* = s_0/|s|$.

4) Define the symmetric matrices $N_s := (O_s + O_s^t)$, $N_s' := (O_s' + O_s'^t)$; then their spectrums are

$$\text{Spec}(N_s) = \{2, -2, -2s_0^*, -2s_0^*\}, \qquad \text{Spec}(N_s') = \{-2s_0^*, -2s_0^*, -2s_0^*, -2s_0^*\}.$$

5) For the product matrices $M_{rst} = M_r \cdot M_s \cdot M_t$, $L_{rst} = L_r \cdot L_s \cdot L_t$, $r, s, t \in \mathbb{R}^4$, we have the characteristic polynomials PM_{rst}, PL_{rst} of M_{rst}, L_{rst}:

$$PM_{rst}(\lambda) = (\lambda^2 - |r|^2|s|^2|t|^2)(\lambda^2 + 2P(r,s,t)\lambda + |r|^2|s|^2|t|^2),$$

$$PL_{rst}(\lambda) = (\lambda^2 + 2P(r,s,t)\lambda + |r|^2|s|^2|t|^2)^2.$$

The characteristic polynomial F of H_0 is calculated in Proposition 4.1.1 and the characteristic polynomial G of H_1 can be calculated in the same way using L_s instead of M_s. □

REMARK 4.2.2. If we do not insist on a computer free proof of the fact, the inclusions $x \in \mathbb{R}$, $y \in \mathbb{R} + e_1\mathbb{R}$, $z \in \mathbb{R} + e_1\mathbb{R} + e_2\mathbb{R}$ will suffice. Indeed, the MAPLE instructions (v being the coordinate vector)

```
H:=hessian(P,v):
X2:=0:  X4:=0:  Y2:=0:  Y4:=0:  Z2:=0:  Z4:=0:  X5:=0:  X6:=0:
X3:=0:
X7:=0:  Y5:=0:  Y6:=0:  Y3:=0:  Y7:=0:  Z5:=0:  Z6:=0:  Z3:=0:
Z7:=0:
G:= factor (charpoly(H,T));
```

return the formula of Proposition 4.2.1 in twenty seconds, < 60 MB of space.

COROLLARY 4.2.1. *Define the angles α, β by*
$$\alpha := \arccos(3\sqrt{3}m), \qquad \beta := \arccos(3\sqrt{3}W).$$
Then
$$\operatorname{Spec}(H(a)) = \left\{ \frac{2}{\sqrt{3}}\cos(\alpha/3 + \pi k/3), 6 \times \left\{ \frac{2}{\sqrt{3}}\cos(\beta/3 + \pi(2l+1)/3) \right\} \right\},$$
for $k = 0, 1, \ldots, 5$, $l = 0, 1, 2$.

PROOF. Indeed, if we put $\lambda = \frac{2}{\sqrt{3}}\cos\gamma$, the equations $\lambda^3 - \lambda + 2m = 0$, $\lambda^3 - \lambda - 2m = 0$, and $\lambda^3 - \lambda + 2W = 0$ become, respectively, $\cos(3\gamma) = -\cos\alpha$, $\cos(3\gamma) = \cos\alpha$, and $\cos(3\gamma) = -\cos\beta$, which implies the result. □

Write the eigenvalues of $H(a)$ in decreasing order:
$$\lambda_1 \geq \lambda_2 \geq \cdots \geq \lambda_{23} \geq \lambda_{24}.$$
Since $|W| \leq m$ and the cosine decreases on $[0, \pi]$, we get

COROLLARY 4.2.2.
$$\begin{array}{llll}
\lambda_1 = \frac{2}{\sqrt{3}}\cos(\frac{\alpha}{3}), & \lambda_8 = l_1, & \lambda_9 = l_2, & \lambda_{16} = -l_2, \quad \lambda_{17} = -l_1, \\
\lambda_2 = \cdots = \lambda_7 = \mu_1, & \lambda_{10} = \cdots = \lambda_{15} = \mu_2, & & \lambda_{18} = \cdots = \lambda_{23} = \mu_3, \\
\lambda_{24} = -\frac{2}{\sqrt{3}}\cos(\frac{\alpha}{3}), & & &
\end{array}$$
for
$$l_1 = \frac{2}{\sqrt{3}}\cos\left(\frac{\alpha + 5\pi}{3}\right) = \frac{2}{\sqrt{3}}\cos\left(\frac{\pi - \alpha}{3}\right),$$
$$l_2 = \frac{2}{\sqrt{3}}\cos\left(\frac{\alpha + \pi}{3}\right),$$
and $\mu_1 \geq \mu_2 \geq \mu_3$ being the roots of $T^3 - T + 2W = 0$.

COROLLARY 4.2.3. *Let $a = (x, y, z) \in \mathbb{S}_1^{23}$, let $H = H_{21} \subset \mathbb{R}^{24} = \mathbb{O}^3$ be a "coordinate" plane, $\dim(H) = 19$, and let*
$$\lambda_1' \geq \lambda_2' \geq \cdots \geq \lambda_{20}' \geq \lambda_{21}'$$
be the eigenvalues of the Hessian $D^2 P_{|H}(a)$ written in decreasing order. Then
$$\lambda_2' = \mu_1, \quad \lambda_{10}' = \mu_2, \quad \lambda_{18}' = \mu_3,$$
$\mu_1 \geq \mu_2 \geq \mu_3$ being the roots of $T^3 - T + 2W = 0$.

PROOF. Taking into account Corollary 4.2.2 it is sufficient to prove that μ_1, μ_2, μ_3 have multiplicity at least 3 in the spectrum of $D^2 P_{|H}(a)$ and that these roots have the right positions in the spectrum. The result on the multiplicity is proved by a normalization similar to that used in the proof of Proposition 4.2.1 and then an explicit MAPLE calculation which gives the result in less than a minute. Let us give the details of positioning μ_1, μ_2, μ_3 for the "coordinate" plane $x_7 = y_7 = z_7 = 0$ (by skew symmetry this is sufficient). One can reduce the calculation to the case
$$x = a \in \mathbb{R}, \quad y = b + di \in \mathbb{R} + i\mathbb{R}, \quad z = c + ei + fj \in \mathbb{R} + i\mathbb{R} + j\mathbb{R}$$
where one can suppose that
$$a^2 + b^2 + c^2 + d^2 + e^2 + f^2 = 1.$$

Then the characteristic polynomial is of the form $P_0^3 \cdot P_1 \cdot P_2$ with

$$P_0 := (T^3 - T + 2bca - 2ead) = (T - \mu_1)(T - \mu_2)(T - \mu_3),$$

$$\begin{aligned}P_1 := \ & T^6 - (c^2 + b^2 + a^2 + 1)T^4 - 2a(de - 2bc)T^3 \\ & - (a^2d^2 - b^2 + b^2d^2 - a^2 + d^4 - c^2 + c^2d^2 + d^2e^2 - d^2)T^2 \\ & - 2a(b^3c + bc^3 - c^2de + bc - a^2de + a^2bc - b^2de)T \\ & + a^2(e^2d^2 - d^2 + 4c^2b^2 - 4dcbe + d^4 + d^2a^2 + d^2c^2 + b^2d^2),\end{aligned}$$

$$P_2 := T^6 - 2T^4 + T^2 - 4b^2a^2 + 8b^2a^2d^2 - 4a^2d^2 + 4b^2a^4 + 4d^4a^2 + 4d^2a^4 + 4b^4a^2.$$

Recall that the *resultant* $\mathrm{Res}(Q_1, Q_2)$ of two monic polynomials Q_1 and Q_2 is a polynomial in the coefficient of Q_1 and Q_2 such that the resultant is zero if and only if Q_1 and Q_2 have a mutual root. Calculating the resultants $\mathrm{Res}(P_1, P_0)$ and $\mathrm{Res}(P_2, P_0)$ using MAPLE, one gets

$$R_1 := \mathrm{Res}(P_1, P_0) = (a^2 + b^2 + c^2 + d^2 + e^2 - 1)^3(1 - 2a^2 + a^4 - 4c^2b^2 - 4e^2d^2 + 8edcb)a^2,$$

$$R_2 := \mathrm{Res}(P_2, P_0) = 64a^6(a^2b^2 + a^2d^2 + b^4 + 2b^2d^2 + d^4 - b^2 - d^2 + c^2b^2 - 2dcbe + e^2d^2)^3.$$

These resultants do not change sign for the variables a, b, c, d, e, f all nonzero. Indeed, for the only nontrivial factor of R_1 one has

$$\begin{aligned}8dcbe - 4e^2d^2 + 1 - 2a^2 + a^4 - 4c^2b^2 &= (1 - a^2 + 2bc - 2de)(1 - a^2 - 2bc + 2de) \\ &\geq (1 - a^2 - b^2 - c^2 - d^2 - e^2)^2 \\ &\geq 0, \quad R_1 \leq 0,\end{aligned}$$

and similarly

$$\begin{aligned}a^2b^2 + a^2d^2 &+ b^4 + 2b^2d^2 + d^4 - b^2 - d^2 + c^2b^2 - 2dcbe + e^2d^2 \\ &= (b^2 + d^2)(a^2 + b^2 + d^2 - 1) + (bc + ed)^2 \\ &\leq -(b^2 + d^2)(e^2 + c^2) + c^2b^2 - 2dcbe + e^2d^2 \\ &= -(be + cd)^2 \leq 0, \quad R_2 \leq 0.\end{aligned}$$

Therefore it is sufficient to prove the positioning of μ_1, μ_2, μ_3 for a single point

$$a = b = c = d = e = f = 1/\sqrt{6}$$

for which $\mu_1 = 1, \mu_2 = 0, \mu_3 = -1$ and the whole spectrum is 1

$$\{-1.14, -1, -1, -1, -.96, -.86, -.74, -.39, -.03, 0, 0, 0, .39, .41, .46, .74, .98, 1, 1, 1, 1.14\}.$$

$\square$

4.2.2. End of proof of Theorem 4.2.1. Recall that a family $\mathcal{A}$ of symmetric matrices is called *uniformly hyperbolic* if

$$\frac{1}{M} < -\frac{\lambda_1(A)}{\lambda_n(A)} < M$$

for a positive constant M and any $A \in \mathcal{A}$. Theorem 4.2.1 is implied by the uniform

hyperbolicity of the family
$$D^2 w_{\delta|H}(a) - O^t \cdot D^2 w_{\delta|H}(b) \cdot O,$$
which we are going to prove.

PROPOSITION 4.2.2. *Let $H \subset \mathbb{R}^{24}$ be a coordinate plane, $\dim H = 21$. Set $M_\delta(u) = D^2 w_{\delta|H}(u)$ for $u \in H$, $1 \le \delta < 2$. Suppose that $a \ne b \in H$ and let $O \in O(21)$ be an orthogonal matrix such that $M_\delta(a,b,O) := M_\delta(a) - O^t \cdot M_\delta(b) \cdot O \ne 0$. Denote by $\Lambda_1 \ge \Lambda_2 \ge \cdots \ge \Lambda_{21}$ the eigenvalues of the matrix $M_\delta(a,b,O)$. Then*
$$\varepsilon \le \frac{\Lambda_1}{-\Lambda_{21}} \le \frac{1}{\varepsilon}$$
for $\varepsilon := \min\{\frac{2-\delta}{4+\delta}, \frac{1}{20}\}$.

PROOF OF PROPOSITION 4.2.2. We can suppose without loss of generality that $|a| \le |b|$; moreover, by homogeneity we can suppose that $a \in \mathbb{S}_1^{20}$ and thus $|b| \ge 1$. Let $\widetilde{b} := b/|b| \in \mathbb{S}_1^{20}$; then $M_\delta(b) = M_\delta(\widetilde{b})|b|^{1-\delta}$. Then one needs the following result for the points $a, \widetilde{b} \in \mathbb{S}_1^{20}$:

LEMMA 4.2.3. *Let $\delta \in [1,2)$, $a, \widetilde{b} \in \mathbb{S}_1^{20}$, $W = W(a)$, $\widetilde{W} = W(\widetilde{b})$, and let $\mu_1(\delta) \ge \mu_2(\delta) \ge \mu_3(\delta)$,*
$$\mu_1(\delta) = \frac{2}{\sqrt{3}} \cos\left(\frac{\arccos(3\sqrt{3}W) - \pi}{3}\right) - W\delta,$$
$$\mu_2(\delta) = \frac{2}{\sqrt{3}} \cos\left(\frac{\arccos(3\sqrt{3}W) + \pi}{3}\right) - W\delta,$$
$$\mu_3(\delta) = -\frac{2}{\sqrt{3}} \cos\left(\frac{\arccos(3\sqrt{3}W)}{3}\right) - W\delta$$

(resp., $\widetilde{\mu}_1(\delta) \ge \widetilde{\mu}_2(\delta) \ge \widetilde{\mu}_3(\delta)$) be the roots of the polynomial
$$P_{1,\delta}(T,W) := Q_1(T + \delta W)$$
$$= T^3 + 3W\delta T^2 + (3W^2\delta^2 - 1)T + W(2-\delta) + W^3\delta^3$$

(resp. of the polynomial
$$\widetilde{P}_{1,\delta}(T,\widetilde{W}) := Q_1(T + \delta\widetilde{W})$$
$$= T^3 + 3\widetilde{W}\delta T^2 + (3\widetilde{W}^2\delta^2 - 1)T + \widetilde{W}(2-\delta) + \widetilde{W}^3\delta^3 \,).$$

Then for any $K > 0$ satisfying $|K-1| + |\widetilde{W} - W| \ne 0$ one has
$$\frac{2-\delta}{4+\delta} =: \varepsilon \le \frac{\mu_+(K)}{-\mu_-(K)} \le \frac{1}{\varepsilon} = \frac{4+\delta}{2-\delta}$$

where
$$\mu_-(K) := \min\{\mu_1(\delta) - K\widetilde{\mu}_1(\delta), \mu_2(\delta) - K\widetilde{\mu}_2(\delta), \mu_3(\delta) - K\widetilde{\mu}_3(\delta)\},$$
$$\mu_+(K) := \max\{\mu_1(\delta) - K\widetilde{\mu}_1(\delta), \mu_2(\delta) - K\widetilde{\mu}_2(\delta), \mu_3(\delta) - K\widetilde{\mu}_3(\delta)\}\,.$$

PROOF. In the proof we will repeatedly use the following easy elementary fact:

CLAIM (∗). Let $l_1 + l_2 + l_3 = t \geq 0$, $l_3 \leq -ht$, with $h > 0$. Denote
$$l := \min\{l_i, i = 1, 2, 3\}, \quad L := \max\{l_i, i = 1, 2, 3\}.$$
Then
$$-L/l \in [h/(2h+1), (2h+1)/h], \ for \ t > 0, \quad -L/l \in [1/2, 2], \ for \ t = 0.$$

If $W = \widetilde{W}$, $K = 1$, there is nothing to prove. If $K = 1$, one can suppose that $\widetilde{W} > W$; we have
$$(\mu_1(\delta) - K\widetilde{\mu}_1(\delta)) + (\mu_2(\delta) - K\widetilde{\mu}_2(\delta)) + (\mu_3(\delta) - K\widetilde{\mu}_3(\delta)) = 3(\widetilde{W} - W)\delta = t$$
and
$$\mu_3(\delta) - K\widetilde{\mu}_3(\delta) = (\widetilde{W} - W)\delta + \frac{2}{\sqrt{3}}\left(\cos\left(\frac{\arccos(3\sqrt{3}\widetilde{W})}{3}\right) - \cos\left(\frac{\arccos(3\sqrt{3}W)}{3}\right)\right)$$
$$\leq (2-\delta)(W - \widetilde{W}) = -(2-\delta)(\widetilde{W} - W) = -\frac{(2-\delta)}{3\delta}t.$$

Therefore, by Claim (∗) with $h = \frac{(2-\delta)}{3\delta}$ one can take
$$\varepsilon = h/(2h+1) = (2-\delta)/(4+\delta)$$
in this case.

We can suppose then $W > \widetilde{W}, K \neq 1$ (if $\widetilde{W} > W$, permute $\widetilde{W}$ with W and replace K by K^{-1}). Using the relations
$$\mu_1(\delta)(-W) = -\mu_3(\delta)(W),$$
$$\mu_2(\delta)(-W) = -\mu_2(\delta)(W),$$
$$\mu_3(\delta)(-W) = -\mu_1(\delta(W),$$
we can suppose without loss of generality that $K < 1$.

We then distinguish three cases corresponding to different signs of $W - K\widetilde{W}$. If $W - K\widetilde{W} = 0$, then one can take $\varepsilon = 1/2$ by Claim (∗) since
$$t = (\mu_1(\delta) - K\widetilde{\mu}_1(\delta)) + (\mu_2(\delta) - K\widetilde{\mu}_2(\delta)) + (\mu_3(\delta) - K\widetilde{\mu}_3(\delta)) = W - K\widetilde{W} = 0.$$
Let $W - K\widetilde{W} = W - \widetilde{W} + (1-K)\widetilde{W} < 0$. Then
$$t = (\mu_1(\delta) - K\widetilde{\mu}_1(\delta)) + (\mu_2(\delta) - K\widetilde{\mu}_2(\delta)) + (\mu_3(\delta) - K\widetilde{\mu}_3(\delta)) = -3(W - K\widetilde{W})\delta > 0$$
and
$$\mu_3(\delta) - K\widetilde{\mu}_3(\delta) = \mu_3(\delta) - \widetilde{\mu}_3(\delta) + (1-K)\widetilde{\mu}_3(\delta) = \mu_3(\delta)(W')(W - \widetilde{W}) + (1-K)\widetilde{\mu}_3(\delta)$$
for $W' \in [W, \widetilde{W}]$. Since
$$\widetilde{\mu}_3(\delta) \leq \frac{\delta - 3}{3\sqrt{3}} < \frac{-1}{3\sqrt{3}} \leq -\widetilde{W}, \ \mu_3(\delta)(W') \leq -2/3 - \delta \leq -5/3 < -1,$$
we get
$$\mu_3(\delta) - K\widetilde{\mu}_3(\delta) < -(W - \widetilde{W} + (1-K)\widetilde{W}) = -(W - K\widetilde{W}) = -ht$$
and one can take $\varepsilon = (2 + 3\delta)^{-1} = 1/(2 + 3\delta) \geq (2-\delta)/(4+\delta)$ applying Claim (∗) with $h = \frac{1}{3\delta}$.

Let then $W - K\widetilde{W} = W - \widetilde{W} + (1-K)\widetilde{W} > 0$. We get
$$t = l_1 + l_2 + l_3 = 3(W - K\widetilde{W})\delta > 0$$

for
$$l_1 = -(\mu_1(\delta) - K\widetilde{\mu}_1(\delta)), \quad l_2 = -(\mu_3(\delta) - K\widetilde{\mu}_3(\delta)), \quad l_3 = -(\mu_2(\delta) - K\widetilde{\mu}_2(\delta)).$$

Note that the function $f(W) := \mu_2(\delta)(W)/W$ decreases for $W \in [\frac{-1}{3\sqrt{3}}, 0]$ and increases for $W \in [0, \frac{1}{3\sqrt{3}}]$ and that $f(0) = 2 - \delta$.

If $W > \widetilde{W} \geq 0$, then
$$\begin{aligned}
l_3 &= -(\mu_2(\delta) - K\widetilde{\mu}_2(\delta)) = \widetilde{\mu}_2(\delta) - \mu_2(\delta) - (1-K)\widetilde{\mu}_2(\delta) \\
&= -\mu_2(\delta)(W')(W - \widetilde{W}) - (1-K)\widetilde{\mu}_2(\delta) \\
&\leq -(2-\delta)(W - \widetilde{W}) - (1-K)(2-\delta)\widetilde{W} \\
&\leq -(2-\delta)(W - K\widetilde{W}) = -ht,
\end{aligned}$$

which gives $\varepsilon = (2-\delta)/(4+\delta)$ again by Claim $(*)$ with $h = \frac{(2-\delta)}{3\delta}$.

Let $\widetilde{W} < 0$, $W \geq 0$. Then
$$l_3 = -(\mu_2(\delta) - K\widetilde{\mu}_2(\delta)) \leq -(2-\delta)W - K(2-\delta)\widetilde{W} = -(2-\delta)(W - K\widetilde{W}) = -ht$$
as well.

Let finally $\widetilde{W} < W < 0$. Then $l_3 \leq -ht$ holds since the function $f(W)$ decreases on $[\frac{-1}{3\sqrt{3}}, 0]$. $\square$

Let us then recall that
$$D^2 w_\delta(a)_{|H} = (D^2 P(a) - \delta P(a))_{|H}$$

for any plane H orthogonal to a unit vector a. Applying Corollary 4.2.3 to $H_{21} = a^\perp \cap b^\perp \cap H'$ we see by Lemma 4.2.2 that $\Lambda_1 \geq \mu_+(K)$, $\Lambda_{21} \leq \mu_-(K)$ for $W := W(a)$, $\widetilde{W} := W(b)$, $K := |b|^{-\delta}$; thus by Lemma 4.2.3 we get the result in all cases except $K = 1$, $W(a) = W(b)$. In this last exceptional case the trace of $M_\delta(a, b, O)$ vanishes and the claim is valid for $\varepsilon = \frac{1}{20}$ (just as for any matrix with vanishing trace). Indeed, for a harmonic cubic polynomial $Q(x)$ on $\mathbb{R}^n$ and any $a \in \mathbb{S}^{n-1} \subset \mathbb{R}^n$ we have $\operatorname{tr}\left(D^2 \frac{Q(x)}{|x|^\delta}(a)\right) = (\delta^2 - 4\delta - n)Q(a)$, and thus
$$\operatorname{tr}(M_\delta(a, b, O)) = (\delta^2 - 4\delta - 21)(W(a) - W(b)) = 0,$$

$P(x)_{|H'}$ being harmonic for a coordinate plane $H' \simeq \mathbb{R}^{21}$. $\square$

Proposition 4.2.2 and Lemma 4.2.3 finish the proof of Theorem 4.2.1. Indeed, we set K to be the dual cone $K := K_\lambda^*$ where
$$K_\lambda = \{(\lambda_1, \ldots, \lambda_n) \in [C/\lambda, C\lambda] : \text{ for some } C > 0\}$$

with $n = 21$, $\lambda = \frac{1}{\varepsilon}$. Then Proposition 4.2.2 gives the K-cone condition in Lemma 4.2.3 on $T_{\sigma_0}\Lambda(B)$ for $\sigma_0 = \operatorname{id} \in \Sigma_{21}$, which implies the same condition on the whole of $M = \bigcup_{\sigma \in \Sigma_{21}} T_\sigma \Lambda(B)$ as well.

4.3. Hessian equations in twelve dimensions

We prove then that $w_{12,\delta}$ is a solution to a Hessian equation.

THEOREM 4.3.1. *For any δ, $1 \leq \delta < 2$, the function $w_{12,\delta}$ is a viscosity solution to a uniformly elliptic Hessian equation $F(D^2 w) = 0$ in a unit ball $B \subset \mathbb{R}^{12}$.*

The proof of this theorem is rather cumbersome and uses extensive MAPLE calculations; it gives a Lipschitz functional F. However, if we drop the Hessian invariance condition, we get

COROLLARY 4.3.1. *For any δ, $1 \leq \delta < 2$, the function $w_{12,\delta}$ is a viscosity solution to a uniformly elliptic (not necessarily Hessian) equation $F(D^2 w) = 0$ in a unit ball $B \subset \mathbb{R}^{12}$ where F is a (C^∞) smooth functional.*

Since the proof of Theorem 4.3.1 in Sections 4.3.1 and 4.3.2 below is somewhat involved and utilizes massive computer (MAPLE) computations, we give here an account of its logical structure and its principal points. First of all, the criterion of ellipticity reduces Theorem 4.3.1 for $\delta = 1$ to the uniform hyperbolicity of $D^2 P(a) - O^t \cdot D^2 P(b) \cdot O$ for a pair $a \neq b$ of unit vectors and an orthogonal matrix O. Lemma 4.2.2 reduces this to the uniform with respect to a and b hyperbolicity of the family of differences $\lambda(D^2 P(a)) - \lambda(D^2 P(b))$. In Section 4.1 above we showed that the characteristic polynomial $\chi(P, a)(T)$ of the Hessian $D^2 P(a)$ of the cubic form P has the form $(T^3 - T + 2m(a))(T^3 - T - 2m(a))(T^3 - T + 2P(a))^2$, which permitted us to describe the structure of the ordered spectrum. The argument of Section 4.3.1 is based on the calculation of the (shifted) characteristic polynomial $\chi(w, a)(T - P(a))$ of the full Hessian $D^2 w(a)$ of the function $w := w_{12,1}$ which is possible thanks to an action of the group $Sp(1) \times Sp(1) \times Sp(1)$ which does not change this polynomial. This action permits us to bring the matrix $D^2 w(a)$ to a simple block form and gives, using a MAPLE calculation, an explicit formula for $\chi(w, a)(T - P(a))$:

$$\chi(w, a)(T - P(a)) = P_6(a, T)(T^3 - T + 2P(a))^2$$

for a certain explicit polynomial $P_6(a, T)$; in fact $P_6(a, T)$ is the (shifted) characteristic polynomial of $D^2 w_6(a')$ for a 6-dimensional version of w and an appropriate 6-dimensional unit vector a'. The crucial point then is that the spectrum in this case is not so different from that of $D^2 P(a)$. In fact, one has the following for this ordered spectrum:

$$\mu_1 = \mu_1' \geq \lambda_1' \geq \lambda_2' \geq \lambda_3' \geq \mu_2 = \mu_2' \geq \lambda_4' \geq \lambda_5' \geq \lambda_6' \geq \lambda_3 \geq \mu_3 = \mu_3'$$

where $\lambda_1' \geq \lambda_2' \geq \lambda_3' \geq \lambda_4' \geq \lambda_5' \geq \lambda_6'$ are the roots of $P_6(a, T)$. To prove this inequality one verifies it for specific points a and then explicitly calculates (using MAPLE) the resultant which (miraculously) vanishes nowhere and thus gives the necessary inequalities. This guarantees the exact formula for the equal 6th and 7th eigenvalues, which permits us to get the necessary uniform hyperbolicity of the differences $\lambda(D^2 P(a)) - \lambda(D^2 P(b))$.

In Section 4.3.2 we generalize this argument to any $\delta \in [1, 2[$. In this situation we need the uniform hyperbolicity of the family $D^2 w_\delta(a) - K O^t \cdot D^2 w_\delta(b) \cdot O$ for a pair $a \neq b$ of unit vectors, any orthogonal matrix O, and any positive constant K. Above in Section 4.2.2 we proved the uniform hyperbolicity of the difference family

$$(\mu_1(a), \mu_2(a), \mu_3(a)) - K(\mu_1(b), \mu_2(b), \mu_3(b)).$$

However, for $\delta \in]1, 2[$ the position of μ_2 in the ordered spectrum of $D^2 w_\delta(a)$ is not fixed anymore, which follows from an explicit calculation of $\chi(w_\delta, a)(T - \delta P(a))$ together with some resultant calculations similar to (but more involved than) those in Section 4.3.1. Nevertheless, the position of the double value $\mu_2 = \mu_2'$ varies from $(5, 6)$ to $(7, 8)$ and an argument using the oddness of w_δ permits us to deduce the uniform hyperbolicity of $\lambda(D^2 w_\delta(a)) - K\lambda(D^2 w_\delta(b))$ and to finish the proof of Theorem 4.3.1.

4.3.1. Nonclassical solution. We begin with a proof of Theorem 4.3.1 in the case of $\delta = 1$, i.e., for a nonclassical, but not singular, solution. We put $w(x) := w_{12,1}(x)$ until the end of the present subsection and denote by $H(a)$ the Hessian $D^2 w(a)$ for $a \in \mathbb{R}^{12} \setminus \{0\}$.

PROPOSITION 4.3.1. *Let $a \neq b \in \mathbb{S}_1^{11}$ and let $O \in O(12)$ be an orthogonal matrix such that $H(a,b,O) := H(a) - O^t \cdot H(b) \cdot O \neq 0$. Denote by $\Lambda_1 \geq \Lambda_2 \geq \cdots \geq \Lambda_{12}$ the eigenvalues of the matrix $H(a,b,O)$. Then*

$$\frac{1}{26} \leq \frac{\Lambda_1}{-\Lambda_{12}} \leq 26.$$

PROOF. We give the proof modulo Lemma 4.3.1 below and consider only the case $\text{tr}(A - B) = 15(P(b) - P(a)) \geq 0$, the proof in the other case being symmetric. Since $\Lambda_1 + \Lambda_2 + \cdots + \Lambda_{12} \geq 0$, one gets $11\Lambda_1 \geq -\Lambda_{12}$. On the other hand,

$$-15\Lambda_{12} \geq \text{tr}(A - B) = \Lambda_1 + \Lambda_2 + \cdots + \Lambda_{12}$$

implies

$$-26\Lambda_{12} \geq -15\Lambda_1 - \Lambda_2 - \Lambda_3 - \cdots - \Lambda_{12} \geq \Lambda_1,$$

which finishes the proof. □

LEMMA 4.3.1. *Let $A := H(a)$, $B := O^t \cdot H(b) \cdot O$.*
  (i) *If $P(a) - P(b) \geq 0$, then $\text{tr}(B - A) = 15(P(a) - P(b)) \leq 15\Lambda_1$.*
  (ii) *If $P(a) - P(b) \leq 0$, then $\text{tr}(B - A) = 15(P(a) - P(b)) \geq 15\Lambda_{12}$.*

PROOF. We need Lemmas 4.3.2 and 4.3.3 below. Let $W = P(a)$, $W' = P(b)$, and $W - W' \geq 0$. By Lemma 4.3.2

$$\Lambda_1 \geq \lambda_6(A) - \lambda_6(B)$$
$$= \frac{2}{\sqrt{3}} \cos\left(\frac{\arccos(3\sqrt{3}W) + \pi}{3}\right) - \frac{2}{\sqrt{3}} \cos\left(\frac{\arccos(3\sqrt{3}W') + \pi}{3}\right) - W + W'.$$

Since $\cos\left(\frac{\arccos(3\sqrt{3}W)+\pi}{3}\right) \geq \sqrt{3}|W|$ and $\cos\left(\frac{\arccos(3\sqrt{3}W')+\pi}{3}\right) \geq \sqrt{3}|W'|$, we get the conclusion. The case $P(a) - P(b) \leq 0$ is studied similarly. □

COROLLARY 4.3.2. *Let $a = (r,s,t) \in \mathbb{S}_1^{11}$. Let $\lambda_1 \geq \lambda_2 \geq \cdots \geq \lambda_{12}$ be the eigenvalues of $A = H(a)$. Then*

$$\lambda_6 = \lambda_7 = \frac{2}{\sqrt{3}} \cos\left(\frac{\arccos(3\sqrt{3}P(a)) + \pi}{3}\right) - P(a).$$

PROOF. By Lemmas 4.3.2 and 4.3.3 below one has $\lambda_6 = \lambda_7 = \mu_2$. One easily verifies that the polynomial $Q_1(X) := P_1(X - W) = X^3 - X + 2W$, in the notation of Lemma 4.3.2. Therefore,

$$\mu_1 = \frac{2}{\sqrt{3}} \cos\left(\frac{\arccos(3\sqrt{3}W) - \pi}{3}\right) - W,$$

$$\mu_2 = \frac{2}{\sqrt{3}} \cos\left(\frac{\arccos(3\sqrt{3}W) + \pi}{3}\right) - W,$$

$$\mu_3 = \frac{2}{\sqrt{3}} \cos\left(\frac{\arccos(3\sqrt{3}W) + 3\pi}{3}\right) - W. \qquad \square$$

Now we give and prove the two lemmas mentioned above.

LEMMA 4.3.2. *Let $a = (r,s,t) \in \mathbb{S}_1^{11}, W = W(a) = P(a), m = m(a) = |s|, n = n(a) = |t|$. Then the characteristic polynomial of the matrix $A := H(a)$ is given by*

$$P_A(\xi) = P_1(\xi)^2 \cdot P_2(\xi)$$

with

$$P_1(\xi) = \xi^3 + 3W\xi^2 + 3W^2\xi - \xi + W + W^3,$$

and

$$\begin{aligned}P_2(\xi) = &\xi^6 + 9W\xi^5 + (21W^2 + 3L - 2)\xi^4 + 2W(7W^2 + 3L - 4)\xi^3 \\ &+ (1 - 6W^2 - 9W^4 - 3L + 9M)\xi^2 \\ &- (15W^4 + 6W^2L - 4W^2 - 6L + 1)W\xi - 5W^6 - 3LW^4 \\ &+ 4W^4 - 3(3M + L)W^2 + W^2 - M\end{aligned}$$

with L and M defined as follows:

$$L := L(m,n) = m^2 + n^2 - n^2m^2 - n^4 - m^4 \in \left[M, \frac{1}{3}\right],$$

$$M := M(m,n) = m^2n^2(1 - n^2 - m^2) \in \left[W^2, \frac{1}{27}\right].$$

PROOF. Note that the function w is invariant under the action of the group $Sp(1) \times Sp(1) \times Sp(1)$ by conjugation on each factor, i.e.,

$$(g_1, g_2, g_3) : (r, s, t) \mapsto (g_1 r g_1^{-1}, g_2 s g_2^{-1}, g_3 t g_3^{-1})$$

for $g_1, g_2, g_3 \in Sp(1) = \{q \in \mathbb{H} : |q| = 1\}$, and hence the spectrum $\mathrm{Spec}(H(a))$ is invariant under this action as well.

Applying this action one can suppose that $r_2 = r_3 = s_2 = s_3 = t_2 = t_3 = 0$, i.e., that $(r, s, t) \in \mathbb{C}^3 \subset \mathbb{H}^3$. In this case the matrix $A = H(a)$ becomes a block matrix

$$A = \begin{pmatrix} A_6 & 0 \\ 0 & M_6 \end{pmatrix}$$

where $A_6 = D^2 w_6(a')$ is the Hessian of the function $w_6(a') = P_6(a') = r_0 s_0 t_0 - r_0 s_1 t_1 - r_1 s_0 t_1 - r_1 s_1 t_0$, $a' = (r_0 + r_1 i, s_0 + s_1 i, t_0 + t_1 i) \in \mathbb{C}^3$, and M_6 is the following matrix:

$$M_6 = \begin{pmatrix} -W & 0 & -t_0 & -t_1 & -s_0 & s_1 \\ 0 & -W & -t_1 & -t_0 & -s_1 & -s_0 \\ -t_0 & -t_1 & -W & 0 & -r_0 & -r_1 \\ -t_1 & -t_0 & 0 & -W & r_1 & -r_0 \\ -s_0 & -s_1 & -r_0 & r_1 & -W & 0 \\ s_1 & -s_0 & -r_1 & -r_0 & 0 & -W \end{pmatrix}.$$

A direct calculation shows that the characteristic polynomial of

$$N_6 = M + W \cdot I_6 = \begin{pmatrix} 0 & 0 & -t_0 & -t_1 & -s_0 & s_1 \\ 0 & 0 & -t_1 & -t_0 & -s_1 & -s_0 \\ -t_0 & -t_1 & 0 & 0 & -r_0 & -r_1 \\ -t_1 & -t_0 & 0 & 0 & r_1 & -r_0 \\ -s_0 & -s_1 & -r_0 & r_1 & 0 & 0 \\ s_1 & -s_0 & -r_1 & -r_0 & 0 & 0 \end{pmatrix}$$

is given by

$$P_{N_6}(\xi) = (\xi^3 - \xi + 2W)^2$$

(one uses that $|a|^2 = |a'|^2 = |r|^2 + |s|^2 + |t|^2 = 1$) which gives the formula for the first factor. To calculate the characteristic polynomial of A_6 one notes an action of the group

$$T^2 = \mathbb{S}^1 \times \mathbb{S}^1 = \{(u_1, u_2, u_3) \in \mathbb{C}^3 \ : |u_1| = |u_2| = |u_3| = 1, u_1 u_2 u_3 = 1\}$$

on $\mathbb{C}^3$ preserving w_6:

$$(u_1, u_2, u_3) : (r, s, t) \mapsto (u_1 r, u_2 s, u_3 t).$$

This action permits us to suppose that $s_1 = t_1 = 0$, $s', t' \in \mathbb{R}^+$, and thus $s' = s_0 = m$, $t' = t_0 = n$, $W = P(r, s, t) = r_0 mn$. In this case a direct calculation gives for $A_6 = (N_{ij})$:

$N_{11} = (3r_0^2 - 3)W$, $\quad N_{12} = (3Wr_0 - mt_0)r_1$, $\quad N_{13} = n(1 - r_0^2 - m^2) + 3Wr_0 m$,

$N_{14} = r_0 n r_1$, $\quad N_{15} = m(1 - r_0^2 - n^2) + 3r_0 n W$, $\quad N_{16} = r_0 m r_1$,

$N_{21} = (3Wr_0 - mn)r_1$, $\quad N_{22} = 3W(r_1^2 - 1)$, $\quad N_{23} = (3Ws_0 - mn)r_1$,

$N_{24} = n(r_1^2 - 1)$, $\quad N_{25} = (3W n r_0 m)r_1$, $\quad N_{26} = m(r_1^2 - 1)$,

$N_{31} = (1 - r_0^2 - m^2)n + 3r_0 m W$, $\quad N_{32} = (3mW - r_0 n)r_1$, $\quad N_{33} = (3m^2 - 3)W$,

$N_{34} = mn r_1$, $\quad N_{35} = (1 - m^2 - n^2)r_0 + 3mt_0 W$, $\quad N_{36} = (m^2 - 1)r_1$,

$N_{41} = r_0 n r_1$, $\quad N_{42} = (r_1^2 - 1)n$, $\quad N_{43} = mn r_1$,

$N_{44} = -W$, $\quad N_{45} = (n^2 - 1)r_1$, $\quad N_{46} = -r_0$, $\quad N_{51} = (1 - r_0^2 - n^2)m + 3r_0 n W$,

$N_{52} = (3nW - mr_0)r_1$, $\quad N_{53} = (1 - m^2 - n^2)r_0 + 3mn W$, $\quad N_{54} = (n^2 - 1)r_1$,

$N_{55} = (3n^2 - 3)W$, $\quad N_{56} = mn r_1$, $\quad N_{61} = m r_0 r_1$, $\quad N_{62} = (r_1^2 - 1)m$,

$N_{63} = (m^2 - 1)r_1$, $\quad N_{64} = -r_0$, $\quad N_{65} = mn r_1$, $\quad N_{66} = -W$.

Applying MAPLE one gets the characteristic polynomial $P_2(T)$. Note that the characteristic polynomial $Q_2(\xi) = P_2(\xi - W)$ of $A_6 + W \cdot I_6$ equals

$$Q_2(\xi) = \xi^6 + 3W\xi^5 - (9W^2 - 3L + 2)\xi^4 - 6WL\xi^3 + (6W^2 - 3L + 9M + 1)\xi^2$$
$$- 3(6M - 4L + 1)W\xi + 3W^2 - 12LW^2 - M. \qquad \square$$

LEMMA 4.3.3. *Let $a = (r, s, t) \in \mathbb{S}_1^{11}$, $A = H(a)$, and let $\mu_1 \geq \mu_2 \geq \mu_3$ be the roots of $P_1(T)$, $\nu_1 \geq \nu_2 \geq \cdots \geq \nu_6$ and the roots of $P_2(T)$. Then*

$$\mu_1 \geq \nu_1 \geq \nu_2 \geq \nu_3 \geq \mu_2 \geq \nu_4 \geq \nu_5 \geq \nu_6 \geq \mu_3.$$

PROOF. Let $\mu_i' = \mu_i + W$, $\nu_j' = \nu_i + W$ for $i = 1, 2, 3$, $j = 1, \ldots, 6$ be the roots of $Q_1(X)$ and $Q_2(X)$, respectively. We have to show that

$$\mu_1' \geq \nu_1' \geq \nu_2' \geq \nu_3' \geq \mu_2' \geq \nu_4' \geq \nu_5' \geq \nu_6' \geq \mu_3'.$$

One notes that $\mu_i'(W) = \mu_i'(-W)$, $\nu_i'(W) = \nu_i'(-W)$. Therefore we can suppose without loss of generality that $W \geq 0$. For $n = 0$ we have $W = mnr_0 = 0$ and

$$Q_2(X) = X^6 - 2X^4 + 3mX^4 - 3mX^4 + X^2 - 3mX^2 + 3mX^2$$
$$= X^2(X - 1)(X + 1)(X^2 - 3m^4 - 1 + 3m^2),$$
$$X^3 - X - 2W = X(X - 1)(X + 1).$$

Thus $\mu_1' = \nu_1' = 1$, $\nu_2' = \sqrt{1 - 3m^2 + 3m^4} \in (0, 1]$, $\nu_3' = \nu_4' = \mu_2' = 0$, $\nu_5' = -\sqrt{1 - 3m^2 + 3m^4} \in [-1, 0)$, $\mu_3' = -1$, and the inequalities hold true. Symmetrically this is true for $m = 0$ as well. We can suppose thus that $m \neq 0$, $n \neq 0$. Let first $r_0^2 + m^2 + n^2 \neq 1$; without loss of generality one supposes also $(m, n, r_0) \in B_1^3 \cap \mathbb{R}_+^3$. We begin with a particular choice: $m = n = r_0 = 1/2$, $W = 1/8$. For that choice easy brute force calculations show that

$$\mu_1' \in [0.83, 0.84], \quad \mu_2' \in [0.26, 0.27], \quad \mu_3' \in [-1.11, -1.1],$$
$$\nu_1' \in [0.7, 0.71], \quad \nu_2' \in [0.54, 0.55], \quad \nu_3' \in [0.42, 0.43],$$
$$\nu_4' \in [-0.39, -0.38], \quad \nu_5' \in [-0.71, 0.7], \quad \nu_6' \in [-0.96, 0.95]$$

and the inequalities hold. Then we consider the resultant $R = R(m, n, r_0)$ of the polynomials $Q_2(X)$ and $X^3 - X + 2W$; brute force (MAPLE) calculations give

$$R = 16n^2m^2(r_0^2 + m^2 + n^2 - 1)^3(27W^2 + 4)(1 - 27W^2) < 0$$

since the condition $W^2 = 1/27$ implies $r_0^2 + m^2 + n^2 = 1$. For any $(m, n, r_0) \in B_1^3 \cap \mathbb{R}_+^3$ there is a line segment joining it to the triple $(1/2, 1/2, 1/2)$, the set $B_1^3 \cap \mathbb{R}_+^3$ being convex. The value of $R(m, n, r_0)$ on the whole segment is strictly negative and thus the order of the roots at (m, n, r_0) is the same as at $(1/2, 1/2, 1/2)$, which finishes the proof of the inequalities for $r_0^2 + m^2 + n^2 \neq 1$. Let finally $m^2 + n^2 + r_0^2 = 1$. Then easy brute force calculations show that

$$Q_2(X) = (X^3 - X + 2W)(X^3 + 3WX^2 - 9W^2X - X + 3LX + W - 6WL).$$

Thus by continuity we get $\lambda_1 = \lambda_2 = \lambda_3 = \mu_1$, $\lambda_6 = \lambda_7 = \mu_2$, $\lambda_{10} = \lambda_{11} = \lambda_{12} = \mu_3$, which is sufficient to conclude the proof. $\qquad \square$

REMARK 4.3.1. Since $|a|^2 = |a'|^2 = |r|^2 + |s|^2 + |t|^2 = r_0^2 + r_1^2 + m^2 + n^2 = 1$, one gets $r_0^2 + m^2 + n^2 \leq 1$ and an application

$$\Phi : S_1^{11} \longrightarrow \bar{B}_{++}^3, \ \ a = (r,s,t) \mapsto \Phi(a) := (r_0, m, n) = (\tfrac{1}{mn}W, m, n)$$

where $\bar{B}_{++}^3 = \bar{B}_1^3 \cap \{m \geq 0, n \geq 0\}$.

REMARK 4.3.2. We use MAPLE calculations extensively in the present section. These calculations concern algebraic identities, do not use any approximation, and are thus completely rigorous. Besides, all of them need only a few seconds on a modest laptop.

Proposition 4.2.2 and Lemma 4.2.3 give a proof of Theorem 4.3.1 in the case of $\delta = 1$. Indeed, we set K to be the dual cone $K := K_\lambda^*$ where

$$K_\lambda = \{(\lambda_1, \ldots, \lambda_n) \in [C/\lambda, C\lambda] : \text{ for some } C > 0 \}$$

with $n = 12$, $\lambda = 26$. Then Proposition 4.2.2 gives the K-cone condition in Lemma 4.2.3 on $T_{\sigma_0}\Lambda(B)$ for $\sigma_0 = \mathrm{id} \in S_{12}$ which implies the same condition on the whole of $M = \bigcup_{\sigma \in S_n} T_\sigma \Lambda(B)$ as well.

REMARK 4.3.3. The ellipticity constant C of the functional F obtained in this way satisfies $C \leq 4 \cdot 26^2 \sqrt{12} < 10^5$; cf. Lemma 4.1.2.

4.3.2. Singular solutions. In this subsection we prove Theorem 4.3.1 for any $\delta \in [1, 2[$. By Lemma 4.2.3 it is sufficient to show that the ellipticity condition (the K-cone condition) valid for the function $w_{12,1}$ remains true for the function $w_{12,\delta}(X) := w(X)|X|^{1-\delta}$. Denote $\delta - 1$ by δ'; hence $0 \leq \delta' < 1$.

For $a \in \mathbb{R}^{12} \setminus \{0\}$ we denote by $H_\delta(a)$ the Hessian $D^2 w_\delta(a)$. The following result is sufficient to prove Theorem 4.3.1:

PROPOSITION 4.3.2. *Let $1 \leq \delta < 2$, i.e., $0 \leq \delta' < 1$. Then for any $a \neq b \in \mathbb{R}^{12} \setminus \{0\}$ and any orthogonal matrix $O \in O(12)$ with $H_\delta(a,b,O) := H_\delta(a) - O^t \cdot H_\delta(b) \cdot O \neq 0$ the eigenvalues $\Lambda_1 \geq \Lambda_2 \geq \cdots \geq \Lambda_{12}$ of $H_\delta(a,b,O)$ verify*

$$\frac{1}{C_\delta} = \frac{1-\delta'}{26+3\delta'-\delta'^2} \leq \frac{\Lambda_1}{-\Lambda_{12}} \leq \frac{26+3\delta'-\delta'^2}{1-\delta'} =: C_\delta \ .$$

Notice that in Proposition 4.3.2 and Lemmas 4.3.4 and 4.3.5 below we use δ' rather than $\delta = 1 + \delta'$ since it is (slightly) more convenient for calculations.

PROOF. We can suppose without loss of generality that $|a| \leq |b|$; moreover, by homogeneity we can suppose that $a \in \mathbb{S}_1^{11}$ and thus $|b| \geq 1$. Let $\bar{b} := b/|b| \in \mathbb{S}_1^{11}$; then $D^2 w_\delta(b) = D^2 w_\delta(\bar{b})|b|^{-\delta}$. One then needs Lemma 4.2.3 applied to the points $a, \bar{b} \in \mathbb{S}^{11}$. □

Lemma 4.2.3 can be applied to our situation thanks to the following formulas generalizing those of Section 4.3.1; the proofs remain essentially the same as for Lemma 4.3.2 (i.e., more or less brute force calculation together with the invariance properties of w).

LEMMA 4.3.4. *Let $\delta' \in [0,1[$, $a = (r,s,t) \in \mathbb{S}_1^{11}$, and let $W = W(a) = P(a), m = m(a) = |s|, n = n(a) = |t|$. Then the characteristic polynomial of the matrix $A = H_\delta(a) := D^2 w_\delta(a)$ is given by*

$$P_{A,\delta}(T) = P_{1,\delta}(T)^2 \cdot P_{2,\delta}(T)$$

where
$$P_{1,\delta}(T) = T^3 + 3W(1+\delta')T^2 + (3W^2(1+\delta')^2 - 1)T + W(1-\delta') + W^3(1+\delta')^3,$$

$$P_{2,\delta}(T) = P_{2,\delta}(T,W) := T^6 + a_{5,\delta}T^5 + a_{4,\delta}T^4 + a_{3,\delta}T^3 + a_{2,\delta}T^2 + a_{1,\delta}T + a_{0,\delta}$$

for

$a_{5,\delta} := W(\delta'+1)(9-\delta'),$

$a_{4,\delta} := W^2(\delta'+1)(21 + 28\delta' - 5\delta'^2) + L(\delta'+1)(3-\delta') - 2,$

$a_{3,\delta} := -2W(1+\delta') \cdot \left(W^2(\delta'+1)(5\delta'^2 - 26\delta' - 7) - L(2\delta'+1)(3-\delta') + 4\right),$

$a_{2,\delta} := -W^4(10\delta'^2 - 53\delta' + 9)(\delta'+1)^3$
$\qquad -2W^2(\delta'+1)(3L\delta'^3 - 6L\delta'^2 - 9L\delta' + 7\delta' + 3),$
$\qquad +L\delta'^2 - 3M\delta'^2 - 2L\delta' + 6M\delta' - 3L + 9M + 1,$

$a_{1,\delta} := -(\delta'+1)\left(W^4(5\delta'-3)(\delta'-5)(\delta'+1)^3\right.$
$\qquad -2(\delta'+1)(-2L\delta'^3 + 5L\delta'^2 + 4L\delta' - 6\delta' - 3L + 2)W^2$
$\qquad \left. +2(3-\delta')(-3\delta'M + L\delta' - L) + 1 - \delta'\right)W,$

$a_{0,\delta} := (1-\delta')\left(W^6(\delta'-5)(\delta'+1)^5 + W^4(\delta'+1)^3(L\delta'^2 - 2L\delta' - 3L + 4)\right.$
$\qquad -W^2(\delta'+1)(L\delta'^2 - 3M\delta'^2 + \delta' + 6M\delta' - 4L\delta' - 1 + 3L + 9M)$
$\qquad \left. -M(1-\delta')\right),$

the quantities $L = m^2 + n^2 - n^2m^2 - n^4 - m^4$, $M = m^2n^2(1 - n^2 - m^2)$ being defined as above.

A MAPLE calculation then gives for the resultant
$$R_\delta(r_0, m, n) := \text{Res}(P_{1,\delta}, P_{2,\delta}) = 16m^4n^4(1 - n^2 - m^2 - r_0^2)^3 \cdot R(W, \delta)$$

where
$$R(W,\delta) = 27(\delta'+1)^3(3-\delta')^3 W^4 + 9(\delta'-1)^2(\delta'-3)^2(\delta'+1)^2 W^2 - (\delta'-1)^2(\delta'^2 - 2\delta' - 2)^2.$$

Denote by $W_0(\delta) \in (0, 1/3\sqrt{3}]$ the unique positive root of $R(W, \delta)$. Recall that the set $\Phi(\mathbb{S}^{11})$ of possible triples $\Phi(a) = (r_0, m, n) : r_0 = r_0(a), m = m(a), n = n(a)$ for $a \in \mathbb{S}_1^{11}$ is a sector $\bar{B}_{++} := B_1 \cap \{m \geq 0, n \geq 0\}$ in the closed unit ball $B = B_1 \subset V$; recall also that $W(a) = r_0mn$. Let $B_+(\delta)$ (resp. $B_-(\delta)$, $B_0(\delta)$) be the subset of $(r_0, m, n) \in \Phi(\mathbb{S}^{11})$ where $R_\delta(W) > 0$ (resp. $R_\delta(W) < 0$, $R_\delta(W) = 0$). Then

$\qquad B_0(\delta) = S_{++}^2 \cup D_{r_0+} \cup D_{m+} \cup D_{n+}$ with $D_{m+} = \bar{B}_{++} \cap \{m = 0\}$, etc.,

$\qquad B_+(\delta) = B_{++} \cap \{r_0mn > W_0(\delta)\}, \quad \bar{B}_+(\delta) = \bar{B}_{++} \cap \{r_0mn \geq W_0(\delta)\},$

$\qquad B_-(\delta) = B_{++} \cap \{0 < r_0mn < W_0(\delta)\}, \quad \bar{B}_-(\delta) = \bar{B}_{++} \cap \{r_0mn \leq W_0(\delta)\}.$

Note that these sets are invariant under the reflection $\sigma : (r_0, m, n) \mapsto (-r_0, m, n)$; $B^0(\delta)$ and $\bar{B}_-(\delta)$ are connected, while $B_-(\delta)$, $\bar{B}_+(\delta)$, and $B_+(\delta)$ have two connected components each.

LEMMA 4.3.5. *Let $a \in \mathbb{S}_1^{11}$, let $\lambda_1(\delta, a) \geq \lambda_2(\delta, a) \geq \cdots \geq \lambda_{12}(\delta, a)$ be the eigenvalues of $D^2 w_\delta(a)$, and let $\mu_1(\delta, a) \geq \mu_2(\delta, a) \geq \mu_3(\delta, a)$ be the roots of $P_{1,\delta}(T, W(a))$. Then:*

   (i) $\lambda_1(\delta, a) = \mu_1(\delta, a)$, $\lambda_{12}(\delta, a) = \mu_3(\delta, a)$.
   (ii) $\lambda_5(\delta, a) = \lambda_6(\delta, a) = \mu_2(\delta, a)$ *for* $\Phi(a) \in \bar{B}_+(\delta)$, $W = W(a) \geq 0$.
   (iii) $\lambda_7(\delta, a) = \lambda_8(\delta, a) = \mu_2(\delta, a)$ *for* $\Phi(a) \in \bar{B}_+(\delta)$, $W = W(a) \leq 0$.
   (iv) $\lambda_6(\delta, a) = \lambda_7(\delta, a) = \mu_2(\delta, a)$ *for* $\Phi(a) \in \bar{B}_-(\delta)$.

PROOF. Since $\lambda_1(\delta, a) = \lambda_{12}(\delta, -a)$, $\lambda_6(\delta, a) = \lambda_7(\delta, -a)$, $\lambda_8(\delta, a) = \lambda_5(\delta, -a)$, $W(-a) = -W(a)$, (iii) is implied by (ii) and, moreover, one can suppose without loss of generality that $\Phi(a) = (r_0, m, n) \in \mathbb{R}^3_+$. Since in the interior of the domain $B_+(\delta) \cap \mathbb{R}^3_+$ (resp. $B_-(\delta) \cap \mathbb{R}^3_+$) the function $R_\delta(r_0, m, n)$ does not vanish, it is sufficient to verify the ordering of the roots at a single point in $B_-(\delta) \cap \mathbb{R}^3_+$ (resp. at a single point in $B_+(\delta) \cap \mathbb{R}^3_+$). We use $a_- := (\varepsilon, \varepsilon, \varepsilon) \in B_-(\delta) \cap \mathbb{R}^3_+$ and $a_+ := (1/\sqrt{3}, 1/\sqrt{3}, 1/\sqrt{3} - \varepsilon) \in B_+(\delta) \cap \mathbb{R}^3_+$ for a sufficiently small $\varepsilon > 0$. Let $\nu_1(\delta, a) \geq \nu_2(\delta, a) \geq \cdots \geq \nu_6(\delta, a)$ be the roots of $P_{2,\delta}(T, W(a))$. Elementary calculations show that for $a = a_-$ one has $W = W(a) = \varepsilon^3$,

$$\mu_1(\delta, a) = 1 + O(\varepsilon^3), \quad \mu_2(\delta, a) = O(\varepsilon^3), \quad \mu_3(\delta, a) = -1 + O(\varepsilon^3),$$

while $P_{2,\delta}(T, W(a)) = F_1(T, \varepsilon) \cdot F_2(T, \varepsilon)$ where

$$F_1(T, \varepsilon) = T^2 - 1 + 2\varepsilon^2 + O(\varepsilon^3)$$

$$F_2(T, \varepsilon) = T^4 + \varepsilon^3(7 + 6\delta' - \delta'^2)T^3$$
$$+ (12\varepsilon^6 \delta'^2 + 3\varepsilon^4 \delta'^2 - 3\varepsilon^6 \delta'^3 + 21\varepsilon^6 \delta' + 6\varepsilon^6 + 4\varepsilon^2 - 2\varepsilon^2 \delta'^2$$
$$- 1 - 6\varepsilon^4 \delta' + 4\varepsilon^2 \delta' - 9\varepsilon^4)T^2$$
$$+ \varepsilon^3(1 - 10\varepsilon^2 - \delta'^2 - 12\varepsilon^4 \delta'^2 + 4\varepsilon^2 \delta' - 18\varepsilon^4 \delta' + 10\varepsilon^2 \delta'^2$$
$$- 4\varepsilon^2 \delta'^3 + 6\varepsilon^4 \delta'^3 + O(\varepsilon^6))T$$
$$+ \varepsilon^4(1 - \delta')^2 - \varepsilon^6(\delta' + 1)(\delta' - 1)^2(2\delta' \varepsilon^2 - 4\varepsilon^2 + 1) + O(\varepsilon^{10}),$$

and thus

$$\mu_1(\delta, a) \geq \nu_1(\delta, a) = 1 - \varepsilon^2 + O(\varepsilon^3) \geq \nu_2(\delta, a) = 1 - \varepsilon^2(2 + 2\delta' - \delta'^2) + O(\varepsilon^3),$$
$$\nu_3(\delta, a) = (1 - \delta')\varepsilon^2 + O(\varepsilon^3) \geq \mu_2(\delta, a) \geq \nu_4(\delta, a) = -(1 - \delta')\varepsilon^2 + O(\varepsilon^3),$$
$$\nu_5(\delta, a) = -1 + \varepsilon^2(2 + 2\delta' - \delta'^2) + O(\varepsilon^3) \geq \nu_6(\delta, a) = -1 + \varepsilon^2 + O(\varepsilon^3) \geq \mu_3(\delta, a),$$

which proves the claim in this case.

For $a = a_+$ one has $W = W(a) = (1/\sqrt{3} - \varepsilon)/3$ and similar calculations give

$$\mu_1(\delta, a) = \frac{2 - \delta'}{3\sqrt{3}} + 3^{-1/4}\sqrt{2}\varepsilon + O(\varepsilon), \quad \mu_2(\delta, a) = \frac{2 - \delta'}{3\sqrt{3}} - 3^{-1/4}\sqrt{2}\varepsilon + O(\varepsilon),$$

$$\mu_3(\delta, a) = \frac{-7 - \delta'}{3\sqrt{3}} + (5/3 + \delta')\varepsilon + O(\varepsilon^2),$$

while $P_{2,\delta}(T, W(a)) = G_1(T, \varepsilon) \cdot G_2(T, \varepsilon)^2 \cdot G_3(T, \varepsilon)^2$ where

$$G_1(T, \varepsilon) := T^2 + \frac{(\delta' + 1)}{3\sqrt{3}}(5 - \delta')(1 - \sqrt{3}\varepsilon)T$$
$$+ \frac{(1 - \delta')}{27}(3\delta'^2\varepsilon^2 - 2\sqrt{3}\delta'^2\varepsilon - 12\delta'\varepsilon^2 + \delta'^2 + 8\sqrt{3}\delta'\varepsilon - 15\varepsilon^2 + 5\delta' + 10\sqrt{3}\varepsilon - 14),$$

$$G_2(T,\varepsilon) := T + \frac{4+\delta'}{3\sqrt{3}} - \frac{\varepsilon(\delta'+1)}{3}, \quad G_3(T,\varepsilon) := T - \frac{2-\delta'}{3\sqrt{3}} - \frac{\varepsilon(\delta'+1)}{3},$$

and thus

$$\mu_1(\delta,a) \geq \nu_1(\delta,a) = \nu_2(\delta,a) = \frac{2-\delta'}{3\sqrt{3}} + O(\varepsilon) \geq \mu_2(\delta,a)$$

$$\geq \nu_3(\delta,a) = \frac{(2-\delta')(1-\delta')}{3\sqrt{3}} + O(\varepsilon) \geq \nu_4(\delta,a) = \nu_5(\delta,a)$$

$$= -\frac{4+\delta'}{3\sqrt{3}} + O(\varepsilon),$$

$$\nu_6(\delta,a) = \frac{-7-\delta'}{3\sqrt{3}} + \frac{\varepsilon(\delta'-5)(\delta'-9)(\delta'+1)}{3(9-2\delta'+\delta'^2)} + O(\varepsilon^2) \geq \mu_3(\delta,a),$$

which finishes the proof of the lemma since

$$\nu_6(\delta,a) - \mu_3(\delta,a) = \frac{2\varepsilon\delta'(7+\delta')(1-\delta')}{3(9-2\delta'+\delta'^2)+O(\varepsilon^2)} \geq 0. \qquad \square$$

END OF PROOF OF PROPOSITION 4.3.2. If $W(a)$ and $W(b)$ are of the same sign, we get the result by applying Lemma 4.2.3 with $K := |b|^{-\delta}$; in the exceptional case $K = 1$, $W(a) = W(b)$, the trace of $H_\delta(a,b,O)$ vanishes and the claim is valid for $C_\delta = 11$. In the case $W(a) \cdot W(b) < 0$ we can suppose without loss of generality that $W(a) > 0$, $W(b) < 0$; if $\Phi(a) \notin B_+$ or $\Phi(\bar{b}) \notin B_+$, then Lemmas 4.2.3 and 4.3.5 work as well. Thus we can suppose $\Phi(a) \in B_+$, $\Phi(\bar{b}) \in B_+$; then

$$\sigma(\Phi(\bar{b})) \in B_+,$$
$$W(-\bar{b}) > 0,$$
$$\lambda_i(-b) = -\lambda_{13-i}(b),$$
$$\lambda_i(-\bar{b}) = -\lambda_{13-i}(\bar{b}),$$
$$\operatorname{tr}(H_\delta(a,b,O)) = -(W(a) + KW(-\bar{b}))(\delta'+1)(15-\delta') < 0,$$

which implies immediately that $11 \geq -\Lambda_1/\Lambda_{12}$. Moreover,

$$\Lambda_1 \geq \lambda_6(a) - K\lambda_6(\bar{b}) = \lambda_6(a) + K\lambda_7(-\bar{b}) = \mu_2(\delta,a) + K\mu_2(\delta,-\bar{b})$$
$$\geq (1-\delta')(W(a) + KW(-\bar{b})) = \frac{(1-\delta')\operatorname{tr}(H_\delta(a,b,O))}{(\delta'+1)(15-\delta')} > 0$$

and thus

$$-\Lambda_1/\Lambda_{12} \geq \left(11 + \frac{(\delta'+1)(15-\delta')}{1-\delta'}\right)^{-1} = \frac{1-\delta'}{26+3\delta'-\delta'^2},$$

which finishes the proof of the proposition. $\qquad \square$

To deduce Corollary 4.3.1 we need the map

$$H_\delta : B_1^{12} \setminus \{0\} \to Q(\mathbb{R}^{12}), \quad a \mapsto D^2 w_\delta(a).$$

The following result is sufficient to conclude using Proposition 4.3.2 and Lemma 4.2.2:

LEMMA 4.3.6. *Let $\delta \in]0,1[$. Then the image $H_\delta\left(B_1^{12} \setminus \{0\}\right) \subset Q(\mathbb{R}^{12})$ is diffeomorphic to the product $V_{11,\delta} \times [1,\infty)$ with a smooth 11-dimensional manifold $V_{11,\delta}$.*

PROOF. Since $D^2 w_\delta(a) = D^2 w_\delta(a/|a|)|a|^{-\delta}$, it is sufficient to show two facts:

(i) $H_\delta{}_{|\mathbb{S}^{11}} : \mathbb{S}^{11} \longrightarrow Q(\mathbb{R}^{12})$ is a smooth embedding.

(ii) If $D^2 w_\delta(a) = D^2 w_\delta(b) \cdot k$ with $k > 0$, then $k = 1$.

Lemmas 4.2.3 and 4.3.5 imply (ii). To prove (i) we fix $a \neq b \in \mathbb{S}^{11}$ and consider $d = \frac{a-b}{|a-b|} \in \mathbb{S}^{11}_1$. Let then $e, f \in \mathbb{S}^{11} \cap a^\perp \cap b^\perp$. Since $e, f \perp a, b$, one has

$$w_{\delta,ee}(a) = P_{ee}(a) - \delta P(a), \quad w_{\delta,ee}(b) = P_{ee}(b) - \delta P(b),$$

$$w_{\delta,ff}(a) = P_{ff}(a) - \delta P(a), \quad w_{\delta,ff}(b) = P_{ff}(b) - \delta P(b)$$

and hence

$$(w_{\delta,ee}(a) - w_{\delta,ee}(b)) - (w_{\delta,ff}(a) - w_{\delta,ff}(b))$$
$$= (P_{ee}(a) - P_{ee}(b)) - (P_{ff}(a) - P_{ff}(b))$$
$$= |a-b|(P_{eed} - P_{ffd}) \geq \frac{2}{\sqrt{3}}|a-b|$$

for suitable vectors e, f as in the proof of Proposition 4.1.2 above. It follows that

$$\max\{|w_{\delta,ee}(a) - w_{\delta,ee}(b)|, |w_{\delta,ff}(a) - w_{\delta,ff}(b)|\} \geq |a-b|/\sqrt{3},$$

which finishes the proof. □

4.4. Isaacs equations

Let us show that the same functions are viscosity solutions to uniformly elliptic Isaacs equations:

THEOREM 4.4.1. *For any δ, $1 \leq \delta < 2$, and any coordinate plane $H' \subset \mathbb{R}^{24}$, $\dim H' = 21$, the functions*

$$w_{24,\delta}(x)_{|H'} = \left(P_{24}(x)/|x|^\delta\right)_{|H'}, \quad w_{12,\delta}(x) = P_{12}(x)/|x|^\delta$$

are viscosity solutions to uniformly elliptic homogeneous Isaacs equations in unit balls $B \subset \mathbb{R}^{21}$ and $B \subset \mathbb{R}^{12}$.

REMARK 4.4.1. The corresponding Isaacs equations have a simple structure and are written explicitly below.

As in Section 4.3.2 we denote by $K_C \subset \mathrm{Sym}_n(\mathbb{R})$ the cone of positive symmetric matrix with the ellipticity constant C; i.e., if $A \in K_C$, $A = \{a_{ij}\}$, then

(4.4.1) $$C^{-1}|\xi|^2 \leq \sum a_{ij}\xi_i\xi_j \leq C|\xi|^2.$$

The proof of the theorem will follow from the lemmas below.

LEMMA 4.4.1. *Let $w \in C^\infty(\mathbb{R}^n \setminus \{0\})$ be a homogeneous order α function for $\alpha \in\ ]1,2]$. Assume that for any $x, y \in \mathbb{S}^n$ there exists a matrix $A \in K_C$ orthogonal to both forms $D^2 w(x), D^2 w(y)$, i.e.,*

$$\mathrm{tr}(AD^2 w(x)) = \mathrm{tr}(AD^2 w(y)) = 0.$$

Then w is a viscosity solution to an Isaacs equation.

PROOF. Define the manifold M by
$$M = \{(a,b) : b \in \mathbb{S}^{n-1},\ a \in D^2 w(b)^\perp \subset \mathrm{Sym}_n(\mathbb{R})\} \subset \mathbb{R}^n \times \mathrm{Sym}_n(\mathbb{R}).$$
Then the projection $\pi(a,b) = b$ gives a vector bundle
$$\pi : M \to \mathbb{S}^{n-1} \tag{4.4.2}$$
over the base $\mathbb{S}^{n-1}$ which is trivial since the quadratic form $D^2 w(b)$ is nonpositive. Let
$$\Sigma = \{a \in K_C, \mathrm{tr}(a) = 1\}.$$
Define then $M' \subset M$ as
$$M' = \{(a,b) : b \in \mathbb{S}^{n-1},\ a \in D^2 w(b)^\perp \cap \Sigma\}.$$
Let $z = (a,b) \in M'$ and let $L_z = L_{ab}$ be the linear uniformly elliptic operator
$$L_{ab}(u) := \sum a_{ij} u_{x_i x_j}$$
where $a = \{a_{ij}\}$. We are going to show that w is a solution of the Isaacs equation (1.1.8) with the set of parameters $(a,b) \in M'$:
$$\sup_{b \in \mathbb{S}^{n-1}} \inf_{a \in \pi^{-1}(b) \cap \Sigma} L_{ab} w(x) = 0 \tag{4.4.3}$$
for $x \in B_1^n \setminus \{0\}$. For a given $x \in B_1^n \setminus \{0\}$ set $b' = x/|x|$. Since w is homogeneous, one has
$$\inf_{a \in \pi^{-1}(b') \cap \Sigma} L_{ab'} w(x) = \inf_{a \in D^2 w(b')^\perp \cap \Sigma} \mathrm{tr}(a D^2 w(x)) = 0.$$
Thus
$$\sup_{b'} \inf_a L_{ab'} w(x) = 0. \tag{4.4.4}$$
Let us suppose that $c \in \mathbb{S}^{n-1} \setminus \{b'\}$. Then $a_0 c = a_0 b' = 0$ for some $a_0 \in D^2 w(b')^\perp \cap D^2 w(c)^\perp \cap \Sigma$ and thus
$$L_{a_0 c} w(x) \leq 0, \tag{4.4.5}$$
which immediately implies (4.4.3). $\square$

Recall that a symmetric matrix A is called *strictly hyperbolic* if
$$\frac{1}{M} < -\frac{\lambda_1(A)}{\lambda_n(A)} < M$$
for some $M > 0$. Let $w = w_{24,\delta}(x)_{|H'}$ or $w = w_{12,\delta}(x)$. To finish the proof of Theorem 4.4.1 we only need to find for any $(x,y) \in \mathbb{S}^{n-1} \times \mathbb{S}^{n-1}$ a positive $Q \in \mathrm{Sym}_n(\mathbb{R})$ orthogonal to both $F_1 := D^2 w(x)$ and $F_2 := D^2 w(y)$,
$$\mathrm{tr}(F_1 Q) = \mathrm{tr}(F_2 Q) = 0,$$
in such a way that Q depends continuously on the pair (F_1, F_2), which guarantees the conditions of Lemma 4.4.1 by the compactness of $\mathbb{S}^{n-1} \times \mathbb{S}^{n-1}$.

Indeed the results of Sections 4.2 and 4.3 imply that $\alpha F_1 - \beta F_2$ is strictly hyperbolic for any $\alpha > 0, \beta > 0$; since the function w is odd, it remains true for any $(\alpha, \beta) \in \mathbb{R}^2 \setminus \{0\}$.

Now we can apply the following result identifying symmetric matrices $\mathrm{Sym}_n(\mathbb{R})$ and quadratic forms $Q(\mathbb{R}^n)$.

LEMMA 4.4.2. *Let $F_1, F_2 \in Q(\mathbb{R}^n)$ and suppose that $\alpha F_1 + \beta F_2$ is strictly hyperbolic for any $(\alpha, \beta) \in \mathbb{R}^2 \setminus \{0\}$. Then there exists a positive quadratic form Q orthogonal to both forms F_1 and F_2,*

$$\operatorname{tr}(F_1 Q) = \operatorname{tr}(F_2 Q) = 0,$$

which depends continuously on the pair (F_1, F_2).

PROOF. We will consider F_1 and F_2 as restricted to $\mathbb{S}^{n-1} \subset \mathbb{R}^n$ and can suppose without loss of generality that the form F_1 is traceless, $\operatorname{tr}(F_1) = 0$. Let $D \subset \mathbb{S}^{n-1}$ be a minimal (with respect to inclusion) domain on which F_1 does not change the sign. We can assume without loss of generality that $F_1|_D > 0$. Our first claim is that F_2 changes the sign on the border ∂D of D. Indeed, if not we assume without loss of generality $F_2|_{\partial D} \geq 0$. Define D_t for $t \in [0, 1]$ as the union of the connected components of the set $\{x \in \mathbb{S}^{n-1} : \cos(\pi t) F_1(x) + \sin(\pi t) F_2(x) > 0\}$ with nonempty intersection with D; thus $D_0 = D$. If for some $s \in [0, 1]$ we get $D_s \cap D \neq D_s$, we are done and we can thus assume that $\forall s \in [0, 1]$, $D_s \subset D$. If for some $s \in [0, 1]$ the set D_s becomes empty, then there exists $s' \in [0, s[$ such that $\bar{D}'_s$ contains an isolated point x_0 with $\cos(\pi s') F_1(x_0) + \sin(\pi s') F_2(x_0) = 0$, which is impossible since then 0 would be a maximal eigenvalue of the strictly hyperbolic form $\cos(\pi s') F_1 + \sin(\pi s') F_2$. In particular, D_1 is nonempty, which is impossible since $F_1 = -F_0$.

Since F_2 changes the sign on the border ∂D of D, there exist two points $a_1, a_2 \in \partial D$ with $F_1(a_1) = F_1(a_2) = 0, F_2(a_1) = a > 0, F_2(a_2) = -a$. Then the form $Q_0(x) := Q_1(x) + Q_2(x)$, with $Q_1(x) := (x, a_1)^2$, $Q_2(x) := (x, a_2)^2$ is clearly orthogonal to both F_1 and F_2 while $Q_1(x)$ and $Q_2(x)$ are orthogonal to F_1. Let $Q_{0,l}(x) := Q_0(x) + l \sum x_i^2$ for some $l > 0$. The form $Q_{0,l}$ is orthogonal to F_1 since F_1 is traceless. Let $m = \operatorname{tr}(F_2)$; then $\operatorname{tr}(F_2 Q_{0,l}) = ml$ and the form $Q := Q_{0,l} - m l Q_1 / a$ being orthogonal to both F_1 and F_2 is positive for sufficiently small $l > 0$. It is also clear that this construction can be made continuously depending on the pair (F_1, F_2). The lemma is proved. $\square$

Let us rewrite the equation (4.4.3) in the following, especially elegant, form

COROLLARY 4.4.1. *The function w is a viscosity solution of the following Isaacs equation:*

$$\sup_{b \in \mathbb{S}^{n-1}} \inf_{L \in D^2 w(b)^\perp \cap \Sigma} \operatorname{tr}(L D^2 w) = 0.$$

CHAPTER 5

Solutions from Isoparametric Forms

In this chapter we describe homogeneous solutions of uniformly elliptic fully nonlinear equations which have Cartan's cubics and other isoparametric polynomials as numerators. Section 5.1 contains a calculation of the spectrum for the Hessian of fractions with Cartan's cubics as numerators and a hyperbolicity result for corresponding Hessian differences. In Section 5.2 some consequences for nonlinear equations in dimensions $5, 8, 14$, and 26, respectively, are given. Section 5.3 contains our most advanced result, namely, the existence of singular solutions in $n = 5$ dimensions, which needs rather elaborate calculations. Section 5.4 treats the case of other isoparametric polynomials (of degrees 4 and 6). In Section 5.5 we give two more applications of isoparametric polynomials to constructing unusual solutions of nonlinear elliptic equations, namely, to singular solutions of the p-Laplace equation and to the Landau-Ginzburg system. These applications do not enter directly into our general framework but are rather close to it in spirit.

5.1. The spectrum

5.1.1. The Hessian of the generic norm. In this section, we establish an explicit representation for the spectrum of the Hessian of certain rational functions associated with Euclidean cubic Jordan algebras $V = \mathrm{Herm}_3(\mathbb{F}_d)$, $d = 1, 2, 4, 8$. Recall that by Theorem 3.1.1 the normalized Jordan norm

(5.1.1) $$u(x) = \sqrt{2}N(x)|_{e^\perp},$$

where $e^\perp$ is the subspace of trace free elements in V, satisfies the eiconal equation

(5.1.2) $$|\nabla u(x)|^2 = 9|x|^4, \quad x \in e^\perp.$$

The cubic form $u(x)$ is rotationally equivalent to the Cartan isoparametric cubic form C_d (3.1.1). For instance, if $d = 1$, one has

$$C_1(x) = x_5^3 + \frac{3x_5}{2}(x_1^2 + x_2^2 - 2x_3^2 - 2x_4^2) + \frac{3\sqrt{3}x_4}{2}(x_2^2 - x_1^2) + 3\sqrt{3}x_1 x_2 x_3.$$

We shall then study the spectrum of the following ratio:

$$w_\alpha(x) = \frac{u(x)}{|x|^\alpha},$$

where $u(x)$ is defined by (5.1.1) and $\alpha \in \mathbb{R}$. This ratio is obviously a homogeneous order $(3 - \alpha)$ function. In Proposition 5.1.1 we give an explicit expression for the spectrum of the Hessian of w_α for $\alpha = 0$, and in Proposition 5.1.2 we establish the general case $\alpha \in \mathbb{R}$. Then we study further analytic properties of the spectrum in Subsection 5.1.4.

First we recall some definitions introduced earlier. Sometimes it is convenient to make computations in local coordinates, so that we shall always assume in this section that the $\{e_i\}_{i=1}^n$ denote an orthonormal basis of V, $n = \dim V = 3d + 3$, with $e_n = e$ chosen to be the unit element in V. We identify a vector $x \in \mathbb{R}^n$ with the element $x = \sum_{i=1}^n e_i x_i \in V$. By our choice of the orthonormal basis, we also have for the bilinear trace form

$$T(x; y) = \frac{1}{3} \operatorname{Tr}(x \bullet y) = \sum_{i=1}^n x_i y_i \equiv \langle x, y \rangle$$

and for the associated Euclidean norm

(5.1.3) $$T(x) \equiv T(x; x) = \langle x, x \rangle = |x|^2 = \sum_{i=1}^n x_i^2.$$

It is convenient for the following to write the minimum polynomial

$$m_x(\lambda) = N(\lambda e - x) \equiv \lambda^3 - \operatorname{Tr}(x)\lambda^2 + S(x)\lambda - N(x)$$

for a generic trace free element $x \in e^\perp$. To this end we note that by (3.2.30)

(5.1.4) $$S(x) = \frac{1}{2} S(x; x) = \frac{1}{2}(\operatorname{Tr}(x)^2 - \operatorname{Tr}(x^{\bullet 2})) = -\frac{1}{2} \operatorname{Tr}(x^{\bullet 2}), \quad x \in e^\perp;$$

hence using (3.2.6), the minimum polynomial for a trace free element x is given by

(5.1.5) $$m_x(\lambda) = \lambda^3 - \frac{1}{2} \operatorname{Tr}(x^{\bullet 2})\lambda - N(x) \equiv \lambda^3 - \frac{1}{2}|x|^2 \lambda - \frac{u(x)}{\sqrt{2}}.$$

In particular, any trace free element satisfies the following cubic identity:

(5.1.6) $$x^{\bullet 3} = \frac{1}{2} \operatorname{Tr}(x^{\bullet 2})x + N(x)e, \quad x \in e^\perp,$$

which yields the following trace representation of the cubic form $u(x)$ defined by (5.1.1):

(5.1.7) $$u(x) = \frac{\sqrt{2}}{3} \operatorname{Tr}(x^{\bullet 3}).$$

PROPOSITION 5.1.1. *In the above notation, the characteristic polynomial of* $\operatorname{Hess}_x(u)$ *is given by*

(5.1.8) $$\det(\lambda \mathbf{1} - \operatorname{Hess}_x(u)) = (\lambda^2 - 36|x|^2) \cdot (\lambda^3 - 27|x|^2\lambda + 9u(x))^d.$$

5.1.2. Proof of Proposition 5.1.1. Since the Hessian of u is a *homogeneous polynomial* map, it suffices to prove Proposition 5.1.1 under the assumption that x belongs to a nonempty Zariski open subset of the unit sphere

(5.1.9) $$\mathbb{S}(e^\perp) = \{\xi \in e^\perp : |\xi| = 1\}.$$

More precisely, in what follows, we shall assume that $x = \xi \in \mathbb{S}(e^\perp)$ is a regular element satisfying the following two conditions:

(5.1.10) $$N(\xi) \ne 0 \quad \text{and} \quad D(m_\xi) \ne 0,$$

where $D(m_x)$ is the discriminant of the minimum polynomial $m_x(\lambda)$ from (5.1.5).

For a regular $\xi \in \mathbb{S}(e^\perp)$ we define the operator $L_e(\xi) : e^\perp \to e^\perp$ by

(5.1.11) $$L_e(\xi)y = (\xi \bullet y)^{e^\perp} = \xi \bullet y - \langle \xi \bullet y, e \rangle e \equiv \xi \bullet y - \langle \xi, y \rangle e, \quad y \in e^\perp,$$

where $z^{e^\perp}$ stands for the projection of $z \in V$ onto $e^\perp$. Observe that for any $y, z \in e^\perp$,

$$\langle L(\xi)y, z \rangle = \langle (L(\xi)y)^{e^\perp}, z \rangle = \langle L_e(\xi)y, z \rangle,$$

and in view of the associativity of the trace form,
$$\langle L_e(\xi)y, z\rangle = \langle L_e(\xi)z, y\rangle,$$
which implies that L_e is a selfadjoint operator.

LEMMA 5.1.1. *In this notation*
$$\mathrm{Hess}_\xi(u) = 6\sqrt{2}\, L_e(\xi).$$

PROOF. Applying the associativity of the generic trace, one obtains from (5.1.7)
$$(5.1.12) \qquad \partial_{x_i} u|_\xi = \frac{\sqrt{2}}{3}\partial_{e_i}\mathrm{Tr}(\xi^{\bullet 3}) = \sqrt{2}\,\mathrm{Tr}(\xi^{\bullet 2}\bullet e_i) = 3\sqrt{2}\langle \xi^{\bullet 2}, e_i\rangle;$$
hence
$$(5.1.13) \qquad \partial_{x_j x_i} u|_\xi = 6\sqrt{2}\langle \xi\bullet e_i, e_j\rangle,$$
which yields for any $y, z \in e^\perp$ that
$$\frac{1}{6\sqrt{2}}\langle\mathrm{Hess}_\xi(u)y, z\rangle = \sum_{i,j=1}^{n-1}\langle \xi\bullet e_i, e_j\rangle y_i z_j = \langle \xi\bullet y, z\rangle = \langle (\xi\bullet y)^{e^\perp}, z\rangle = \langle L_e(\xi)y, z\rangle.$$
By the nonsingularity of the inner product we derive the desired identity. □

LEMMA 5.1.2. *The space*
$$V_\xi := \mathbb{R}[\xi] \cap e^\perp$$
is an invariant vector subspace of $L_e(\xi)$ and
$$(5.1.14) \qquad \det(t\mathbf{1} - L_e(\xi))\big|_{V_\xi} = t^2 - \frac{1}{2}.$$

PROOF. We have for any $k \geq 0$
$$L_e(\xi)\xi^{\bullet k} = (\xi^{\bullet(k+1)})^{e^\perp} = \xi^{\bullet(k+1)} - \langle \xi^{\bullet(k+1)}, e\rangle e,$$
which shows that $L_e(\xi)\xi^{\bullet k} \in \mathbb{R}[\xi]\cap e^\perp$; hence $L_e(x) : \mathbb{R}[\xi] \to V_\xi$. Since $V_\xi \subset \mathbb{R}[\xi]$, we conclude that V_ξ is an eigenspace of $L_e(\xi)$.

Furthermore, we have by (5.1.6)
$$(5.1.15) \qquad \xi^{\bullet 3} = \frac{3}{2}\xi + N(\xi)e, \qquad \forall \xi \in \mathbb{S}(e^\perp).$$
Define $v_\pm = \xi^{\bullet 2} \pm \frac{1}{\sqrt{2}}\xi - e$. Then using (5.1.15) and the fact that $(e)^{e^\perp} = 0$ we find
$$L_e(\xi)v_\pm = \left(\xi^{\bullet 3} \pm \frac{\xi^{\bullet 2}}{\sqrt{2}} - \xi\right)^{e^\perp} = \left(\frac{1}{2}\xi \pm \frac{\xi^{\bullet 2}}{\sqrt{2}} + N(x)e\right)^{e^\perp} \equiv \pm\frac{1}{\sqrt{2}}v_\pm,$$
which proves that $v_\pm$ is an eigenvector with eigenvalue $\pm\frac{1}{\sqrt{2}}$. Since these vectors are linearly independent and $\dim V_\xi = 2$, they form an eigenbasis of V_ξ; hence (5.1.14) follows. □

Now, using the definition $\mathbb{R}[\xi] = V_\xi \oplus \mathbb{R}e$, we obtain
$$(5.1.16) \qquad \mathbb{R}[\xi]^\perp = V_\xi^\perp \cap (\mathbb{R}e)^\perp \equiv V_\xi^\perp \cap e^\perp,$$
which shows that the orthogonal complement of V_ξ in $e^\perp$ is $\mathbb{R}[\xi]^\perp$. By Lemma 5.1.2,
$$(5.1.17) \qquad e^\perp = V_\xi \oplus \mathbb{R}[\xi]^\perp$$

is an orthogonal decomposition of $L_c(\xi)$ into invariant subspaces; thus, it suffices to determine the spectrum of the restriction $L_e(\xi)$ onto $\mathbb{R}[\xi]^\perp$.

LEMMA 5.1.3. *It is true that*

(5.1.18) $$\operatorname{tr} L_e(\xi)|_{\mathbb{R}[\xi]^\perp} = \operatorname{tr} L_e(\xi) = 0$$

and

(5.1.19) $$\det(\lambda \mathbf{1} - L_e(\xi))|_{\mathbb{R}[\xi]^\perp} = (-2)^{-3d} m_\xi(-2\lambda)^d,$$

where m_ξ is defined by (5.1.5).

PROOF. Since $e \in \mathbb{R}[\xi]$, we have

(5.1.20) $$L_e(\xi)|_{\mathbb{R}[\xi]^\perp} = L(\xi)|_{\mathbb{R}[\xi]^\perp}.$$

On the other hand, $\mathbb{R}[\xi]^\perp$ also is an invariant subspace of $L(\xi)$ for if $y \in \mathbb{R}[\xi]^\perp$ and $z \in \mathbb{R}[\xi]$, then $\xi \bullet z \in \mathbb{R}[\xi]$ and

$$\langle L(\xi) y, z \rangle = \langle \xi \bullet y, z \rangle = \langle y, \xi \bullet z \rangle = 0;$$

hence $L(\xi) y \in \mathbb{R}[\xi]^\perp$. Thus, it suffices to determine the spectrum of $L(\xi)$ on $\mathbb{R}[\xi]^\perp$.

Since V is a simple Euclidean Jordan algebra, one can find a Jordan frame $\{c_1, c_2, c_3\}$ of ξ such that

(5.1.21) $$\xi = \sum_{k=1}^{3} \theta_k c_k, \qquad c_i \bullet c_j = \delta_{ij} c_i, \ c_1 + c_2 + c_3 = e,$$

where each idempotent $c_i \ne 0$ is primitive and $\theta_i \in \mathbb{R}$ are the roots of

(5.1.22) $$m_\xi(t) = \prod_{k=1}^{3} (t - \theta_k);$$

see [**85**, p. 44]. Let $V = \bigoplus_{i \le j} V_{ij}$ be the corresponding Peirce decomposition associated with c_1, c_2, c_3, where

(5.1.23) $$L(c_k)|_{V_{ij}} = c_{kij} \mathbf{1}_{V_{jk}}, \qquad c_{kij} = \begin{cases} 1, & \text{if } i = j = k, \\ 0, & \text{if } i, j, k \text{ are pairwise distinct}, \\ \frac{1}{2}, & \text{otherwise.} \end{cases}$$

By virtue of the regularity of ξ, $\bigoplus_{i=1}^{3} V_{ii} \equiv \operatorname{span}(c_1, c_2, c_3) = \mathbb{R}[\xi]$; hence

$$\bigoplus_{1 \le i, j \le 3} V_{ij} = \mathbb{R}[\xi]^\perp.$$

Now observe that in view of $\operatorname{Tr} \xi = 0$, we see that $\sum_{i=1}^{3} \theta_i = 0$; hence for any $i \ne j$,

$$L(\xi)|_{V_{ij}} = \frac{1}{2} (\theta_i + \theta_j) \mathbf{1}_{V_{ij}} = -\frac{1}{2} \theta_k \mathbf{1}_{V_{ij}},$$

where (i, j, k) is a permutation of $(1, 2, 3)$. Since V is simple, $V_{ii} = \operatorname{span}(c_i)$, and $\dim V_{ij} = d \equiv \dim_\mathbb{R} \mathbb{F}_d$ (see Corollary IV.2.6 in [**85**]); hence

$$\det(L(\xi) - \mathbf{1})|_{\mathbb{R}[\xi]^\perp} = \prod_{k=1}^{3} (\lambda + \frac{1}{2} \theta_k)^d = (-2)^{3d} \prod_{k=1}^{3} (-2\lambda - \theta_k)^d = (-2)^{3d} m_\xi(-2\lambda),$$

which yields (5.1.19).

To finish the proof, it only remains to notice that (5.1.18) follows from the Viète theorem applied to the coefficient of λ^{3d-1} of the right-hand side of (5.1.19) and the relation (5.1.20) □

Now, Proposition 5.1.1 follows from Lemmas 5.1.3, 5.1.2, and 5.1.1.

5.1.3. The case of general α.

PROPOSITION 5.1.2. *Let $\alpha \neq 2$. Then for any $\xi \in \mathbb{S}(e^\perp)$ the following holds:*

$$(5.1.24) \qquad \det((\mu - \alpha u)\mathbf{1}_{e^\perp} - \mathrm{Hess}_\xi(w_\alpha)) = Q_1(\mu)^d Q_2(\mu),$$

with

$$(5.1.25) \quad Q_1(\mu) = \mu^3 - 27\mu + 54u, \quad Q_2(\mu) = \mu^2 + \beta u\,\mu + 3(3\beta - \beta u^2 - 12),$$

where $\beta = \alpha(4-\alpha)$ and $u = u(\xi)$.

PROOF. We have

$$|x|^{\alpha+4}\partial^2_{x_i x_j} w_\alpha = ((\alpha+2)x_i x_j - \delta_{ij}|x|^2)\alpha u + |x|^4 \partial^2_{x_i x_j} u - \alpha(x_i \partial_{x_j} u + x_j \partial_{x_i} u).$$

Setting $x = \xi \in \mathbb{S}(e^\perp)$ in the latter identity, multiplying it by y_i, z_j, and summing the resulting identities over all $1 \leq i, j \leq n-1$, we find using (5.1.12) and (5.1.13),

$$(5.1.26) \quad \begin{aligned} \langle \mathrm{Hess}_\xi(w_\alpha)y, z\rangle &= \big((\alpha+2)\langle \xi, y\rangle\langle \xi, z\rangle - \langle y, z\rangle\big)\alpha u + 6\sqrt{2}\,\langle L_e(\xi)y, z\rangle \\ &\quad - 3\sqrt{2}\,\alpha\big(\langle y, \xi\rangle\langle \xi^{\bullet 2}, z\rangle + \langle z, \xi\rangle\langle \xi^{\bullet 2}, y\rangle\big) \end{aligned}$$

where we denote for short $u = u(\xi)$. Let $y \otimes z$ denote the rank 1 operator $x \mapsto \langle z, x\rangle y$. Then by the nonsingularity of the inner product we find from (5.1.26),

$$(5.1.27) \qquad H := \mathrm{Hess}_\xi(w_\alpha) + \alpha u \mathbf{1}_V = 6\sqrt{2}\,L_e(\xi) + \alpha B,$$

where

$$B = (\alpha+2)u \cdot \xi \otimes \xi - 3\sqrt{2}\big(\xi \otimes \eta + \eta \otimes \xi\big)$$

and $\eta = \xi^{\bullet 2} - e \in e^\perp$.

Observe that $V_\xi = \mathbb{R}[\xi] \cap e^\perp$ and $\mathbb{R}[\xi]^\perp$ are invariant subspaces of H. Indeed, B is obviously an operator with values in V_ξ; hence V_ξ is an invariant subspace of B. On the other hand, by Lemma 5.1.2, V_ξ is an invariant subspace of $L_e(\xi)$; hence $H : V_\xi \to V_\xi$ by (5.1.27). Next, observe that H is selfadjoint because $L_e(\xi)$ and B are; hence (5.1.17) implies that $\mathbb{R}[\xi]^\perp$ is also an invariant subspace of $\mathrm{Hess}_\xi(w_\alpha)$, which proves our claim.

In particular, we have

$$(5.1.28) \qquad \det(\mu \mathbf{1}_{e^\perp} - H) = \det(\mu\mathbf{1} - H)\big|_{\mathbb{R}[\xi]^\perp} \cdot \det(\mu\mathbf{1} - H)\big|_{V_\xi}.$$

Furthermore, observe that $\mathbb{R}[\xi]^\perp \subset \ker B$; hence $H|_{\mathbb{R}[\xi]^\perp} = 6\sqrt{2}\,L_e(\xi)|_{\mathbb{R}[\xi]^\perp}$, which yields by (5.1.19)

$$\det(\mu\mathbf{1} - H)\big|_{\mathbb{R}[\xi]^\perp} = \det(\mu\mathbf{1} - 6\sqrt{2}\,L_e(\xi))\big|_{\mathbb{R}[\xi]^\perp} = (\mu^3 - 27\mu + 54u)^d.$$

The first factor in (5.1.28) and, thereby, the explicit expression for Q_1 in (5.1.24) are established.

In order to establish the second factor in (5.1.28) we observe that ξ and $\eta = \xi^{\bullet 2} - e \in e^\perp$ form a basis in V_ξ because they are linearly independent (recall that by the regularity assumption ξ is an element of order 3) and $\dim V_\xi = 2$. Since $\langle \xi^{\bullet 2}, e\rangle = \langle \xi, \xi\rangle = 1$, we have

$$L_e(\xi)\xi = \xi^{\bullet 2} - \langle \xi^{\bullet 2}, e\rangle e = \xi^{\bullet 2} - e = \eta.$$

Furthermore, by (5.1.15) and the definition of $u(\xi) = \sqrt{2}N(\xi)$, we have $\xi^{\bullet 3} = \frac{3}{2}\xi + \frac{u}{\sqrt{2}}e$ and $\eta \bullet \xi = \xi^{\bullet 3} - \xi = \frac{\xi}{2} + \frac{u}{\sqrt{2}}e$; hence

$$L_e(\xi)\eta = \eta \bullet \xi - \langle \eta \bullet \xi, e\rangle e = \frac{1}{2}\xi.$$

Similarly, we find

$$\langle \eta, \xi\rangle = \langle \xi^3, e\rangle = \frac{u}{\sqrt{2}}, \qquad \langle \eta, \eta\rangle = \langle \xi^{\bullet 4} - 2\xi^{\bullet 2} + e, e\rangle = \frac{1}{2}.$$

Then (5.1.27) yields

$$H\xi = \alpha(\alpha - 1)u \cdot \xi - 3\sqrt{2}(\alpha - 2) \cdot \eta,$$

$$H\eta = \frac{\sqrt{2}}{2}\left(6 - 3\alpha + \alpha(\alpha + 2)u^2\right) \cdot \xi - 3\alpha u \cdot \eta,$$

with the following matrix in the basis $\{\xi, \eta\}$:

(5.1.29) $$H = \begin{pmatrix} \alpha(\alpha - 1)u & -3\sqrt{2}(\alpha - 2) \\ 3\sqrt{2}\left(1 - \frac{\alpha}{2} + \frac{\alpha(\alpha+2)}{6}u^2\right) & -3\alpha u \end{pmatrix}.$$

This immediately yields

$$\det(\mu\mathbf{1} - H)\big|_{V_\xi} = \mu^2 + \beta u\,\mu + (9\beta - 3\beta u^2 - 36)$$

with $\beta = 4\alpha - \alpha^2$. The proposition is proved. $\square$

5.1.4. Hyperbolicity. First we make some elementary observations.

LEMMA 5.1.4. *If $u(x)$ is defined by (5.1.1) and $\xi \in \mathbb{S}(e^\perp)$, then $|u(\xi)| \le 1$. If ξ is also regular, then $|u(\xi)| < 1$.*

PROOF. By (5.1.2), $|\nabla u(x)| \le 3|x|$ for all $x \in V$. Hence

$$|u(\xi)| = |u(\xi) - u(0)| = \left|\int_0^1 \frac{\partial u(\xi t)}{\partial t}\,dt\right| = \left|\int_0^1 \langle \nabla u(\xi t), \xi\rangle t\,dt\right| \le \int_0^1 3t^2\,dt = 1.$$

Now assume that ξ is a regular element. By (5.1.5), $m_\xi(t) = t^3 - \frac{3}{2}t - \frac{u}{\sqrt{2}}$; hence (5.1.10) yields $D(m_\xi) \equiv \frac{27}{2}(1 - u(\xi)^2) \ne 0$, which implies the required conclusion. $\square$

For a more detailed look at the Hessian's eigenvalues we observe that the roots of the polynomials Q_1 and Q_2 are real because each Q_i is a factor of the characteristic polynomial of a selfadjoint operator.

LEMMA 5.1.5. *For any regular $\xi \in \mathbb{S}(e^\perp)$, the cubic polynomial $Q_1(\mu)$ has three distinct real roots $\mu_3 < \mu_2 < \mu_1$ which satisfy the following properties:*
  (i) $\mu_1 \mu_3 < 0$.
  (ii) $|\mu_2| < 3$.
  (iii) $3 < |\mu_k| < 6$, $k \in \{1, 3\}$.

PROOF. We have $Q_1'(\mu) = 3(\mu^2 - 9)$; hence $\mu = \pm 3$ are the critical points of the cubic polynomial Q_1. We have the following for the critical values:

$$(5.1.30) \qquad Q_1(\pm 3) = \mp 54(1 - u(\xi));$$

hence $Q_1(-3) > 0$ and $Q_1(3) < 0$, which easily implies that $Q(\mu)$ has distinct real roots satisfying $\mu_3 < -3 < \mu_2 < 3 < \mu_1$. In particular, this proves (i), (ii), and the lower bound in (iii). The upper bound in (iii) follows similarly from

$$(5.1.31) \qquad Q_1(\pm 6) = \pm 54(1 - u(\xi)). \qquad \square$$

LEMMA 5.1.6. *Let $\xi \in \mathbb{S}(e^\perp)$ be regular, let $\mu_3 < \mu_2 < \mu_1$ be the roots of $Q_1(\mu)$, and let $\rho_2 \leq \rho_1$ be the roots of $Q_2(\mu)$. Then for any $\alpha \in [1, 2)$,*

$$(5.1.32) \qquad \mu_1 < \rho_2 \leq \rho_1 < \mu_3.$$

Moreover, if $\alpha = 1$, then one has a stronger inequality:

$$(5.1.33) \qquad \mu_3 < \rho_2 < \mu_2 < \rho_1 < \mu_1.$$

PROOF. We have $\beta = 4\alpha - \alpha^2 \in [3; 4)$ for $\alpha \in [1, 2)$. This yields by Lemma 5.1.4 that $3\beta - \beta u^2 \leq 3\beta < 12$ and

$$Q_2(0) = 3(3\beta - \beta u^2 - 12) < 0;$$

hence $\rho_2 < 0 < \rho_1$. Thus, in order to verify (5.1.32) it suffices to show that Q_2 is positive at the roots μ_1 and μ_3. To this end, we denote by y one of the roots μ_1 or μ_3 of the polynomial Q_1. Observe that $Q_1(y) = 0$ yields $u = \frac{1}{54}(27y - y^3)$, and substituting this into (5.1.25), we obtain

$$Q_2(y) = y^2 + \beta u\, y + 9\beta - 3\beta u^2 - 36 = \frac{36 - y^2}{972}\left(\beta y^4 - 243(4 - \beta)\right).$$

By the upper bound in (iii) in Lemma 5.1.5, we have $36 - y^2 > 0$, and using the lower bound in (iii) in Lemma 5.1.5, we obtain

$$\beta y^4 - 243(4 - \beta) \geq 81\beta - 243(4 - \beta) = 324(\beta - 3) > 0,$$

which shows $Q_2(y) > 0$ and thereby proves (5.1.32).

In order to establish (5.1.33), we note that $\alpha = 1$ yields $\beta = 3$ and the above argument together with (ii) in Lemma 5.1.5 immediately shows

$$Q_2(\mu_2) = \frac{36 - \mu_2^2}{324}\left(\mu_2^4 - 81\right) < 0.$$

This yields that μ_2 lies between the roots of Q_2, and the proposition is proved completely. $\qquad \square$

Now we are ready to establish the hyperbolicity result.

PROPOSITION 5.1.3. *Let $\alpha = 1$ and*

$$M(\xi, \xi', O) = \operatorname{Hess}_\xi(w_1) - O^t \operatorname{Hess}_{\xi'}(w_1) O,$$

where O is an orthogonal endomorphism of V, and let

$$\Lambda_n \leq \Lambda_{n-1} \leq \cdots \leq \Lambda_2 \leq \Lambda_1, \quad n = 3d + 2,$$

be the eigenvalues of $M(\xi, \xi', O)$. Then

$$(5.1.34) \qquad \frac{1}{c} \leq -\frac{\Lambda_1}{\Lambda_n} \leq c, \qquad \text{with } c = 2(n + 1).$$

PROOF. Since $\Delta u(x) = 0$, we find by homogeneity of u for any $\xi \in \mathbb{S}(e^\perp)$
$$\Delta w_1(\xi) = |\xi|^{-1}\Delta u - 2|\xi|^{-3}\langle \nabla u(\xi), \xi\rangle + u\Delta(|\xi|^{-1})$$
$$= -(n+3)|\xi|^{-3}u(\xi) \equiv -(n+3)u(\xi).$$

By density it suffices to prove the proposition under the assumption that ξ and ξ' are regular. Observe that

(5.1.35) $$\operatorname{tr} M(\xi, \xi', O) = \Delta w_1(\xi) - \Delta w_1(\xi') = -(n+3)\omega,$$

where $\omega := u(\xi) - u(\xi')$. If $\omega = 0$, then (5.1.34) trivially holds with $c = (n-1)$ as for any traceless matrix in dimension n. By symmetry, we may assume that $\omega > 0$. Then

(5.1.36) $$-(n+3)\omega = \operatorname{tr} M(\xi, \xi', O) \geq \Lambda_1 + (n-1)\Lambda_n.$$

This yields by the assumption $\omega > 0$ that

(5.1.37) $$\Lambda_1 \leq -(n-1)\Lambda_5.$$

Denote by
$$\lambda_n \leq \cdots \leq \lambda_2 \leq \lambda_1, \qquad \lambda'_n \leq \cdots \leq \lambda'_2 \leq \lambda'_1$$
the eigenvalues of $\operatorname{Hess}_\xi(w_1)$ and $\operatorname{Hess}_{\xi'}(w_1)$, respectively. Then by Weyl's inequality,

(5.1.38) $$\Lambda_1 \geq \max_{1\leq i \leq n}(\lambda_i - \lambda'_i) \geq \lambda_\gamma - \lambda'_\gamma,$$

$\gamma = [n/2]$, i.e., λ_γ is the middle term. Observe that in our case $\alpha = 1$; hence by Proposition 5.1.1 a real number λ is an eigenvalue of $\operatorname{Hess}_\xi(w_1)$ if and only if $\mu = \lambda + u(\xi)$ is a root of $Q_1(\mu)Q_2(\mu)$. Applying (5.1.33), we obtain for the middle term
$$\lambda_\gamma = \mu_2 - u(\xi), \quad \lambda'_\gamma = \mu'_2 - u(\xi');$$
hence (5.1.38) yields

(5.1.39) $$\Lambda_1 \geq \mu_2 - \mu'_2 - \omega.$$

On the other hand, $Q_1(\mu_2) = 0$ yields $\mu_2^3 - \mu_2'^3 - 27(\mu_2 - \mu'_2) + 54\omega = 0$; hence
$$\mu_2 - \mu'_2 = 54\frac{\omega}{\sigma_2},$$
where $\sigma_2 := 27 - (\mu_2^2 + \mu_2\mu'_2 + \mu_2'^2) > 0$ by (ii) in Lemma 5.1.5. On the other hand, $\mu_2^2 + \mu_2\mu'_2 + \mu_2'^2 \geq 0$; hence $\sigma_2 \leq 27$ and
$$\mu_2 - \mu'_2 = 54\frac{\omega}{\sigma_2} > 2\omega,$$
which yields $\Lambda_1 \geq \omega$ by (5.1.39). Combining this with

(5.1.40) $$-(n+3)\omega = \operatorname{tr} M(\xi, \xi', O) \leq (n-1)\Lambda_1 + \Lambda_n,$$

we find

(5.1.41) $$-\Lambda_5 \leq 2(n+1)\Lambda_1,$$

which together with (5.1.37) yields the required two-sided estimate. □

5.2. Nonclassical and singular solutions

We will then construct nonclassical and singular solutions using the Cartan cubics in dimensions 5, 8, 4, and 26. To do that we recall that the argument in the proof of Theorems 4.2.1 and 4.3.1 gives that if $w = w_n$ is an odd homogeneous function of order $2 - \delta$, $0 \leq \delta < 1$, defined on a unit ball $B = B_1 \subset \mathbb{R}^n$, smooth in $B \setminus \{0\}$ and with

$$\Lambda : B \longrightarrow \lambda(S) \in \mathbb{R}^n ,$$

where $\lambda(S) = \{\lambda_i : \lambda_1 \leq \cdots \leq \lambda_n\} \in \mathbb{R}^n$ is the corresponding eigenvalue map, then the uniform hyperbolicity of the family of nonzero differences

(5.2.1) $$D^2(w)(a) - O^t \cdot D^2(w)(a) \cdot O$$

with $a \neq b \in \mathbb{R}^n \setminus \{0\}$ and an orthogonal matrix O is sufficient for the validity of Lemma 4.2.3. Therefore, w is a viscosity solution in B of a uniformly elliptic Hessian equation of the form (0.1) as well as of a uniformly elliptic Isaacs equation.

Applying Proposition 5.1.3 one deduces that the functions $w_n(x) = P_n(x)/|x|$ for $n = 5, 8, 14$, and 26 are nonclassical solutions of uniformly elliptic Hessian equations and uniformly elliptic Isaacs equations. However, this fact can be deduced directly, not using the theory of Jordan algebras in dimensions 5 and 8, and in dimensions 14 and 26 much more is true. We give these developments below.

5.2.1. Five dimensions.
The form $C_1(x)$ admits a 3-dimensional automorphism group. Indeed, one easily verifies that the orthogonal transformations

$$A_1(\phi) := \begin{pmatrix} \frac{3\cos(\phi)^2-1}{2} & \frac{\sqrt{3}\sin(\phi)^2}{2} & 0 & 0 & \frac{\sqrt{3}\sin(2\phi)}{2} \\ \frac{\sqrt{3}\sin(\phi)^2}{2} & \frac{1+\cos(\phi)^2}{2} & 0 & 0 & \frac{-\sin(2\phi)}{2} \\ 0 & 0 & \cos(\phi) & \sin(\phi) & 0 \\ 0 & 0 & -\sin(\phi) & \cos(\phi) & 0 \\ \frac{-\sqrt{3}\sin(2\phi)}{2} & \frac{\sin(2\phi)}{2} & 0 & 0 & \cos(2\phi) \end{pmatrix},$$

$$A_2(\psi) := \begin{pmatrix} 1 & 0 & 0 & 0 & 0 \\ 0 & \cos(2\psi) & 0 & 0 & -\sin(2\psi) \\ 0 & 0 & \cos(\psi) & -\sin(\psi) & 0 \\ 0 & 0 & \sin(\psi) & \cos(\psi) & 0 \\ 0 & \sin(2\psi) & 0 & 0 & \cos(2\psi) \end{pmatrix},$$

$$A_3(\theta) := \begin{pmatrix} \frac{3\cos(\theta)^2-1}{2} & \frac{-\sqrt{3}\sin(\theta)^2}{2} & 0 & \frac{-\sqrt{3}\sin(2\theta)}{2} & 0 \\ \frac{-\sqrt{3}\sin(\theta)^2}{2} & \frac{1+\cos(\theta)^2}{2} & 0 & \frac{-\sin(2\theta)}{2} & 0 \\ 0 & 0 & \cos(\theta) & 0 & -\sin(\theta) \\ \frac{\sqrt{3}\sin(2\theta)}{2} & \frac{\sin(2\theta)}{2} & 0 & \cos(2\theta) & 0 \\ 0 & 0 & \sin(\theta) & 0 & \cos(\theta) \end{pmatrix}$$

do not change the value of $C_1(x)$.

Moreover, one easily gets

LEMMA 5.2.1. *Let G_P be the subgroup of $SO(5)$ generated by $\{A_1(\phi), A_2(\psi), A_3(\theta) : (\phi, \psi, \theta) \in \mathbb{R}^3\}$. Then the orbit $G_P \mathbb{S}^1$ of the circle*

$$\mathbb{S}^1 = \{(\cos(\chi), 0, \sin(\chi), 0, 0) : \chi \in \mathbb{R}\} \subset \mathbb{S}^4$$

under the natural action of G_P is the whole of $\mathbb{S}^4$.

PROOF. Indeed, calculating the differential of the action
$$(\mathbb{S}^1)^4 \longrightarrow \mathbb{S}^4, \quad (\phi, \psi, \theta, \chi) \mapsto (\cos(\chi), 0, \sin(\chi), 0, 0) A_1(-\phi) A_2(-\psi) A_3(-\theta)$$
at $(\phi, \psi, \theta, \chi) = (0, 0, 0, 0)$ one sees that its rank is 4, which implies surjectivity. □

REMARK 5.2.1. One notes that from the general theory of the isoparametric forms it follows immediately that $C_1(x)$ has a 3-dimensional automorphism group (in fact, a large majority of the isoparametric forms are homogeneous), but we prefer to give the above explicit generators to be as elementary as possible. Let us briefly comment about how to find those generators: one considers skew-symmetric matrices M which do not change $C_1(x)$ infinitesimally, i.e., $C_1(x + Mx\varepsilon) = C_1(x) + O(\varepsilon^2)$, which gives homogeneous linear conditions on M. One easily finds three linearly independent solutions, e.g.,

$$M_1(\phi) := \phi \begin{pmatrix} 0 & 0 & 0 & 0 & \sqrt{3} \\ 0 & 0 & 0 & 0 & -1 \\ 0 & 0 & 0 & 1 & 0 \\ 0 & 0 & -1 & 0 & 0 \\ -\sqrt{3} & 1 & 0 & 0 & 0 \end{pmatrix}, \quad M_2(\psi) := \psi \begin{pmatrix} 0 & 0 & 0 & 0 & 0 \\ 0 & 0 & 0 & 0 & -2 \\ 0 & 0 & 0 & 1 & 0 \\ 0 & 0 & -1 & 0 & 0 \\ 0 & 2 & 0 & 0 & 0 \end{pmatrix},$$

$$M_3(\theta) := \theta \begin{pmatrix} 0 & 0 & \sqrt{3} & 0 & 0 \\ 0 & 0 & -1 & 0 & 0 \\ -\sqrt{3} & 1 & 0 & 1 & 0 \\ 0 & 0 & 0 & 0 & -1 \\ 0 & 0 & 0 & 1 & 0 \end{pmatrix}$$

and then sets $A_i := \exp(M_i)$. Notice also that in [**174**] the matrices A_2 and A_3 are erroneous and should be changed (permute the last two lines and columns).

REMARK 5.2.2. It is clear that C_1 is a specialization of C_2 which is in turn a specialization of C_3, the last being a specialization of C_4. One can also compare the form C_1 with a natural specialization of $P_9(X, Y, Z)$ and thus of P_{12}. Recall that (Remark 4.1.5) the determinant function
$$P_9(X, Y, Z) = X_1 Y_3 Z_2 - X_1 Y_2 Z_3 + X_2 Y_1 Z_3 - X_2 Y_3 Z_1 + X_3 Y_2 Z_1 - X_3 Y_1 Z_2$$
is a specialization of P_{12}. One can specialize further, making the matrix M_{XYZ} symmetric and traceless and thus a traceless element of $\mathrm{Sym}_3(\mathbb{R}) = \mathrm{Herm}_3(\mathbb{R})$ that corresponds to the equations $Y_1 := X_2, Z_1 := X_3, Z_2 := Y_3, Z_3 := -X_1 - Y_2$. The determinant of the matrix $N(X, Y)$ obtained in this way becomes the cubic
$$Q_5(X_1, X_2, X_3, Y_2, Y_3) := (X_2^2 - X_1 Y_2)(X_1 + Y_2) + 2 X_2 X_3 Y_3 - X_1 Y_3^2 - Y_2 X_3^2$$
in five variables.

It is important to notice that Q_5 is *not* proportional to $P_5 = C_1$ rotated. The reason is that C_1 is the determinant of a traceless symmetric matrix M under the condition that the matrix norm of M is proportional to the norm of the coordinate vector. This last condition does not hold for $N(X, Y)$:
$$\|N(X, Y)\|^2 = \mathrm{tr}(N(X, Y)^2) = 2(X_1^2 + X_2^2 + X_3^2 + Y_2^2 + Y_3^2 + X_1 Y_2).$$

Brute force (MAPLE) calculations show that Q_5 does not verify the ellipticity criterion in its weakest form, that of Subsection 4.1.1. Therefore, one cannot replace P_5 by Q_5 in the main theorems of the present section and Section 5.3; it does not give solutions of elliptic equations (even degenerate). Note also that, naturally, Q_5 is not a solution to the eiconal equation.

5.2. NONCLASSICAL AND SINGULAR SOLUTIONS

Let $w_5(x) = C_1(x)/|x|$. By Lemma 4.2.2 it is sufficient to prove the uniform hyperbolicity of
$$M_5(x,y,O) := D^2 w_5(x) - O^t D^2 w_5(y) O.$$
We begin with calculating the eigenvalues of $D^2 w_5(x)$.

More precisely, we need

LEMMA 5.2.2. *Let $x \in \mathbb{S}^4$, let $\lambda_1 \geq \lambda_2 \geq \cdots \geq \lambda_5$ be the eigenvalues of $D^2 w_5(x)$, and let $x \in G_P(p,0,q,0,0)$ with $p^2 + q^2 = 1$. Then*
$$\lambda_1 = \frac{p^3 - 6p + 3\sqrt{12 - 3p^2}}{2}, \quad \lambda_3 = \frac{p^3 + 3p}{2}, \quad \lambda_5 = \frac{p^3 - 6p - 3\sqrt{12 - 3p^2}}{2}.$$

PROOF. Since w_5 is invariant under G_P, we can suppose that $x = (p,0,q,0,0)$. Then $w_5(x) = \frac{p(3-p^2)}{2}$ and we get by a brute force calculation
$$D^2 w_5(x) := \begin{pmatrix} M_1 & 0 \\ 0 & M_2 \end{pmatrix},$$
a block matrix with
$$M_1 := \frac{1}{2} \begin{pmatrix} p(1 + 2p^2 - 3p^4) & 3\sqrt{3}p(p^2 - 1) & 3q(1 - p^4) \\ 3\sqrt{3}p(p^2 - 1) & p^3 - 15p & 3\sqrt{3}q(p^2 + 1) \\ 3q(1 - p^4) & 3\sqrt{3}q(p^2 + 1) & p^3 + 3p^5 \end{pmatrix},$$
$$M_2 := \frac{1}{2} \begin{pmatrix} p^3 + 3p & 6\sqrt{3}q \\ 6\sqrt{3}q & p^3 - 15p \end{pmatrix},$$
which gives $F(S) = F_1(S) \cdot F_2(S)$ for the characteristic polynomial where
$$F_1(S) = \left(S - \frac{3p}{2} - \frac{p^3}{2}\right)\left(S^2 + 6pS - p^3 S + \frac{63p^2}{4} - 3p^4 + \frac{p^6}{4} - 27\right),$$
$$F_2(S) = \left(S^2 + \frac{15pS}{2} - \frac{5p^3 S}{2} - \frac{45p^2}{4} + \frac{15p^4}{2} - \frac{5p^6}{4} - 9\right)$$
have the roots
$$\lambda_1 = \frac{p^3 - 6p + 3\sqrt{12 - 3p^2}}{2}, \quad \lambda_3 = \frac{p^3 + 3p}{2},$$
$$\lambda_5 = \frac{p^3 - 6p - 3\sqrt{12 - 3p^2}}{2},$$
$$\lambda_2 = \frac{5p^3 - 15p + 3\sqrt{5p^6 - 30p^4 + 45p^2 + 16}}{4},$$
$$\lambda_4 = \frac{5p^3 - 15p - 3\sqrt{5p^6 - 30p^4 + 45p^2 + 16}}{4}.$$
One only needs to verify that indeed
$$\lambda_1 \geq \lambda_2 \geq \cdots \geq \lambda_5,$$
which is elementary. For example, let us verify for $p \in [-1, 1]$ the inequality
$$\lambda_1 = \frac{p^3 - 6p + 3\sqrt{12 - 3p^2}}{2} \geq \lambda_2 = \frac{5p^3 - 15p + 3\sqrt{5p^6 - 30p^4 + 45p^2 + 16}}{4}.$$

Indeed, $4(\lambda_1 - \lambda_2)/3 = r_1 - r_2$ with

$$r_1 := p - p^3 + 2\sqrt{12 - 3p^2} \geq 0, \qquad r_2 := \sqrt{5p^6 - 30p^4 + 45p^2 + 16},$$

$$(r_1^2 - r_2^2)/4 = (1 - p^2)(p^4 - 6p^2 + 8 + p\sqrt{12 - 3p^2}) \geq 0,$$

since

$$s_1^2 - s_2^2 := (p^4 - 6p^2 + 8)^2 - p^2(12 - 3p^2)$$
$$= (1 - p)(2 - p)(p + 2)(p^5 + p^4 - 7p^3 - 7p^2 + 13p + 16) \geq 0$$

because

$$p^5 + p^4 - 7p^3 - 7p^2 + 13p + 16 = (p + 1)(p^4 - 7p^2 + 13) + 3 \geq 3. \qquad \square$$

PROPOSITION 5.2.1. *Let* $O \in O(5), x \neq y \in \mathbb{S}^4$. *Suppose* $M_5(x, y, O) \neq 0$ *and let* $\Lambda_1 \geq \Lambda_1 \geq \cdots \geq \Lambda_5$ *be its eigenvalues. Then*

$$\frac{1}{20} \leq -\frac{\Lambda_1}{\Lambda_5} \leq 20.$$

PROOF. Let now $y \in G_P(\bar{p}, 0, \bar{q}, 0, 0)$. If $p = \bar{p}$ but $M_5(x, y, O) \neq 0$, then $\operatorname{tr}(M_5(x, y, O)) = 0$ and the conclusion follows as for any traceless matrix in dimension 5. Let then $p > \bar{p}$. Then by Lemma 5.2.2

$$\Lambda_1 \geq \lambda_2(p) - \lambda_2(\bar{p}) = \frac{(p - \bar{p})(p^2 + p\bar{p} + \bar{p}^2 + 3)}{2} \geq \frac{3(p - \bar{p})}{2},$$
$$-\Lambda_5 \geq \max\{\lambda_3(\bar{p}) - \lambda_3(p), \lambda_1(\bar{p}) - \lambda_1(p)\}$$
$$\geq (p - \bar{p}) \inf_{p \in [-1,1]} \max\{|\lambda_1'(p)|, |\lambda_3'(p)|\} = 3(p - \bar{p}),$$
$$0 \geq \operatorname{tr} M_5(x, y, O) = 8w_5(y) - 8w_5(x)$$
$$= -8(p - \bar{p})(3 - p^2 - p\bar{p} - \bar{p}^2) \geq -24(p - \bar{p}).$$

Therefore,
$$-\Lambda_5 \geq 4\Lambda_1$$
since $\operatorname{tr} M_5(x, y, O) \leq 0$ and

$$4\Lambda_1 + \Lambda_5 \geq \operatorname{tr} M_5(x, y, O) \geq -24(p - \bar{p}),$$

$$-\Lambda_5 \leq 4\Lambda_1 + 24(p - \bar{p}) \leq 4\Lambda_1 + 16\Lambda_1 = 20\Lambda_1.$$

The case $p < \bar{p}$ is completely parallel, which finishes the proof of our results. $\square$

Thus we get another proof of

THEOREM 5.2.1. *The function* $w_5(x) = C_1(x)/|x|$ *is a viscosity solution of a Hessian uniformly elliptic equation and of a uniformly elliptic Isaacs equation in the unit ball.*

5.2.2. Smooth equation.

Let us note that the uniformly elliptic functional F in all preceding constructions is, in general, only Lipschitz. Let us now show that in five dimensions we can make F (C^∞)-smooth; moreover, near the solution, F can be constructed to coincide with a simple explicit polynomial.

We begin with this explicit polynomial equation.

PROPOSITION 5.2.2. *The function $w = w_5(x)$ satisfies the Hessian equation*

$$g_5(w) = 0, \tag{5.2.2}$$

where

$$g_5(w) := \sigma_1(w)\left(\sigma_1(w)^2 + 1728\right)\left(5\sigma_1(w)^2 + 576\right) + 2^{15}\sigma_5(w)$$
$$= (\Delta w)^5 + 2^8 3^2 (\Delta w)^3 + 2^{12} 3^5 \Delta w + 2^{15}\sigma_5(w)$$

where $\sigma_1(w) = \Delta(w)$ is the Laplacian of w and $\sigma_5(w) := \det(D^2 w)$ is its Monge-Ampère operator.

PROOF. Indeed, the calculations above give for the characteristic polynomial

$$S^5 + a_1 S^4 + a_2 S^3 + a_3 S^2 + a_4 S + a_5$$

of $D^2 w$: $a_1 = -\Delta(w) = 4p(3 - p^2)$, $a_5 = -\sigma_5(w) = -\det(D^2 w)$,

$$a_5 = \frac{p}{32}(p^2 - 3)(p^2 + 3)(5p^6 - 30p^4 + 45p^2 + 36)(p^4 - 9p^2 + 36)$$

and a brute force calculation shows that $g_5(w)$ is identically zero as a polynomial in p. □

Let us now show that the equation $g_5(w)$ is uniformly elliptic in a neighborhood of the solution.

LEMMA 5.2.3. *The five partial derivatives $\frac{\partial g_5}{\partial \lambda_i}$, $i = 1, \ldots, 5$, are strictly positive (and bounded, which is automatic due to compactness) on the symmetrized image*

$$M_5 := \bigcup_{\sigma \in \Sigma_5} T_\sigma \Lambda(B) \subset \mathbb{R}^5$$

of the unit ball under the map Λ.

PROOF. Let us note first that M_5 is an algebraic curve, the union of 120 curves $T_\sigma \Lambda(B)$, and that it is sufficient, by symmetry, to verify the condition on the curve $\Lambda(B)$ only. Denote $h_i(p) := 2^{-15} \frac{\partial g_5}{\partial \lambda_i}$ as a function of p. One easily verifies by a brute force calculation that $h_2(p) = h_1(-p)$, $h_4(p) = h_3(-p)$,

$$h_1 := \frac{45}{128}p^{12} - \frac{135}{32}p^{10} + \frac{1215}{64}p^8 - \frac{243}{32}p^6 - \frac{19683}{128}p^4 + \frac{2187}{8}p^2 + \frac{243}{8}$$
$$+ p\sqrt{5p^6 - 30p^4 + 45p^2 + 16}\left(-\frac{3}{32}p^8 + \frac{27}{32}p^6 - \frac{81}{32}p^4 - \frac{243}{32}p^2 + \frac{243}{8}\right),$$

$$h_3 := -\frac{15}{128}p^{12} + \frac{15}{32}p^{10} + 495/64 p^8 \frac{495}{64}p^8 - \frac{1125}{32}p^6 - \frac{999}{128}p^4 + 162p^2 + \frac{243}{8}$$
$$+ \sqrt{12 - 3p^2}\left(\frac{15}{16}p^8 - \frac{45}{16}p^6 - \frac{135}{16}p^4 + \frac{513}{16}p^2 + \frac{81}{4}\right)p,$$

$$h_5 := -\frac{15}{128}p^{12} + \frac{105}{128}p^{10} - \frac{2205}{64}p^8 + \frac{5625}{32}p^6 - \frac{53487}{128}p^4 + \frac{567}{2}p^2 + \frac{2187}{8}.$$

A simple calculation shows that $\forall i = 1, \ldots, 5$, $\min\{h_i(p) : p \in [-1, 1]\} = h_i(0) > 30$, which proves the result. □

Let us then use the following result:

LEMMA 5.2.4. *Let w be a homogeneous order 2 function which is a viscosity solution of a Hessian uniformly elliptic equation in $\mathbb{R}^n$, and let $g(w) : U \to \mathbb{R}$ be a smooth function of the symmetric functions $\sigma_1(w), \sigma_2(w), \ldots, \sigma_n(w)$ of $D^2 w$ where U is a domain containing M for*

$$M := \bigcup_{\sigma \in \Sigma_n} T_\sigma \Lambda(B) \subset \mathbb{R}^n$$

and $g_{|M} = 0$, which satisfies the condition

(5.2.3) $$\min_{1 \le i \le n} \inf_{x \in M} \frac{\partial g}{\partial \lambda_i}(\lambda) > 0.$$

Let $\mu_1(a, b, O) \le \cdots \le \mu_n(a, b, O)$ be the eigenvalues of $D^2 w(a) - O^t D^2 w(b) O$ for some $a, b \in \mathbb{S}^{n-1}$ and orthogonal O. Assume that for any $a, b \in B$ either

$$\mu_1(a, b, O) = \cdots = \mu_n(a, b, O) = 0$$

or

$$C^{-1} \le -\frac{\mu_1}{\mu_n} \le C,$$

where C is a positive constant independent of a and b. If $\delta > 0$, we assume additionally that w changes sign in B. Then w is a viscosity solution in B of a uniformly elliptic Hessian equation $F(D^2 w) = 0$ with a smooth F.

PROOF. For $t > 0$ let us denote by $K_t \subset \mathbb{R}^n$ the cone $\{\lambda \in \mathbb{R}^n,\ \lambda_i/|\lambda| > t\}$, and let K_t^* be its dual cone. Let (x, y) be orthogonal coordinates in $\mathbb{R}^n$ such that $x = \lambda_1 + \cdots + \lambda_n$ and y are coordinates on the orthogonal complement of x. Denote by p the orthogonal projection of $\mathbb{R}^n$ onto the subspace $x^\perp$. Denote

$$\Gamma = \{g = 0\} \subset U, \quad G = p(\Gamma), \quad m = p(M).$$

It follows from (5.2.3) above and (5.2.4) below that the surface Γ is a graph of a smooth function h defined on G. By k_t we denote the function on $x^\perp$ with the graph ∂K_t^*. We define a function $H(y)$ by the formula

$$H(y) = \inf_{z \in G} \{h(z) + k_t(y - z)\}.$$

Let us fix a sufficiently small $t > 0$. Then from (5.2.3) and (5.2.4) it follows that $H = h$ on G. Denote by J the graph of H. It is easy to show, see similar arguments in Sections 4.1 and 4.2, that for any $a, b \in J$, $a \ne b$,

(5.2.4) $$1/C \le -\min_i(a_i - b_i)/\max_i(a_i - b_i) \le C.$$

Let E be a smooth function in $\mathbb{R}^{n-1}$ with support in a unit ball and integral 1. For $c > 0$ set $E_c(y) = c^{1-n} E(y/c)$, and define

$$H_c := H * E_c.$$

Then H_c is a smooth function such that any two points a, b on its graph satisfy (5.2.5). Moreover $H_c \to H$ in $C(\mathbb{R}^n)$ as c goes to 0, and $H_c \to h$ in C^∞ on compact subdomains of G. Thus for a sufficiently small $c > 0$ we can easily modify the function H_c to a function $\widetilde{H}$ such that $\widetilde{H}$ will coincide with h in a neighborhood of m and coincide with H in the complement of G and the points on the graph of $\widetilde{H}$

will still satisfy (5.2.5) possibly with a larger constant C. Define the function F in $\mathbb{R}^n$ by
$$F := x - \widetilde{H}(y).$$
Then w is a solution in $\mathbb{R}^n \setminus \{0\}$ of a uniformly elliptic Hessian equation $F(D^2 w) = 0$ with this F. As in Subsection 4.2.2 it follows that w is a viscosity solution of $F(D^2 w) = 0$ in the whole space $\mathbb{R}^n$. $\square$

We get immediately

THEOREM 5.2.2. *The function w_5 is a viscosity solution in the unit ball of a smooth uniformly elliptic equation $F(D^2 w) = 0$ with smooth Hessian F which coincides with the polynomial $g_5(w)$ in a neighborhood of the curve M_5.*

Moreover, a continuity argument permits us to prove the existence of singular solutions in five dimensions:

THEOREM 5.2.3. *The function $w_5/|x|^\varepsilon$ is a viscosity solution in the unit ball of a smooth uniformly elliptic equation $F(D^2 w) = 0$ with smooth Hessian F for some (unspecified) $\varepsilon > 0$.*

5.2.3. Dimensions $8, 14$, and 26. In eight dimensions we automatically get

THEOREM 5.2.4. *The function $w_8(x) = C_2(x)/|x|$ is a viscosity solution of a Hessian uniformly elliptic equation and of a uniformly elliptic Isaacs equation in the unit ball.*

Moreover, calculations parallel (albeit more cumbersome) to those in five dimensions permit us to prove the following result:

THEOREM 5.2.5. *The function $w_8(x)$ is a viscosity solution in the unit ball of a smooth uniformly elliptic equation $F(D^2 w) = 0$ with smooth Hessian F which coincides with the polynomial*
(5.2.5)
$$g_8(w) := 5\left(\frac{\Delta w}{11}\right)^8 + 3^2 31 \left(\frac{\Delta w}{11}\right)^6 + 3^5 17 \left(\frac{\Delta w}{11}\right)^4 + 3^8 \left(\frac{\Delta w}{11}\right)^2 - \det(D^2 w)$$
in a neighborhood of the curve M_8. For the function $u = w_8/11$ the polynomial is
$$5(\Delta u)^8 + 3^2 31 (\Delta u)^6 + 3^5 17 (\Delta u)^4 + 3^8 (\Delta u)^2 - 11^8 \det(D^2 u).$$

In $n = 14$ and 26 dimensions the multiplicity of the cubic factor of the Hessian characteristic polynomial is 4 and 8, respectively. Since this multiplicity is ≥ 3, one can apply the argument of Section 4.2 to the functions $w_{14,\delta} = C_3(x)/|x|^\delta$ and $w_{26,\delta} = C_4(x)/|x|^\delta$. Repeating this argument (and one of Section 4.4) one gets the following results.

THEOREM 5.2.6. *Let $1 \leq \delta < 2$. Then there exists a uniformly elliptic Hessian equation of the form (0.1) such that the function*
$$w_{14,\delta} = C_3(x)/|x|^\delta$$
is its viscosity solution in the unit ball $B \subset \mathbb{R}^{14}$. The same function is also a solution to a uniformly elliptic Isaacs equation.

One should compare this result with that of Section 4.3; it is completely human-controlled, does not use computer computations, and is valid in a dimension rather close to $n = 12$.

For $n = 26$ and the Cartan cubic C_4 one has the following result.

THEOREM 5.2.7. *Let $1 \leq \delta < 2$. Then there exists a uniformly elliptic Hessian equation of the form (0.1) such that the function*
$$w_{26,\delta} = C_4(x)/|x|^\delta$$
is its viscosity solution in the unit ball $B \subset \mathbb{R}^{26}$. The same function is also a solution to a uniformly elliptic Isaacs equation.

The analogues of Theorems 5.2.5 and 5.2.6 are also valid; moreover, one can write the corresponding polynomial equations in a uniform manner: The function
$$u_l(x) := \frac{C_l(x)}{(3l+5)|x|} = \frac{w_{3l+2}(x)}{(3l+5)}$$
satisfies the Hessian uniformly elliptic, in a neighborhood of the curve $M_{3l+2} \subset \mathbb{R}^{3l+2}$, polynomial equation

(5.2.6) $\qquad (3l+5)^{3l+2} \det D^2 u_l = (\Delta u)^l (5(\Delta u)^2 + 9)((\Delta u)^2 + 27)^l$

for $l = 1, 2, 4, 8$. One can also write down the corresponding polynomial equations for the singular solutions in dimensions $n = 14$ and 26. Note, however, that verifying their local ellipticity for any $\delta \in]1, 2]$ needs very long computer calculations.

REMARK 5.2.3. One notes that constructions of Sections 4.1–4.3 are parallel to those of Sections 5.2–5.3. In fact, there exists a common algebraic framework for considering the trialities and Cartan's cubics; see Subsection 6.11.5.

5.3. Singular solutions in five dimensions

For the reader's convenience we formulate Theorem 1.6.11 below.

THEOREM 5.3.1. *The function $w_5/|x|^\delta$ is a viscosity solution in the unit ball of a smooth uniformly elliptic equation $F(D^2 w) = 0$ with smooth Hessian F for any $\delta \in [0, 1[$.*

In order to prove this result one needs more elaborate techniques, which we are going to explain. As above we also get that $w_{5,\delta}(x)$, $\delta \in [0, 1[$, is a viscosity solution to a uniformly elliptic Isaacs equation:

COROLLARY 5.3.1. *The function*
$$w_{5,\delta}(x) = P_5(x)/|x|^{1+\delta}, \quad \delta \in [0, 1[,$$
is a viscosity solution to a uniformly elliptic Isaacs equation in a unit ball $B \subset \mathbb{R}^5$.

Let then $w = w_n$ be an odd homogeneous function of order $2 - \delta$, $0 \leq \delta < 1$, defined on a unit ball $B = B_1 \subset \mathbb{R}^n$ and smooth in $B \setminus \{0\}$. Then the Hessian of w is homogeneous of order $-\delta$. One notes that Lemma 5.2.4 with its proof remains valid for w; we will apply this result to the function $w_{5,\delta}(x) = P_5(x)/|x|^{1+\delta}$.

We give then the proof of the theorem.

5.3. SINGULAR SOLUTIONS IN FIVE DIMENSIONS

It is sufficient to prove the existence of a smooth function g satisfying the conditions (5.2.3) and (5.2.4). We begin with calculating the eigenvalues of $D^2 w_{5,\delta}(x)$. More precisely, we need

LEMMA 5.3.1. *Let $x \in \mathbb{S}^4$, and let $x \in G_P(p, 0, q, 0, 0)$ with $p^2 + q^2 = 1$. Then*
$$\mathrm{Spec}(D^2 w_{5,\delta}(x)) = \{\mu_{1,\delta}, \mu_{2,\delta}, \mu_{3,\delta}, \mu_{4,\delta}, \mu_{5,\delta}\}$$

for

$$\mu_{1,\delta} = \frac{p(p^2\delta + 6 - 3\delta)}{2},$$

$$\mu_{2,\delta} = \frac{p(p^2\delta - 3 - 3\delta) + 3\sqrt{12 - 3p^2}}{2},$$

$$\mu_{3,\delta} = \frac{p(p^2\delta - 3 - 3\delta) - 3\sqrt{12 - 3p^2}}{2},$$

$$\mu_{4,\delta} = -\frac{p\delta(6-\delta)(3-p^2) + \sqrt{D(p,\delta)}}{4},$$

$$\mu_{5,\delta} = -\frac{p\delta(6-\delta)(3-p^2) - \sqrt{D(p,\delta)}}{4},$$

and
$$D(p, \delta) := (6-\delta)(4-\delta)(2-\delta)\delta(p^2 - 3)^2 p^2 + 144(\delta - 2)^2 > 0.$$

The characteristic polynomial $F(S)$ of $D^2 w$ is given by
$$F(S) = S^5 + a_{1,\delta} S^4 + a_{2,\delta} S^3 + a_{3,\delta} S^2 + a_{4,\delta} S + a_{5,\delta}$$

for

$$a_{1,\delta} = \frac{(\delta+1)(\delta-8)b}{2},$$

$$a_{2,\delta} = \frac{(\delta+1)(21\delta + 13 - 4\delta^2)b^2}{4} + 9(2\delta - \delta^2 - 4),$$

$$a_{3,\delta} = \frac{(6\delta^2 - 31\delta - 1)(\delta+1)^2 b^3}{8} + \frac{27(4\delta - 2\delta^2 + 5 + \delta^3)}{2},$$

$$a_{4,\delta} = \frac{(2\delta - 1)(5 - \delta)(\delta+1)^2 b^4}{8} + \frac{9(\delta - 1)(\delta^2 - 2\delta + 9)}{2},$$

$$a_{5,\delta} = \frac{b(1-\delta)\left(b^2(\delta+1)^3 + 108(1-\delta)\right)\left(b^2(\delta+1)(\delta-5) + 36(\delta-1)\right)}{32},$$

where $b := p(p^2 - 3)$.

Note that the spectrum in this lemma is an unordered one.

PROOF. Since $w_{5,\delta}$ is invariant under G_P, we can suppose that $x = (p, 0, q, 0, 0)$. Then $w_{5,\delta}(x) = \frac{p(3-p^2)}{2}$ and we get by a brute force calculation:
$$D^2 w_{5,\delta}(x) := \begin{pmatrix} M_{1,\delta} & 0 \\ 0 & M_{2,\delta} \end{pmatrix},$$

being a block matrix with
$$M_{1,\delta} := \frac{1}{2}\begin{pmatrix} m_{1,1} & m_{1,2} & m_{1,3} \\ m_{1,2} & m_{2,2} & m_{2,3} \\ m_{1,3} & m_{2,3} & m_{3,3} \end{pmatrix},$$

where
$$m_{1,1} := -(\delta + 2)\delta p^5 + (\delta + 3)\delta p^3 + (12 - 9\delta)p,$$
$$m_{1,2} := 3\sqrt{3}p(p^2 - 1)\delta,$$
$$m_{1,3} := -q\left((\delta + 2)\delta p^4 + 3\delta(1 - \delta)p^2 + (3\delta - 6)\right),$$
$$m_{2,2} := \delta p^3 - 3(\delta + 4)p,$$
$$m_{2,3} := 3\sqrt{3}q(\delta p^2 + 2 - \delta),$$
$$m_{3,3} := (\delta + 2)\delta p^5 + (5 - 4\delta)\delta p^3 - 3(\delta - 1)(2 - \delta)p,$$

and
$$M_{2,\delta} := \frac{1}{2}\begin{pmatrix} \delta p^3 + 3(2-\delta)p & 6\sqrt{3}q \\ 6\sqrt{3}q & \delta p^3 - 3(4+\delta)p \end{pmatrix},$$

which gives $F(S) = F_1(S) \cdot F_2(S) \cdot F_3(S)$ for the characteristic polynomial, where
$$F_1(S) := S - \frac{p(p^2\delta + 6 - 3\delta)}{2},$$
$$F_2(S) = S^2 + \frac{\delta p(p^2 - 3)(\delta - 6)S}{2}$$
$$+ \frac{(2-\delta)\left((\delta-6)\delta p^6 + 6(6-\delta)\delta^2 p^4 + 9(\delta^2 - 6\delta)p^2 + 36(\delta - 2)\right)}{4},$$
$$F_3(S) := S^2 + (3 + 3\delta - \delta p^2)pS + \frac{(p^2 - 3)(\delta^2 p^4 - 3\delta^2 p^2 - 6\delta p^2 + 36)}{4},$$

and one gets the spectrum. Developing $F(S)$ we get the last formulas. □

COROLLARY 5.3.2. *Denote $\varepsilon = 1 - \delta$. The function w satisfies the following Hessian equation:*
$$\det(D^2 w) = e_5(\Delta w)^5 + e_3(\Delta w)^3 \sigma_2(w) + e_1 \Delta w \sigma_4(w)$$
where
$$e_5 = \frac{\varepsilon^2(168 - 5\varepsilon^4 - 24\varepsilon^3 - 56\varepsilon)}{(\varepsilon^2 + 3)(\varepsilon + 7)^5(\varepsilon - 2)^3},$$
$$e_3 = \frac{\varepsilon^2(2\varepsilon^2 + \varepsilon + 8)}{(\varepsilon - 2)^2(\varepsilon + 7)^3(\varepsilon^2 + 3)}, \quad e_1 = \frac{\varepsilon}{(2 - \varepsilon)(\varepsilon + 7)},$$
$\Delta w = \mathrm{tr}(D^2 w)$ *being the Laplacian, $\sigma_2(w)$ and $\sigma_4(w)$ being, respectively, the second and the fourth symmetric functions of the eigenvalues of $D^2 w$.*

PROOF. This follows immediately from Lemma 5.3.1 and a simple calculation since
$$\Delta(w) = -a_{1,\delta}, \quad \sigma_2(w) = a_{2,\delta}, \quad \sigma_4(w) = a_{4,\delta}, \quad \det(D^2 w) = -a_{5,\delta}. \quad \square$$

REMARK 5.3.1. The equation becomes (after multiplying by 2^{17} to get integer coefficients)
$$2^{17} \det(D^2 w) = -83(\Delta w)^5 + 2^6 11 \sigma_2(w)(\Delta w)^3 + 2^{14} \sigma_4(w)\Delta w$$
for $\varepsilon = 1$, i.e., for $\delta = 0$, and therefore is different from the equation (5.2.2). This is not a contradiction since the function w_5 verifies (infinitely) many different uniformly elliptic equations.

Let us then determine the ordered spectrum $\{\lambda_1, \lambda_2, \ldots, \lambda_5\}, \lambda_1 \geq \lambda_2 \geq \cdots \geq \lambda_5$ of $D^2 w$.

LEMMA 5.3.2. *Let $\lambda_1 \geq \lambda_2 \geq \cdots \geq \lambda_5$ be the eigenvalues of $D^2 w_{5,\delta}(x)$. Then*

$$\lambda_1 = \mu_{2,\delta}, \quad \lambda_5 = \mu_{3,\delta},$$

$$\lambda_2 = \begin{cases} \mu_{4,\delta} & \text{for } p \in [-1, p_0(\delta)], \\ \mu_{1,\delta} & \text{for } p \in [p_0(\delta), 1], \end{cases}$$

$$\lambda_3 = \begin{cases} \mu_{5,\delta} & \text{for } p \in [-1, -p_0(\delta)], \\ \mu_{1,\delta} & \text{for } p \in [-p_0(\delta), p_0(\delta)], \\ \mu_{4,\delta} & \text{for } p \in [p_0(\delta), 1], \end{cases}$$

$$\lambda_4 = \begin{cases} \mu_{1,\delta} & \text{for } p \in [-1, -p_0(\delta)], \\ \mu_{5,\delta} & \text{for } p \in [-p_0(\delta), 1], \end{cases}$$

where

$$p_0(\delta) := \frac{3^{1/4}\sqrt{1-\delta}}{(3+2\delta-\delta^2)^{1/4}} = \frac{3^{1/4}\sqrt{\varepsilon}}{(4-\varepsilon^2)^{1/4}} \in]0,1].$$

PROOF. The inequalities $\mu_{2,\delta}(p) \geq \mu_{1,\delta}(p) \geq \mu_{3,\delta}(p)$ are obvious since $\mu_{2,\delta}(p)$ and $\mu_{3,\delta}(p)$ are decreasing at p, $\mu_{1,\delta}(p)$ is increasing in p, $\mu_{3,\delta}(-1) = \mu_{1,\delta}(-1)$, and $\mu_{2,\delta}(1) = \mu_{1,\delta}(1)$.

The resultant

$$R(\delta, p) = \text{Res}(F_2, F_3) = 144(p-1)^2(p+1)^2 \left(r_8 p^8 - r_6 p^6 + r_4 p^4 - r_2 p^2 + r_0\right),$$

where

$$r_8 = (\varepsilon^2 - 4)^2, \quad r_6 = 12(\varepsilon^2 - 4)^2, \quad r_4 = 3(4-\varepsilon^2)(72 - 17\varepsilon^2),$$
$$r_2 = 108(\varepsilon^2 - 4)^2, \quad r_0 = 144(3-\varepsilon^2)^2,$$

is strictly positive for $(\varepsilon, p) \in]0,1[\times]-1,1[$. Indeed, let

$$r := \frac{R}{144(p-1)^2(p+1)^2} = r_8 p^8 - r_6 p^6 + r_4 p^4 - r_2 p^2 + r_0.$$

Then

$$d := \frac{\partial r}{4\varepsilon \partial \varepsilon} = (\varepsilon^2 - 4)p^8 + 12p^6(4-\varepsilon^2) + 3(17\varepsilon^2 - 70)p^4 + 108(4-\varepsilon^2)p^2 + 144(\varepsilon^2 - 3) < 0$$

for $(\varepsilon, p) \in]0,1[\times[0,1[$ since

$$\frac{1}{4p}\frac{\partial d}{\partial p} = (4-\varepsilon^2)(-2p^6 + 18p^4 - 51p^2 + 54) - 6p^2 \geq (4-\varepsilon^2) \cdot 19 - 6 \geq 51,$$

and for $p = 1$ one has $d = -166 + 76\varepsilon^2 \leq -90$. For $\delta = 0$, $\varepsilon = 1$ we get

$$R(\delta, p) \geq R(1, p) = 9((1-p^2)(4-p^2))(p^4 - 7p^2 + 16),$$

which proves positivity. Using then the inequalities

$$\mu_{2,\delta}(-1) = \mu_{4,\delta}(-1) > \mu_{5,\delta}(-1) > \mu_{3,\delta}(-1),$$
$$\mu_{2,\delta}(1) > \mu_{4,\delta}(1) > \mu_{5,\delta}(-1) = \mu_{3,\delta}(-1)$$

and the postivity of the resultant, we get

$$\mu_{2,\delta}(p) \geq \mu_{4,\delta}(p) \geq \mu_{5,\delta}(p) \geq \mu_{3,\delta}(p)$$

for any $p \in [-1, 1]$.

Calculating the resultant,
$$R_1(\delta, p) = \text{Res}(F_1, F_3) = 12(p^2 - 3)\left(p^4(\varepsilon^2 - 4) + 3\varepsilon^2\right)$$
and taking into account the equalities
$$\mu_{4,\delta}(p_0(\delta)) = \mu_{1,\delta}(p_0(\delta)), \quad \mu_{5,\delta}(-p_0(\delta)) = \mu_{1,\delta}(-p_0(\delta)),$$
we get the result. $\square$

Note the oddness property of the spectrum:
$$\lambda_{1,\delta}(-p) = -\lambda_{5,\delta}(p), \quad \lambda_{2,\delta}(-p) = -\lambda_{4,\delta}(p), \quad \lambda_{3,\delta}(-p) = -\lambda_{3,\delta}(p).$$

Let us now verify the uniform hyperbolicity of $M_\delta(a,b,O)$.

PROPOSITION 5.3.1. *Let $M_\delta(x) = D^2 w_\delta(x)$, $0 \leq \delta < 1$. Suppose that $a \neq b \in B_1 \setminus \{0\}$ and let $O \in O(5)$ be an orthogonal matrix such that*
$$M_\delta(a,b,O) := M_\delta(a) - O^t \cdot M_\delta(b) \cdot O \neq 0.$$
Denote by $\Lambda_1 \geq \Lambda_2 \geq \cdots \geq \Lambda_5$ the eigenvalues of the matrix $M_\delta(a,b,O)$. Then
$$\frac{1}{C} \leq -\frac{\Lambda_1}{\Lambda_5} \leq C$$
for $C := \frac{1000(\delta+1)(3-\delta)}{3(1-\delta)^2}$.

PROOF. The proof depends on the value of
$$k := p_0(\delta) := \frac{3^{1/4}\sqrt{1-\delta}}{(3 + 2\delta - \delta^2)^{1/4}} = \frac{3^{1/4}\sqrt{\varepsilon}}{(4 - \varepsilon^2)^{1/4}}.$$
Note that $C = \frac{1000(\delta+1)(3-\delta)}{3(1-\delta)^2} = \frac{1000}{k^4}$. We shall give the proof for $k \in]0, \frac{1}{2}]$; the proof for $k \in [\frac{1}{2}, 1]$ is similar and simpler and uses $C = 10^4$.

Suppose that the conclusion does not hold, that is, for some $a \neq b$ and some $O \in O(5)$ one has
$$M_\delta(a,b,O) := M_\delta(a) - O^t \cdot M_\delta(b) \cdot O \neq 0,$$
but
$$\frac{1}{C} > -\frac{\Lambda_1}{\Lambda_5} \quad \text{or} \quad -\frac{\Lambda_1}{\Lambda_5} > C.$$
We can suppose without loss of generality that $|b| \leq 1 = |a| \in \mathbb{S}_1^4$. Let $\bar{b} := b/|b| \in \mathbb{S}_1^4$, $W := W(a)$, $\overline{W} := W(\bar{b})$, $K := |b|^{-1-\delta}$. Note that since for any harmonic cubic polynomial $Q(x)$ on $\mathbb{R}^n$ and any $a \in \mathbb{S}_1^{n-1} \subset \mathbb{R}^n$ one has
$$\text{tr}\, D^2(|x|^{-1-\delta} Q(x))|_{x=a} = (\delta^2 - 2\delta - 3 - n) Q(a),$$
we get $\text{tr}\,(M_\delta(a,b,O)) = (2\delta + 8 - \delta^2)(K\overline{W} - W)$, P_5 being harmonic. Let us prove the claim for $(K\overline{W} - W) \geq 0$, the proof for $(K\overline{W} - W) \leq 0$ being the same while permuting a with b and Λ_1 with Λ_5. Since
$$\text{tr}\,(M_\delta(a,b,O)) = (2\delta + 8 - \delta^2)(K\overline{W} - W) \geq 0,$$
we get $4\Lambda_1 + \Lambda_5 \geq 0$ and $-\Lambda_5/\Lambda_1 \leq 4$. Therefore, we have only to rule out the inequality $\frac{1}{C} > -\frac{\Lambda_1}{\Lambda_5}$, i.e., $-\Lambda_5 > C\Lambda_1$. Recall that
$$W = \frac{3p - p^3}{2}, \quad \overline{W} = \frac{3\bar{p} - \bar{p}^3}{2}$$
for some $p, \bar{p} \in [-1, 1]$.

We have then three possibilities:
1) $p, \overline{p} \in [-k, k]$.
2) $p \in [-k, k]$, $\overline{p} \notin [-k, k]$.
3) $p, \overline{p} \notin [-k, k]$.

In cases 1) and 3) applying the Weyl inequalty, we get $\Lambda_1 \geq \mu_+(K)$, $\Lambda_5 \leq \mu_-(K)$, which permits us to finish the proof as above. We thus have to treat the more difficult case 2). Then we get

$$-\min_{i=1,\cdots,5}\{K\lambda_{i,\delta}(\overline{p}) - \lambda_{i,\delta}(p)\} > C \max_{i=1,\cdots,5}\{K\lambda_{i,\delta}(\overline{p}) - \lambda_{i,\delta}(p)\}$$

since $-\Lambda_5 > C\Lambda_1$.

Thanks to the oddness of the spectrum we suppose without loss of generality that $\overline{p} > k$. Recall that then by Lemma 5.3.2 one has

$$\lambda_{1,\delta}(\overline{p}) = \mu_{2,\delta}(\overline{p}), \quad \lambda_{1,\delta}(p) = \mu_{2,\delta}(p), \quad \lambda_{2,\delta}(\overline{p}) = \mu_{1,\delta}(\overline{p}), \quad \lambda_{1,\delta}(p) = \mu_{4,\delta}(p),$$
$$\lambda_{3,\delta}(\overline{p}) = \mu_{4,\delta}(\overline{p}), \quad \lambda_{3,\delta}(p) = \mu_{1,\delta}(p), \quad \lambda_{4,\delta}(\overline{p}) = \mu_{5,\delta}(\overline{p}), \quad \lambda_{4,\delta}(p) = \mu_{5,\delta}(p),$$
$$\lambda_{5,\delta}(\overline{p}) = \mu_{3,\delta}(\overline{p}), \quad \lambda_{5,\delta}(p) = \mu_{3,\delta}(p).$$

We have then two possibilities for p:
2a) $p \in [-k, 0]$.
2b) $p \in]0, k]$.

Let $p \in [-k, 0]$; then $\mu_{1,\delta}(p) \leq 0$ and thus

$$C \max_{i=1,\ldots,5}\{K\lambda_{i,\delta}(\overline{p}) - \lambda_{i,\delta}(p)\} \geq CK\lambda_{3,\delta}(\overline{p}) = CK\mu_{4,\delta}(\overline{p}) \geq CK\mu_{4,\delta}(p_0(\delta))$$
$$= CK\mu_{1,\delta}(k) = CK\left(k^3\left(\sqrt{k^4+3} - k^2 + 3\right)/\sqrt{k^4+3}\right) \geq 2CKk^3$$

since one verifies that the function $\mu_{4,\delta}(p)$ is increasing on $[k, 1]$.

On the other hand,

$$\left|\min_{i=1,\ldots,5}\{K\lambda_{i,\delta}(\overline{p}) - \lambda_{i,\delta}(p)\}\right|$$
$$\leq K\max_{i=1,\ldots,5,p}|\{\lambda_{i,\delta}(p)\}| + \max_{i=1,\ldots,5,p}|\{\lambda_{i,\delta}(p)\}| \leq 8(K+1).$$

Therefore one gets $8(K+1) \geq 2CKk^3$, which clearly is a contradiction for, say, $K \geq 1/4$. For $0 < K \leq 1/4$ we get

$$C \max_{i=1,\ldots,5}\{K\lambda_{i,\delta}(\overline{p}) - \lambda_{i,\delta}(p)\} \geq C(K\lambda_{5,\delta}(\overline{p}) - \lambda_{5,\delta}(p)) \geq C(K(-8) - (-5)) \geq 3C,$$

which cannot be less than $8(K+1) \leq 10$.

Let finally $p \in]0, k]$. We consider then two possibilities for K:

(i) $K \leq 20/31 = (1.55)^{-1}$.
(ii) $K > 20/31 = (1.55)^{-1}$.

In case (i) one has

$$C \max_{i=1,\ldots,5}\{K\lambda_{i,\delta}(\overline{p}) - \lambda_{i,\delta}(p)\} \geq C(K\lambda_{5,\delta}(\overline{p}) - \lambda_{5,\delta}(p))$$
$$\geq C(K\mu_{3,\delta}(1) - \mu_{3,\delta}(0)) \geq C(3\sqrt{3} + 20(\varepsilon - 8)/31) > C/30 > 8(K+1)$$

since $\lambda_{5,\delta}(p) = \mu_{3,\delta}(p)$ is decreasing, $\mu_{3,\delta}(0) = -3\sqrt{3}$, $\mu_{3,\delta}(1) = \varepsilon - 8$.

We suppose then that $K > 20/31 = (1.55)^{-1}$. Then if $p \leq 3k/4$, one has

$$C \max_{i=1,\ldots,5}\{K\lambda_{i,\delta}(\overline{p}) - \lambda_{i,\delta}(p)\}$$
$$\geq CK(\lambda_{3,\delta}(\overline{p}) - \lambda_{3,\delta}(p)/K) \geq CK(\mu_{4,\delta}(\overline{p}) - \mu_{1,\delta}(p)/K)$$
$$\geq CK(\mu_{4,\delta}(k) - \mu_{1,\delta}(3k/4)/K) = CK(\mu_{1,\delta}(k) - \mu_{1,\delta}(3k/4)/K)$$
$$> CK\mu_{1,\delta}\left(\frac{3k}{4}\right)\left(\frac{\mu_{4,\delta}(k))}{\mu_{1,\delta}\left(\frac{3k}{4}\right)} - \frac{31}{20}\right)$$
$$\geq \frac{CK}{100}\mu_{1,\delta}\left(\frac{3k}{4}\right) \geq CK\frac{2k^3}{100} > 20K > 8(K+1),$$

which is a contradiction for $k \in [0, 1/2]$ since $\frac{\mu_{4,\delta}(k))}{\mu_{1,\delta}\left(\frac{3k}{4}\right)} > 1.56, \mu_{1,\delta}\left(\frac{3k}{4}\right) > 2k^3$ there. Thus we can suppose that $p \in]\frac{3k}{4}, k]$. One notes then that $\mu_{4,\delta}(p) \leq \mu_{4,\delta}(k)$ for $p \in [\frac{k}{4}, 1]$. This permits us to rule out the case $\overline{p} \geq \frac{3k}{2}$. Indeed, one has in this case

$$C \max_{i=1,\ldots,5}\{K\lambda_{i,\delta}(\overline{p}) - \lambda_{i,\delta}(p)\} \geq CK(\lambda_{2,\delta}(\overline{p}) - \lambda_{2,\delta}(p)/K)$$
$$\geq CK(\mu_{1,\delta}(\overline{p}) - \mu_{4,\delta}(p)/K)$$
$$\geq CK(\mu_{1,\delta}(3k/2) - \mu_{1,\delta}(k)/K)$$
$$= CK(\mu_{1,\delta}(3k/2) - \mu_{1,\delta}(k)/K),$$

and one gets a contradiction as above since

$$\frac{\mu_{1,\delta}\left(\frac{3k}{2}\right)}{\mu_{1,\delta}(k)} > 2$$

for $k \in [0, 1/2]$.

The last case to rule out is thus $K \geq 20/31, p \in [\frac{3k}{4}, k], \overline{p} \in [k, \frac{3k}{2}]$. Let then

$$\alpha := k - p \in \left[0, \frac{k}{4}\right] \subset \left[0, \frac{1}{8}\right], \quad \beta := \overline{p} - k \in \left[0, \frac{k}{2}\right] \subset \left[0, \frac{1}{4}\right], \quad a := \max\{\alpha, \beta\}.$$

It is easy to verify that on $\left[\frac{3k}{4}, \frac{3k}{2}\right]$ one has the following inequalities:

$$\frac{\partial \mu_{1,\delta}(p)}{\partial p} \geq 3k^2, \qquad \frac{11k^3}{4} \geq \mu_{1,\delta}(k) \geq \frac{5k^3}{2},$$
$$\frac{\partial \mu_{2,\delta}(p)}{\partial p} \geq -5, \qquad 4k - 5 \geq \mu_{2,\delta}(k) \geq 4k - \frac{11}{2},$$
$$\frac{\partial \mu_{3,\delta}(p)}{\partial p} \geq -\frac{9}{2}, \qquad -5 - 3k \geq \mu_{3,\delta}(k) \geq -\frac{11+7k}{2},$$
$$\frac{\partial \mu_{4,\delta}(p)}{\partial p} \geq -\frac{k}{29}, \qquad \frac{11k^3}{4} \geq \mu_{4,\delta}(k) \geq \frac{5k^3}{2},$$
$$\frac{\partial \mu_{5,\delta}(p)}{\partial p} \geq 10k^2 - 12, \qquad -10k \geq \mu_{5,\delta}(k) \geq -12k.$$

Let then $K \in \left[\frac{20}{31}, 1\right]$. Therefore,

$$C \max_{i=1,\ldots,5}\{K\lambda_{i,\delta}(\overline{p}) - \lambda_{i,\delta}(p)\}$$
$$\geq C \max\{K\mu_{1,\delta}(k+\alpha) - \mu_{1,\delta}(k-\beta), K\mu_{3,\delta}(k+\alpha) - \mu_{3,\delta}(k-\beta)\}$$
$$\geq \max\{M_1, M_2\}$$
$$= C \max\left\{3(K-1)k^3 a + 3(K+1)k^2, (1-K)(5+3k)a - \frac{(56+29k)(K+1)}{10}\right\}$$

for linear forms M_1, M_2 in K. Note that the minimal value of $\max\{M_1, M_2\}$ as a function of K equals (recall that our $C = 1000/k^4$)

$$\frac{1500a(40 - 9k)}{k^2(12k^2a + 18a + 11k^3 + 20 + 12k)} > \frac{1250a}{k^2} > 0,$$

which is attained for $K = K_0 := (11k^3 + 20 + 12k)/(12k^2a + 18a + 11k^3 + 20 + 12k) < 1$.

On the other hand,

$$-\Lambda_5 \leq -\min_{i=1,\ldots,5}\{K\lambda_{i,\delta}(\overline{p}) - \lambda_{i,\delta}(p)\} \leq \max\{l_1, l_2, l_3, l_4, l_5\}$$

for the following linear forms (in K):

$$l_1 := -k^2\left(\frac{11a}{4} + 3k\right)K - \frac{11a}{4}k^2 + 3k^3,$$

$$l_2 := \left(5a + 4k - \frac{11}{2}\right)K + 5a + \frac{11}{2} - 4k,$$

$$l_3 := (5a + 5 + 3k)K + 5a - 5 - 3k,$$

$$l_4 := \left(\frac{ak}{29} - \frac{11k^3}{2}\right)K + \frac{ak}{29} + \frac{11k^3}{2},$$

$$l_5 := ((12 - 10k^2)a + 12k)K + (12 - 10k^2)a - 12k.$$

To refute our inequality it is sufficient to prove that $M_i(K_{j,k}) > 0$ for any triple (i, j, k) with $i, j \in \{1, 2\}$, $i \neq j$, $k \in \{1, 2, 3, 4, 5\}$ where $l_k(K_{j,k}) = M_j(K_{j,k})$.

Explicit calculations give (for the values $m_{ijk} := \frac{M_i(K_{j,k})}{500ak^2}$)

$$\frac{m_{121}}{3} = \frac{(9k^4 + 6k^6)a + 10000 + 5k^7 + 20k^4 + 3k^5 - 2250k}{k^2((3k^4 + 3000)a + 3k^5 + 2750k)} > 0,$$

$$\frac{m_{211}}{3} = \frac{(9k^4 + 6k^6)a + 10000 + 5k^7 + 20k^4 + 3k^5 - 2250k}{(4500 - 3k^6)a + 5000 + 3000k - 3k^7} > 0,$$

$$m_{122} = \frac{60000 - (60k^4 + 90k^2)a - 192k^3 + 66k^4 - 13500k - 101k^2 - 158k^5}{k^2((6000 - 10k^2)a - 11k^2 + 8k^3 + 5500k)} > 0,$$

$$m_{212} = \frac{60000 - (60k^4 + 90k^2)a - 192k^3 + 66k^4 - 13500k - 101k^2 - 158k^5}{(10k^4 + 9000)a + 11k^4 - 8k^5 + 10000 + 6000k} > 0,$$

$$m_{123} = \frac{30000 - (45k^2 + 30k^4)a - 55k^2 - 6750k - 33k^3 + 30k^4 - 37k^5}{k^2((3000 - 5k^2)a + 2750k - 5k^2 - 3k^3)} > 0,$$

$$m_{213} = \frac{30000 - (45k^2 + 30k^4)a - 55k^2 - 6750k - 33k^3 + 30k^4 - 37k^5}{(5k^4 + 4500)a + 5k^4 + 3k^5 + 5000 + 3000k} > 0,$$

$$m_{124} = \frac{3480000 - (36k^3 + 24k^5)a - t(k)}{k^2((348000 - 4k^3)a + 319000k + 319k^5)} > 0,$$

$$m_{214} = \frac{3480000 - (36k^3 + 24k^5)a - t(k)}{(522000 + 4k^5)a + 580000 + 348000k - 319k^7} > 0,$$

$$m_{215} = \frac{(9k^4 + 30k^6 - 54k^2)a + 15000 - u(k)}{5k^4a + 1500a - 6k^3 + 1375k - 6k^2a} > 0,$$

$$m_{125} = \frac{(9k^4 + 30k^6 - 54k^2)a + 15000 - u(k)}{(2250 + 6k^4 - 5k^6)a + 2500 + 1500k + 6k^5} > 0,$$

for the following functions of k: $t(k) := 783000k + 80k^3 + 48k^4 + 44k^6 + 2871k^5 + 1914k^7 < 4 \cdot 10^5$, $u(k) := 120k^2 - 30k^5 + 3375k + 18k^3 - 100k^4 - 55k^7 < 2000$.

Finally, let $K \geq 1$. Then
$$C\Lambda_1 \geq C(K\mu_{1,\delta}(k+\alpha) - \mu_{1,\delta}(k-\beta)) \geq \frac{5C}{2}(K-1)k^3 a + 3(K+1)Ck^2$$
$$= L_1 := \frac{500(5kK + 6a - 5k)}{k^2} = \frac{2500(K-1)}{k} + \frac{3000a}{k^2},$$
and
$$-\Lambda_5 \leq -\min_{i=2,\ldots,5}\{K\lambda_{i,\delta}(\overline{p}) - \lambda_{i,\delta}(p)\} \leq \max\{L_2, L_3, L_4, L_5\}$$
for the following linear forms in K:
$$L_2 := 5(K+1)a + (1-K)(5-4k) = (5a - 5 + 4k)(K-1) + 10a,$$
$$L_3 := \frac{9}{2}(K+1)a + \frac{11+7k}{2}(K-1) = \frac{9a + 11 + 7k}{2}(K-1) + 9a,$$
$$L_4 := (K+1)a\frac{k}{29} - \frac{5k^3}{2}(K-1) = \left(\frac{ak}{29} - \frac{5k^3}{2}\right)(K-1) + \frac{2ak}{29},$$
$$L_5 := 12k(K+1)a + \frac{9}{2}(K-1) = 3\left(4ak + \frac{3}{2}\right)(K-1) + 24ak.$$

One immediately sees that both the slope and the value at $K=1$ of L_1 are (much) bigger than those of $L_i, i = 2, 3, 4, 5$, which finishes the proof. $\square$

PROOF. To prove Theorem 5.3.1 it is sufficient to verify the condition (5.2.3), namely, that the five partial derivatives $\frac{\partial g}{\partial \lambda_i}, i = 1, \ldots, 5$, are strictly positive (and bounded, which is automatic due to compactness) on the symmetrized image
$$M := \bigcup_{\sigma \in \Sigma_n} T_\sigma \Lambda(B) \subset \mathbb{R}^n$$
of the unit ball under the map Λ,
$$g(\lambda_1, \ldots, \lambda_5) = \det(D^2 w) - e_5(\Delta w)^5 - e_3(\Delta w)^3 \sigma_2(w) - e_1 \sigma_4(w) \Delta w$$
being our equation. By homogeneity it is sufficient to show this on $M' := \Lambda(\mathbb{S}_1^4)$, which is an algebraic curve, the union of 120 curves $T_\sigma \Lambda(\mathbb{S}_1^4)$, and it is sufficient, by symmetry, to verify the condition on the curve $\Lambda(\mathbb{S}_1^4)$ only. A brute force calculation shows then that
$$g_1(p, \varepsilon) := \frac{\partial g}{\partial \lambda_1} = \sum_{i=0}^{12} m_i p^i$$
$$= m_{12} p^8 (p^4 - 12p^2 + 54) + m_9 b^3 + m_6 p^4 \left(p^2 - \frac{3}{4}\right) + m_2 + m_0,$$
with
$$m_{12} = 3(\varepsilon^4 + 3\varepsilon^3 - 20\varepsilon^2 + 12\varepsilon - 56)(\varepsilon - 2)^2(\varepsilon + 2)^2,$$
$$m_9 = 3D(p, \varepsilon)(\varepsilon + 7)(\varepsilon + 2)(\varepsilon^2 + 2)(\varepsilon - 2)^2,$$
$$m_6 = 108(2 - \varepsilon)(3\varepsilon^7 + 17\varepsilon^6 - 54\varepsilon^5 - 152\varepsilon^4 + 72\varepsilon^3 - 42\varepsilon^2 + 384\varepsilon + 1344),$$
$$m_2 = 1944\varepsilon^2(2 - \varepsilon)(\varepsilon^2 - 7)(\varepsilon^2 + 3),$$
$$m_0 = -7776\varepsilon^2(\varepsilon^2 + 3)$$
for $D(p, \varepsilon) := \sqrt{(16 - \varepsilon^2)(4 - \varepsilon^2)b^2 + 144\varepsilon^2}$, $b = (p^2 - 3)p$;

$$g_2(p,\varepsilon) := \frac{\partial g}{\partial \lambda_2} = \sum_{i=0}^{12} n_i p^i,$$

with

$$n_{12} = (\varepsilon+4)(\varepsilon+1)(4-\varepsilon^2)^2,$$
$$n_{10} = -(\varepsilon+2)(\varepsilon^4 + 19\varepsilon^3 + 86\varepsilon^2 + 182\varepsilon + 96)(2-\varepsilon)^2,$$
$$n_{11} = n_1 = 0, \quad n_8 = 9(\varepsilon+2)(\varepsilon^2 + 10\varepsilon + 6)(\varepsilon^2 + 3\varepsilon + 8)(2-\varepsilon)^2,$$
$$n_9 = \varepsilon(\varepsilon+7)(\varepsilon+2)(\varepsilon^2+2)(2-\varepsilon)^2\sqrt{3(4-p^2)}, \quad n_7 = -9n_9, \quad n_5 = 27n_9,$$
$$n_6 = 3(2-\varepsilon)(13\varepsilon^6 + 115\varepsilon^5 + 218\varepsilon^4 + 170\varepsilon^3 - 876\varepsilon^2 - 2856\varepsilon - 1152),$$
$$n_4 = 9(\varepsilon-2)(11\varepsilon^6 + 62\varepsilon^5 + 33\varepsilon^4 - 24\varepsilon^3 - 348\varepsilon^2 - 1176\varepsilon - 288),$$
$$n_3 = \varepsilon^3(2-\varepsilon)(\varepsilon+7)(3\varepsilon^2+2)\sqrt{3(4-p^2)},$$
$$n_2 = 108\varepsilon^2(2-\varepsilon)(\varepsilon^2+3)(\varepsilon^2+4\varepsilon-3), \quad n_0 = 1296\varepsilon^2(\varepsilon^2+3);$$

$$g_3(p,\varepsilon) := \frac{\partial g}{\partial \lambda_3} = \sum_{i=0}^{6} h_{2i} p^{2i},$$

with

$$h_{12} = (\varepsilon+4)(\varepsilon+1)(4-\varepsilon^2)^2,$$
$$h_{10} = 2(\varepsilon+2)(\varepsilon^4 + \varepsilon^3 - 40\varepsilon^2 - 70\varepsilon - 48)(2-\varepsilon)^2,$$
$$h_8 = -18(\varepsilon+2)(\varepsilon^4 + 4\varepsilon^3 - 19\varepsilon^2 - 28\varepsilon - 24)(2-\varepsilon)^2,$$
$$h_6 = 6(\varepsilon-2)(7\varepsilon^6 + 37\varepsilon^5 - 136\varepsilon^4 - 274\varepsilon^3 + 330\varepsilon^2 + 672\varepsilon + 576),$$
$$h_4 = 9(\varepsilon-2)(2\varepsilon^6 - \varepsilon^5 + 27\varepsilon^4 - 66\varepsilon^3 - 348\varepsilon^2 - 1176\varepsilon - 288),$$
$$h_2 = 108\varepsilon(2-\varepsilon)(\varepsilon^3 + 4\varepsilon^2 - 15\varepsilon - 84)(\varepsilon^2+3),$$
$$h_0 = -1296\varepsilon(\varepsilon^2+3)(\varepsilon^2+4\varepsilon-14);$$

$$g_4(p,\varepsilon) := \frac{\partial g}{\partial \lambda_4} = g_1(-p,\varepsilon);$$

$$g_5(p,\varepsilon) := \frac{\partial g}{\partial \lambda_5} = g_2(-p,\varepsilon);$$

and thus we need to consider only the functions $g_1(p,\varepsilon), g_2(p,\varepsilon), g_3(p,\varepsilon)$ on the set $[-1,1] \times]0,1]$. We have to prove that for any fixed $\varepsilon \in]0,1]$ they are strictly positive.

The technique of the proof is identical for all three derivatives, and we begin with g_3, which is slightly simpler since it is a polynomial in two variables. One can rearrange it in the form

$$g_3(p,\varepsilon) = g_{37}\varepsilon^7 + g_{36}\varepsilon^6 + g_{35}\varepsilon^5 + g_{34}\varepsilon^4 + g_{33}\varepsilon^3 + g_{32}\varepsilon^2 + g_{31}\varepsilon + g_{30}$$

with

$$g_{37} = 2q^5 + 42q^3 - 18q^4 - 108q + 18q^2, \quad g_{36} = 138q^3 - 2q^5 - 216q + q^6 - 36q^4 - 45q^2,$$
$$g_{35} = -92q^5 + 558q^4 + 2160q + 5q^6 - 1296 - 1260q^3 + 261q^2,$$
$$g_{34} = -4q^6 + 5184q - 12q^3 - 5184 - 1080q^2 + 28q^5 - 36q^4,$$
$$g_{33} = -1944q^2 - 2520q^4 - 40q^6 + 14256 - 10692q + 5268q^3 + 520q^5,$$
$$g_{32} = -144q^4 + 72q^3 + 112q^5 + 17496q - 4320q^2 - 16q^6 - 15552,$$
$$g_{31} = -736q^5 + 80q^6 + 2304q^4 - 54432q + 54432 - 4608q^3 + 18576q^2,$$
$$g_{30} = 64q^2(q-3)^4 \geq 0$$

for $q = p^2 \in [0,1]$. Therefore,
$$g_3(p,\varepsilon) \geq \varepsilon(\bar{g}_{37}\varepsilon^6 + \bar{g}_{36}\varepsilon^5 + \bar{g}_{35}\varepsilon^4 + \bar{g}_{34}\varepsilon^3 + \bar{g}_{33}\varepsilon^2 + \bar{g}_{32}\varepsilon + \bar{g}_{31}),$$
where $\bar{g}_{3i} := \min_{q \in [0,1]} g_{3i}(q)$, and elementary calculations give
$$\frac{g_3(p,\varepsilon)}{\varepsilon} \geq -64\varepsilon^5 - 160\varepsilon^4 - 5184\varepsilon^3 + 4848\varepsilon^2 - 15552\varepsilon + 15616 > 1620$$
for $\varepsilon \in (0, \frac{9}{10}]$. For $\varepsilon \in [\frac{9}{10}, 1]$ we have $g_3(p,\varepsilon) \geq \sum_{i=0}^{6} \bar{h}_{2i} q^i$ where $\bar{h}_{2i} := \min_{\varepsilon \in [\frac{9}{10},1]} h_{2i}$ and thus
$$g_3(p,\varepsilon) \geq -736q^5 + 80q^6 + 2304q^4 - 54432q + 54432 - 4608q^3 + 18576q^2 > 4840.$$
Thus, finally, $g_3(p,\varepsilon) \geq \min\{1620\varepsilon, 4840\} = 1620\varepsilon$.

The function $g_1(p,\varepsilon) = s_1 - t_1$ with
$$\begin{aligned}s_1 := \, &(q^3 - 6q^2 + 9q)\varepsilon^7 + (5q^3 - 30q^2 + 45q - 72)\varepsilon^6 + (108q^2 - 18q^3 - 162q)\varepsilon^5 \\ &+ (-432q + 288 - 48q^3 + 288q^2)\varepsilon^4 + (24q^3 - 144q^2 + 216q)\varepsilon^3 \\ &+ 1512\varepsilon^2 + (1152q - 768q^2 + 128q^3)\varepsilon + 448q^2(q-3)^4 \\ \geq \, &{-72\varepsilon^6} - 72\varepsilon^5 + 96\varepsilon^4 + 1512\varepsilon^2 \geq 1440\varepsilon^2 > 0\end{aligned}$$
and
$$t_1 := (\varepsilon^2 - 4)(\varepsilon + 7)(\varepsilon^2 + 2)bD(p,\varepsilon);$$
simplifying $s_1^2 - t_1^2 = (s_1 - t_1)(s_1 + t_1) = g_1(p,\varepsilon)(s_1 + t_1)$ one finds
$$\begin{aligned}&(540q^2 - 1296q - 216q^4 - 4q^6 + 288q^3 + 48q^5)\varepsilon^{13} \\ &+ (-1944q^4 + 3024q^3 + 432q^5 - 7776q + 5184 + 2268q^2 - 36q^6)\varepsilon^{12} \\ &+ (-2052q^2 + 3240q^4 - 5328q^3 - 720q^5 + 60q^6 + 10368q)\varepsilon^{11} \\ &+ (936q^6 + 50544q^4 + 55944q^2 - 11232q^5 - 41472 - 97776q^3 + 29808q)\varepsilon^{10} \\ &+ (26136q^2 + 120q^6 - 15696q^3 + 6480q^4 - 24624q - 1440q^5)\varepsilon^9 \\ &+ (57456q^5 + 156816q - 492372q^2 - 4788q^6 - 134784 + 534528q^3 - 258552q^4)\varepsilon^8 \\ &+ (180576q^2 - 352q^6 - 313632q + 3168q^3 + 4224q^5 - 19008q^4)\varepsilon^7 \\ &+ (54432q^4 - 238464q^3 + 859248q^2 - 1166400q + 1008q^6 + 870912 - 12096q^5)\varepsilon^6 \\ &+ (1118592q^3 - 518400q^4 - 9600q^6 - 1268352q^2 + 736128q + 115200q^5)\varepsilon^5 \\ &+ (1524096q^4 + 207360q + 2286144 + 28224q^6 - 338688q^5 + 2147904q^2 - 3025152q^3)\varepsilon^4 \\ &+ (10752q^6 + 580608q^4 - 903168q^3 - 677376q^2 + 2322432q - 129024q^5)\varepsilon^3 \\ &+ (3640320q^3 - 25344q^6 + 304128q^5 - 1368576q^4 + 8128512q - 7471872q^2)\varepsilon^2 \\ &+ (3096576q^4 + 4644864q^2 + 57344q^6 - 688128q^5 - 6193152q^3)\varepsilon \\ &\geq 64\varepsilon^4(-10\varepsilon^9 + 18\varepsilon^8 - 648\varepsilon^6 - 141\varepsilon^5 - 2214\varepsilon^4 - 2266\varepsilon^3 + 5760\varepsilon^2 + 35721) \\ &\geq 2.2 \cdot 10^6 \varepsilon^4.\end{aligned}$$
Since $s_1 + t_1 \leq 10^6$, one gets $g_1(p,\varepsilon) > 2\varepsilon^4$.

Similarly, $g_2(p,\varepsilon) = s_2 - t_2$ with a polynomial $s_2 \geq 3000\varepsilon^2$ and
$$t_2 = \varepsilon(\varepsilon + 7)(\varepsilon - 2)t(\varepsilon, q)p^3\sqrt{3(4-q)},$$

where

$$t(\varepsilon, q) := (\varepsilon^4 - 2\varepsilon^2 - 8)q^3 + 9(-\varepsilon^4 + 2\varepsilon^2 + 8)q^2 + 27(\varepsilon^4 - 2\varepsilon^2 - 8)p^2 - 9(3\varepsilon^2 + 2)\varepsilon^2.$$

Simplifying $s_2^2 - t_2^2 = g_2(p, \varepsilon)(s_2 + t_2)$ one gets a polynomial

$$\geq (-2560\varepsilon^9 - 18176\varepsilon^8 - 325632\varepsilon^6 - 1254656\varepsilon^4 + 2202112\varepsilon^2 + 15116544)\varepsilon^4$$
$$\geq 1.5 \cdot 10^7 \varepsilon^4$$

and $g_2(p, \varepsilon) \geq 15\varepsilon^4$, which finishes the proof. □

5.4. Other isoparametric forms

Let us then show that all homogeneous isoparametric polynomials give rise to nonclassical solutions of uniformly elliptic Hessian equations and uniformly elliptic Isaacs equations, which is a consequence of the corresponding hyperbolicity results.

5.4.1. Examples of isoparametric polynomials. Recall that isoparametric functions are defined in Section 3.1, which treats the case of cubics.

Let us summarize some well-known facts about isoparametric functions below (we refer to [**58**], [**245**], and [**67**] for a more detailed account).

(**F1**) [**54**], [**170**], [**171**]. Any isoparametric hypersurface with g distinct principal curvatures in the standard sphere arises as a level set $M_t := u^{-1}(t) \cap \mathbb{S}^{n-1}$ of a homogenous order g polynomial solution of the system

(5.4.1) $\qquad |\nabla u(x)|^2 = g^2 |x|^{2g-2}, \quad \Delta u(x) = \dfrac{m_2 - m_1}{2} g^2 |x|^{g-2}, \qquad x \in V = \mathbb{R}^n.$

The restriction $u(x)|_{\mathbb{S}^{n-1}}$ takes its values in $[-1, 1]$, and the number g of distinct principal curvatures

(5.4.2) $\qquad\qquad\qquad \varkappa_1 > \cdots > \varkappa_g,$

of M_t must be $1, 2, 3, 4,$ or 6.

(**F2**) [**170**], [**171**], [**1**]. If $m_1, m_2, \ldots, m_g$ are the multiplicities of the principal curvatures (5.4.2), then

$$n = \dim V = \dfrac{(m_1 + m_2)g}{2} + 2$$

and

(5.4.3) $\qquad\qquad m_{i+2} = m_i, \quad i \geq 1 \ \text{(indices modulo } g\text{)}.$

Moreover, if $g = 3$, then $m_i = m_1$ for all i, and if $g = 6$, then either $(m_1, m_2) = (1, 1)$ or $(m_1, m_2) = (2, 2)$.

(**F3**) [**170**], [**171**]. If $t = \cos g\theta$ with $0 < \theta < \frac{\pi}{g}$, then one has for the principal curvatures of M_t

(5.4.4) $\qquad\qquad \varkappa_j = \cot\left(\theta + \dfrac{\pi j}{g}\right), \qquad \text{where } 0 \leq j \leq g-1.$

Examples for $g = 3$ are described in Section 3.1.

EXAMPLE 5.4.1 ($g = 4$, a homogeneous isoparametric hypersurface). Consider the quartic form

$$u(x) = \dfrac{1}{2}\left(\operatorname{tr} X^4 - \dfrac{3}{8}(\operatorname{tr} X^2)^2\right),$$

where X is a skew-symmetric matrix

$$X = \begin{pmatrix} 0 & x_1 & x_2 & x_3 & x_4 \\ -x_1 & 0 & x_5 & x_6 & x_7 \\ -x_2 & -x_5 & 0 & x_8 & x_9 \\ -x_3 & -x_6 & -x_8 & 0 & x_{10} \\ -x_4 & -x_7 & -x_9 & -x_{10} & 0 \end{pmatrix}.$$

Then $u(x)$ is an isoparametric function in $\mathbb{R}^{10}$ with $g = 4$, $m_1 = m_2 = 2$. All level sets $M_t = u^{-1}(t) \cap \mathbb{S}^9$, $t \in (-1, 1)$, are *homogeneous* isoparametric hypersurfaces.

EXAMPLE 5.4.2 ($g = 4$, the OT-FKM type). By using representations of Clifford algebras Ozeki and Takeuchi [**193**] constructed two infinite series of isoparametric families with four principal curvatures and this construction was generalized later by Ferus, Karcher, and Münzner in [**88**]. More precisely, a quartic form u is said to be of OT-FKM type if it is congruent to

$$u(x) = |x|^4 - 2\sum_{i=0}^{s}(x^t A_i x)^2,$$

where $\{A_i\}_{0 \leq i \leq s}$ is a system of symmetric endomorphisms of $\mathbb{R}^n$ satisfying $A_i A_j + A_j A_i = 2\delta_{ij} \cdot \mathbf{1}_{\mathbb{R}^n}$. Observe that the latter condition imposes several restrictions on the possible pairs (s, n). In particular, n must be an even number and moreover $\delta(s)|m$, where $\delta(s)$ is defined by Table 5.4.1

TABLE 5.4.1. The degree of an irreducible representation of symmetric Clifford systems.

s	1	2	3	4	5	6	7	8	...	q
$\delta(s)$	1	2	4	4	8	8	8	8	...	$16\,\delta(s-8)$

In this notation, the multiplicities of the principal curvatures are given by

(5.4.5) $$m_1 = s, \quad m_2 = \frac{n}{2} - s - 1.$$

Recent progress in the classification of isoparametric hypersurfaces with four principal curvatures [**59**], [**66**], [**67**] resulted in the following remarkable observation: Except for the case of the multiplicity pair $(m_1, m_2) = (7, 8)$ the only anomalous isoparametric hypersurfaces with four principal curvatures (i.e., those *not* of OT-FKM type) are the homogeneous hypersurfaces with $(m_1, m_2) = (2, 2)$ (see Example 5.4.1) or $(m_1, m_2) = (4, 5)$ (the case of multiplicity pair $(7, 8)$ is still open).

EXAMPLE 5.4.3 ($g = 6$). If $g = 6$, U. Abresch [**1**] showed that the multiplicity of each principal curvature is the same number, $m = m_1 = m_2$, which takes only the values 1 or 2. In the former case, [**77**] proves the homogeneity of such hypersurfaces and conjectures that it is true for the case $m = 2$. Recently, homogeneity in the case $m = 2$ was established by R. Miyaoka [**167**].

5.4.2. Spectrum of the Hessian.

LEMMA 5.4.1. *Given $\alpha \in \mathbb{R}$, we define the following ratio:*

$$w_\alpha(x) = \frac{u(x)}{|x|^\alpha},$$

where u is an isoparametric polynomial of order $g \geq 3$ satisfying (5.4.1). Let $\xi \in \mathbb{S}^{n-1}$ and $u(\xi) = \cos g\theta$, where $\theta \in (0, \frac{\pi}{g})$. Then
(5.4.6)
$$\lambda_\pm(\theta) = \frac{\alpha(\alpha - 2g)u}{2} \pm \frac{\sqrt{\alpha(\alpha+2)(\alpha+2-2g)(\alpha-2g)u^2 + 4g^2(1+\alpha-g)^2}}{2}$$

and
(5.4.7)
$$\lambda_j(\theta) = \frac{g\sin(\frac{\pi j}{g} - (g-1)\theta)}{\sin(\frac{\pi j}{g} + \theta)} - \alpha \cos g\theta \equiv (-1)^j \lambda_0(\theta + \frac{\pi j}{g}), \qquad 0 \leq j \leq g-1,$$

are the eigenvalues of the characteristic polynomial of the Hessian of $\mathrm{Hess}_\xi(w_\alpha)$ at the point ξ. Moreover, the $\mu_\pm$ are single eigenvalues whereas the eigenvalue λ_j has multiplicity m_s, where $s = 1, 2$ is chosen such that $s \equiv j \mod 2$.

PROOF. Let us denote for short $H_w = \mathrm{Hess}_\xi(w_\alpha)$ and $H_u = \mathrm{Hess}_\xi(u)$. Without loss of generality one can assume that $\xi \in \mathbb{S}^{n-1}$ is a regular point, i.e., $\nabla u(\xi)$ is noncollinear to ξ. A direct computation shows that

(5.4.8) $$H_w = H_u - \alpha u \mathbf{1}_V + B,$$

where
$$B = \alpha(\alpha+2)u\,\xi \otimes \xi - \alpha(\eta \otimes \xi + \xi \otimes \eta),$$

$u = u(\xi)$ and $\eta = \nabla u(\xi)$. Observe that by Euler's homogeneous function theorem $\langle \eta, \xi \rangle = \langle \nabla u(\xi), \xi \rangle = gu(\xi)$, which yields

$$B\xi = \alpha(\alpha+2-g)u\,\xi - \alpha\eta, \quad B\eta = \alpha g((\alpha+2)u^2 - g)\,\xi - \alpha gu\,\eta.$$

This shows that $V_1 = \mathrm{span}(\xi, \eta)$ is an invariant subspace of B.

Similarly, applying homogeneity again, $H_u\xi = (g-1)\eta$, and polarizing the first equation in (5.4.1) one readily finds that

(5.4.9) $$H_u\eta = (g-1)g^2\xi.$$

This shows that V_1 is an invariant subspace of $H_u(\xi)$ too, and thus by (5.4.8), V_1 is an eigenspace of H_w. We also have the following matrix representation of $H_w|_{V_1}$ in the basis ξ and η:

(5.4.10) $$H_w|_{V_1} = \begin{pmatrix} \alpha(\alpha+1-g)u & g-1-\alpha \\ (g-1-\alpha)g^2 + \alpha(\alpha+2)gu^2 & -\alpha(g+1)u \end{pmatrix},$$

and the characteristic polynomial is found to be

$$\det(H_w - \lambda\mathbf{1})|_{V_1} = \lambda^2 + \alpha(2g-\alpha)u\lambda - (g-\alpha-1)(\alpha(2g-\alpha)u^2 + g^2(g-\alpha-1)).$$

This yields the two single eigenvalues of (5.4.6). The orthogonal complement is given by $V_2 = V \ominus V_1$, which is obviously an invariant subspace of H_w. By the definition, $B|_{V_2} = 0$; hence

(5.4.11) $$H_w|_{V_2} = (H_u - \alpha u\mathbf{1})|_{V_2}.$$

Now, let us consider the isoparametric submanifold $M = u^{-1}(t) \subset \mathbb{S}^{n-1}$, where $t = u(\xi) \in (-1,1)$. Then $V_2 = T_\xi M$ is the tangent space at the point $\xi \in M$. Observe that

$$\nabla u(\xi) - \xi \langle \nabla u(\xi), \xi \rangle = \nabla u(\xi) - gu(\xi)\xi \equiv \eta - gu\xi$$

is a normal vector field to $M \subset \mathbb{S}^{n-1}$; hence, normalizing it, we obtain a unit normal

$$\nu(\xi) = \frac{\nabla u(\xi) - gu(\xi)\xi}{g\sqrt{1-u^2(\xi)}}.$$

Therefore, we have for the shape operator

$$A^\nu(E) \equiv -(\nabla_E \nu)^{V_2} = -\frac{1}{g\sqrt{1-u^2}}(H_u(E)|_{V_2} - gu\,E)$$

where $E \in T_\xi M \equiv V_2$. This shows that $H_u(E)|_{V_2}$ is diagonalizable in the same basis as the shape operator A^ν; hence λ is an eigenvalue of $H_u(E)|_{V_2}$ if and only if $\varkappa = \frac{gu-\lambda}{g\sqrt{1-u^2}}$ is a principal curvature of M. Let us write $u(\xi) = \cos g\theta$ for some $\theta \in (0, \frac{\pi}{g})$. Using (5.4.4), we obtain

$$(5.4.12) \qquad \lambda = g\left(\cos g\theta - \sin g\theta \cot\left(\theta + \frac{\pi j}{g}\right)\right) = \frac{g\sin(\theta + \frac{\pi j}{g} - g\theta)}{\sin(\theta + \frac{\pi j}{g})},$$

where $\theta_j = \theta + \frac{\pi j}{g}$. Then (5.4.11) finishes the proof. $\square$

We shall need the following elementary corollary of Lemma 5.4.1.

COROLLARY 5.4.1. *For any $\theta \in (0, \frac{\pi}{g})$ the following holds:*

$$(5.4.13) \qquad \lambda_j(\theta) - \lambda_{j-1}(\theta) = \frac{\sin g\theta \, \sin \frac{\pi}{g}}{\sin(\frac{\pi j}{g} + \theta)\sin(\frac{\pi(j-1)}{g} + \theta)};$$

thus $\lambda_j(\theta) - \lambda_{j-1}(\theta) > 0$ for all $1 \leq j \leq g-1$. In particular,

$$\lambda_0(\theta) < \lambda_1(\theta) < \cdots < \lambda_{g-1}(\theta), \qquad \theta \in \left(0, \frac{\pi}{g}\right).$$

Moreover, if additionally $\alpha = g - 2$ and $g \geq 2$, then

$$(5.4.14) \qquad \lambda_-(\theta) < -2, \qquad \lambda_+(\theta) > 2.$$

PROOF. The first claim follows immediately from (5.4.7). In order to show (5.4.14) it suffices to note that for $\alpha = g-2$, $\lambda_{pm}(\theta)$ are the roots of

$$P(\lambda) := \lambda^2 + (g^2-4)u(\theta)\lambda - (g^2-4)u^2 - g^2 = 0$$

and

$$P(\pm 2) = -(g^2-4)(1 \pm u(\theta))^2 < 0. \qquad \square$$

5.4.3. The case $\alpha = g-2$. A closer look at the eigenvalues reveals that the case $\alpha = g-2$ is distinguished in many relations. In fact, exactly for this value of α, the Hessian $H_w(x)$ becomes homogeneous of order 0 and, thus, independent of scale changes.

First we consider the case with $g = 3$. Recall that by (F2) in Subsection 5.4.1 $m_1 = m_2 = d$ in the cubic case, where d is the dimension of the associated real division algebra $\mathbb{F}_d$.

COROLLARY 5.4.2 (The case $g = 3$). *Let $\mu_1 > \mu_2 > \cdots > \mu_5$ be the distinct eigenvalues of $\operatorname{Hess}_\xi(w_1)$ evaluated at $\xi \in \mathbb{S}^{n-1}$ counted with the multiplicities $k_1, k_2, \ldots, k_5$, respectively. Then $n = 3d + 2$, $d \in \{1, 2, 4, 8\}$,*

$$\mu_1 = \lambda_2, \quad \mu_2 = \lambda_+, \quad \mu_3 = \lambda_1, \quad \mu_4 = \lambda_-, \quad \mu_5 = \lambda_0, \tag{5.4.15}$$

where $k_1 = k_3 = k_5 = d$, $k_2 = k_4 = 1$, and λ_j and $\lambda_\pm$ are defined as in Lemma 5.4.1 above.

PROOF. Setting $g = 3$ and $\alpha = g - 2 = 1$ in (5.4.7) and (5.4.6), we have

$$\lambda_j(\theta) = \lambda_0\left(\theta + \frac{\pi j}{3}\right), \quad 0 \le j \le 2, \quad 0 < \theta < \frac{\pi}{3}, \tag{5.4.16}$$

where $\lambda_0(\theta) = -6\cos\theta - \cos 3\theta$, and since

$$\lambda_0'(\theta) = 3(2\sin\theta + \sin 3\theta) = 3\sin\theta(1 + 4\cos^2\theta),$$

we conclude that $\lambda_0(\theta)$ is strictly increasing in $(0, \pi)$. On the other hand, $\lambda_1(0) = -7$, $\lambda_1(\pi/3) = -2$, $\lambda_1(2\pi/3) = 2$, and $\lambda_1(\pi) = 7$. Thus we get by Corollary 5.4.1 that

$$-7 < \lambda_0(\theta) < -2 < \lambda_1(\theta) < 2 < \lambda_2(\theta) < 7, \qquad \theta \in \left(0, \frac{\pi}{3}\right).$$

Furthermore, since $\lambda_\pm$ are the roots of the quadratic polynomial

$$P(t) = t^2 + 5t\cos 3\theta - 5\cos^2 3\theta - 9 = 0$$

and since $P(\lambda_0(\theta)) = 9\sin^2\theta(16\cos^4\theta - 1)$, it follows from $P(\lambda_1(\theta)) > 0$, $P(\lambda_2(\theta)) < 0$, $P(\lambda_3(\theta)) > 0$ for $\theta \in (0, \frac{\pi}{3})$ and (5.4.16) that (5.4.15) holds. □

COROLLARY 5.4.3 (The case $g = 4$). *Let u be an isoparametric polynomial of order $g = 4$ satisfying (5.4.1), and let $\mu_1 > \mu_2 > \cdots > \mu_6$ be the distinct eigenvalues of $\operatorname{Hess}_\xi(w_2)$ evaluated at $\xi \in \mathbb{S}^{n-1}$ counted with the multiplicities $k_1, k_2, \ldots, k_6$, respectively. Then $n = 2(m_1 + m_2 + 1)$, and*

$$\mu_1 = \lambda_3, \quad \mu_2 = \lambda_+, \quad \mu_3 = \lambda_2, \quad \mu_4 = \lambda_1, \quad \mu_5 = \lambda_-, \quad \mu_6 = \lambda_0, \tag{5.4.17}$$

where $k_1 = k_4 = m_1$, $k_2 = k_5 = 1$, $k_3 = k_6 = m_2$, and λ_j and $\lambda_\pm$ are defined as in Lemma 5.4.1 above.

PROOF. As above, we obtain from (5.4.7) and (5.4.6) for $g = 4$ and $\alpha = g - 2 = 2$ that $\lambda_j(\theta) = \lambda_0(\theta + \frac{\pi j}{4})$, $0 \le j \le 3$, which yields for $0 < \theta < \frac{\pi}{4}$

$$-14 < \lambda_1(\theta) < -2 < \lambda_2(\theta) < \lambda_3(\theta) < 2 < \lambda_4(\theta) < 14.$$

Furthermore, since $\lambda_\pm$ are the roots of the quadratic polynomial

$$P(t) = t^2 + 12t\cos 3\theta - 12\cos^2 3\theta - 16 = 0$$

and since $P(\lambda_1(\theta)) = 1024\cos^4\theta\sin^2\theta(2\cos^2\theta - 1)$, we obtain

$$P(\lambda_0(\theta)) > 0, \quad P(\lambda_1(\theta)) < 0, \quad P(\lambda_2(\theta)) < 0, \quad P(\lambda_3(\theta)) > 0$$

for $\theta \in (0, \frac{\pi}{3})$. This proves the order in (5.4.17). □

The case $g \ge 5$ is somewhat more involved because the $\lambda_\pm$-curves have now common points with the λ_j-curves and we formulate here the only relevant case: $g = 6$.

COROLLARY 5.4.4 (The case $g = 6$). *Let u be an isoparametric polynomial of order $g = 6$ satisfying (5.4.1), and let $\mu_1 > \mu_2 > \cdots > \mu_8$ be the distinct eigenvalues of $\mathrm{Hess}_\xi(w_2)$ evaluated at $\xi \in \mathbb{S}^{n-1}$ counted with the multiplicities $k_1, k_2, \ldots, k_8$, respectively. Then $n = 6m + 2$, and*

$$(5.4.18) \qquad \mu_4 = \lambda_3, \quad \mu_5 = \lambda_2,$$

and λ_j and $\lambda_\pm$ are defined as in Lemma 5.4.1 above.

5.4.4. Hyperbolicity.

PROPOSITION 5.4.1. *Let $u(x)$ be an isoparametric polynomial of order $g \geq 3$ and let $w_{g-2}(x) = u(x)|x|^{2-g}$, $x \in \mathbb{R}^n$. Define*

$$M(\xi, \xi', O) = \mathrm{Hess}_\xi(w_1) - O^t \, \mathrm{Hess}_{\xi'}(w_1) O,$$

where O is an orthogonal endomorphism of V, and let

$$\Lambda_n \leq \Lambda_{n-1} \leq \cdots \leq \Lambda_2 \leq \Lambda_1$$

be the eigenvalues of $M(\xi, \xi', O)$. Set $\Lambda_n^* = -\Lambda_n$; then

$$(5.4.19) \qquad \frac{1}{c} \leq \frac{\Lambda_1}{\Lambda_n^*} \leq c,$$

with $c = c(g, n)$.

PROOF. Since $\Delta u(x) = 0$, we find by homogeneity of u for any $\xi \in \mathbb{S}(e^\perp)$

$$\Delta w_1(\xi) = |\xi|^{2-g} \Delta u - 2(g-2)|\xi|^{-g} \langle \nabla u(\xi), \xi \rangle + u \Delta |\xi|^{2-g}$$
$$= \frac{(m_2 - m_1)g^2}{2} - (g-2)(n+g) u(\xi).$$

By a density argument, it suffices to prove the proposition assuming that ξ and ξ' are regular. Observe that

$$(5.4.20) \qquad \mathrm{tr}\, M(\xi, \xi', O) = \Delta w_1(\xi) - \Delta w_1(\xi') = -(g-2)(n+g)\omega,$$

where $\omega := u(\xi) - u(\xi')$. If $\omega = 0$, then (5.4.19) holds with $c = (n-1)$ as for any traceless matrix in dimension n. Let us now assume that $\omega \neq 0$. Then by symmetry we can assume that

$$(5.4.21) \qquad \omega = u(\xi) - u(\xi') > 0,$$

which yields, in particular, that

$$(5.4.22) \qquad \begin{aligned} -(g-2)(n+g)\omega = \mathrm{tr}\, M(\xi, \xi', O) &\geq \Lambda_1 + (n-1)\Lambda_n \\ &= \Lambda_1 - (n-1)\Lambda_n^*, \end{aligned}$$

and by the negativity of ω,

$$(5.4.23) \qquad (n-1)\Lambda_n^* \geq \Lambda_1.$$

In order to establish a lower estimate on Λ_1 we write $u(\xi) = \cos g\theta$ and $u(\xi') = \cos g\theta'$, where by (5.4.21),

$$(5.4.24) \qquad 0 < \theta < \theta' < \frac{\pi}{g},$$

and we observe that by Weyl's inequality,

$$\Lambda_1 \geq \max_{1 \leq i \leq n} (\lambda_i(\theta) - \lambda_i(\theta')).$$

We consider the three possible cases separately.

The case $g = 3$. In this situation

(5.4.25) $$\Lambda_1 \geq \lambda_1(\theta) - \lambda_1(\theta').$$

By (5.4.7), $\lambda_1(\theta) = -\lambda_0(\theta + \frac{\pi}{3})$ is a decreasing function and

$$\frac{d\lambda_1}{du} = \frac{\lambda_1'(\theta)}{u'(\theta)} = \frac{\sin\theta + \sqrt{3}\cos\theta - \sin 3\theta}{\sin 3\theta} = 2\frac{\sin(\theta + \frac{\pi}{3})}{\sin 3\theta} - 1 \geq 1, \quad \theta \in \left(0, \frac{\pi}{3}\right),$$

which by (5.4.25) yields $\Lambda_1 \geq \omega$. Combining this with

(5.4.26) $$-(n+3)\omega = \operatorname{tr} M(\xi, \xi', O) \leq (n-1)\Lambda_1 + \Lambda_n = (n-1)\Lambda_1 - \Lambda_n^*,$$

we find

(5.4.27) $$\Lambda_n^* \leq 2(n+1)\Lambda_1,$$

which together with (5.4.23) yields the required estimate with $c_3 = 2(n+1)$.

The case $g = 4$. In this situation

(5.4.28) $$\Lambda_1 \geq \max_{i=1,2}\{\lambda_i(\theta) - \lambda_i(\theta')\} \geq f(\theta) - f(\theta'),$$

where

$$f(\theta) = \frac{1}{2}(\lambda_1(\theta) + \lambda_2(\theta)) = -2\cos 4\theta - 4\sin 2\theta + 4\cos 2\theta$$

is a decreasing function and arguing as above we find

$$\frac{df}{du} = -\frac{f'(\theta)}{4\sin 4\theta} = -\frac{2(\sin 4\theta - \sin 2\theta - \cos 2\theta)}{\sin 4\theta}$$
$$= -2 + \frac{1}{\sin 2\theta} + \frac{1}{\cos 2\theta} \geq 2(\sqrt{2} - 1), \quad \theta \in \left(0, \frac{\pi}{4}\right).$$

Thus, (5.4.28) yields $\Lambda_1 \geq (2\sqrt{2} - 2)\omega$ and combining this with (5.4.20) and

(5.4.29) $$-2(n+4)\omega = \operatorname{tr} M(\xi, \xi', O) \leq (n-1)\Lambda_1 + \Lambda_n = (n-1)\Lambda_1 - \Lambda_n^*,$$

we find

(5.4.30) $$\Lambda_n^* \leq (n - 1 + (\sqrt{2} + 1)(n+4))\Lambda_1,$$

which together with (5.4.23) yields the required estimate.

The case $g = 6$ is established similarly to the case $g = 4$. $\square$

COROLLARY 5.4.5. *Let $u(x)$ be an isoparametric polynomial of order $g \geq 3$ and let $w_{g-2}(x) = u(x)|x|^{2-g}$, $x \in \mathbb{R}^n$. Then u is a solution of a uniformly elliptic Hessian equation as well as of a uniformly elliptic Isaacs equation.*

Notice that while isoparametric functions with $g = 3$ or 6 exist only in a finite number of dimensions (5, 8, 14, 26, and 8, 14, respectively), quartic functions exist in every even dimension ≥ 6. Note also that for $g = 4$ and 6 the corresponding solutions are rational functions. Let us give an explicit solution for quartics in six dimensions.

EXAMPLE 5.4.4. There exists the unique (up to a scalar factor) isoparametric quartic in six dimensions, that of OT-FKM type, namely

(5.4.31) $$P_6(x, y) := 6|x|^2|y|^2 - |x|^4 - |y|^4 - 2\langle x, y\rangle^2, \quad x, y \in \mathbb{R}^3.$$

The corresponding solution to a Hessian uniformly elliptic equation is

$$u_6(x,y) = \frac{P_6(x,y)}{|x|^2+|y|^2} = \frac{6|x|^2|y|^2 - |x|^4 - |y|^4 - 2\langle x,y\rangle^2}{|x|^2+|y|^2}, \quad x,y \in \mathbb{R}^3;$$

even simpler solutions are

$$v_6(x,y) = \frac{2|x|^4 + 2|y|^4 + \langle x,y\rangle^2}{|x|^2+|y|^2},$$

$$w_6(x,y) = \frac{4|x|^2|y|^2 - \langle x,y\rangle^2}{|x|^2+|y|^2},$$

related to $u_6(x,y)$ in an obvious way,

$$v_6(x,y) = \frac{3(|x|^2+|y|^2) - u_6(x,y)}{2}, \quad w_6(x,y) = \frac{|x|^2+|y|^2 + u_6(x,y)}{2}.$$

One can easily write down a polynomial Hessian, locally uniformly elliptic equation for u_6, v_6, or w_6, which leads to a smooth uniformly elliptic Hessian equation. For instance, for $w = w_6$ one can take

$$\begin{aligned}F(D^2 w) = &-2^8 3^2 5^6 \det(D^2 w) + 2^5 3^2 5^6 \,\sigma_4(w) - 2^5 5^5 \,\sigma_2(w)\sigma_3(w)\\&+ 2^4 3^2 5^5 \,\sigma_2(w)^2 - 2^2 3^3 (\Delta w)^6 + 2^2 3^3 179 (\Delta w)^5\\&- 5^3 10687 (\Delta w)^4 + 2^3 5^2 238723 (\Delta w)^3\\&- 2^3 \cdot 3 \cdot 5 \cdot 13 \cdot 17 \cdot 336187 (\Delta w)^2 + 2^5 3^4 37 \cdot 271 \cdot 379 \,\Delta w\\&- 2^4 3^6 109 \cdot 31799\end{aligned}$$

with all derivatives $> 10^6$ on the corresponding image of the spectrum map.

There is no doubt that one can also obtain singular solutions of the form

$$\frac{6|x|^2|y|^2 - |x|^4 - |y|^4 - 2\langle x,y\rangle^2}{(|x|^2+|y|^2)^\delta}, \quad \delta \in \left]1, \frac{3}{2}\right[.$$

for smooth uniformly elliptic Hessian equations, but to prove that, one will need some rather long and cumbersome computations.

5.5. Two more applications

In this section we give two more constructions of unusual solutions from isoparametric polynomials, the first one concerning the p-Laplacian equation and the second concerning the Landau-Ginzburg system. These constructions do not fit exactly into our main framework; however, they are rather close in spirit to our principal applications.

5.5.1. Singular solutions of the p-Laplacian equation.

We consider solutions to the p-Laplacian equation, $p > 1$, defined in the punctured unit ball $B \setminus \{0\}$.

First we recall the simplest radial solution

(5.5.1) $\qquad v_0(x) := |x|^{\frac{p-n}{p-1}}, \quad p \neq n, \qquad v_0(x) := (-\log|x|), \quad p = n,$

which is a fundamental solution of the equation in B.

The following famous result [**222**] by Serrin (which holds for any quasilinear divergence form equation) says that a positive singular solution of the p-Laplacian equation should behave as v_0:

THEOREM 5.5.1. *Let u be a continuos positive solution of $\Delta_p u = 0$ in $B \setminus \{0\}$. Then either 0 is a remouvable singularity (i.e., u can be extened to the whole of B) or $u \sim v_0$ for $x \to 0$.*

In the perspective of this theorem the following result could be of interest:

PROPOSITION 5.5.1. *Let $w(x)$ be a homogeneous harmonic isoparametric polynomial with*
$$|\nabla w(x)|^2 = m^2 |x|^{2m-2} \quad (m = 2, 3, 4, \text{ or } 6),$$
and let
$$p = \frac{2(m+n-1)}{m+1}.$$
Then
$$\Delta_p \left(\frac{w(x)}{|x|^{2m}} \right) = 0$$
in $\mathbb{R}^n \setminus \{0\}$.

PROOF. This is a straightforward calculation. □

5.5.2. Radial solutions of the Ginzburg-Landau system. Recall first some properties of the Ginzburg-Landau system.

DEFINITION 5.5.1. The system
$$(5.5.2) \qquad \Delta u = u(|u|^2 - 1)$$
on a function $u : \mathbb{R}^N \to \mathbb{R}^N, N \geq 2$, is called the Ginzburg-Landau system.

This system describes many natural phenomena and is of exceptional importance. One is concerned with solutions of (5.5.2) satisfying
$$(5.5.3) \qquad |u(x)| \to 1 \text{ as } |x| \to \infty.$$
For $N = 2$ it is easy to construct solutions of (5.5.2)–(5.5.3) in polar coordinates, using separation of variables. Given any integer $q \in \mathbb{Z}$ the function
$$u = u(r, \theta) = e^{iq\theta} f(r)$$
is a solution of (5.5.2)–(5.5.3), $N = 2$, provided the real-valued function f satisfies the ordinary differential equation
$$f'' + \frac{f'}{r} - \frac{q^2}{r^2} f = f(f^2 - 1)$$
on $(0, \infty)$ with $f(0) = 0$ and $f(\infty) = 1$.

It is not difficult to see that for every integer q it has a unique solution f_q and thus one obtains a family of special solutions $u_q = u_q(r, \theta) = e^{iq\theta} f_q(r)$. An outstanding open problem is whether these are the only solutions of (5.5.2)–(5.5.3), $N = 2$, modulo translations and rotations.

This is (almost) trivially true for $q = 0$, which is a version of Liouville's theorem. It is also the case for $q = 1$ [**165**], [**34**] under an additional condition of a topological nature:

THEOREM 5.5.2. *Let u be a solution of (5.5.2), $N = 2$, satisfying* $\deg(u, \infty) = 1$. *Then u has radial symmetry, $u = e^{i\theta} f_1(r)$ modulo rotation, translation, and complex conjugation.*

Here $\deg(u, \infty)$ is the degree of u at infinity = the degree of the map
$$x \in \mathbb{S}^1 \mapsto \frac{u(Rx)}{|u(Rx)|} \in \mathbb{S}^1 \text{ for large } R.$$

In view of these results Brézis [**33**] posed the following:

QUESTION (Open Problem 3 in [**33**]). Let $u : \mathbb{R}^N \to \mathbb{R}^N$ be a solution of
$$\Delta u = u(|u|^2 - 1)$$
on $\mathbb{R}^N, N \geq 3$, with $|u(x)| \to 1$ as $|x| \to \infty$ (possibly with a good rate of convergence). Assume $\deg(u, \infty) = \pm 1$. Does u have the form

(5.5.4)
$$u(x) = \frac{x}{|x|} f(|x|)$$

modulo translation and isometry, where $f : \mathbb{R}_+ \to \mathbb{R}_+$ is a smooth function, such that $f(0) = 0$ and $f(\infty) = 1$?

Isoparametric polynomials permit us to give counterexamples in the dimensions $N = 6$ [**93**] and $N = 8$ [**86**], respectively:

THEOREM 5.5.3. *There exists a solution $u : \mathbb{R}^6 \to \mathbb{R}^6$ of the Ginzburg-Landau system in $N = 6$ dimensions which can be written as*

(5.5.5)
$$u(x) = \Phi\left(\frac{x}{|x|}\right) h(|x|), \quad \Phi := \frac{\nabla P_6}{4},$$

the gradient map of the isoparametric quartic (5.4.31), $h \in C^2(\mathbb{R}_+, \mathbb{R})$ being the unique solution of
$$h'' + 5\frac{h'}{r} - 21\frac{h}{r^2} = h(h^2 - 1)$$
on $(0, \infty)$ with $h(0) = 0$ and $h(\infty) = 1$ satisfying
  (i) $|u(x)| \to 1$ as $|x| \to \infty$,
  (ii) $\deg(u, \infty) = 1$.

THEOREM 5.5.4. *There exists a solution $u : \mathbb{R}^8 \to \mathbb{R}^8$ of the Ginzburg-Landau system in 8 dimensions which can be written as*

(5.5.6)
$$u(x) = G\left(\frac{x}{|x|}\right) f(|x|), \quad G := \frac{\nabla P_8}{6},$$

the gradient map of the harmonic isoparametric sextic P_8 in $\mathbb{R}^8$,
$$G \in C^2(\mathbb{S}^7, \mathbb{S}^7) \cap (\mathcal{SH}_{5,8})^8,$$

$\mathcal{SH}_{k,l}$ *being the space of spherical harmonics of degree k in $\mathbb{R}^l$, $f \in C^2(\mathbb{R}_+, \mathbb{R})$ being the unique solution of*
$$f'' + 7\frac{f'}{r} - 55\frac{f}{r^2} = f(f^2 - 1)$$

on $(0,\infty)$ with $f(0) = 0$ and $f(\infty) = 1$. Furthermore, u satisfies
  (i) $|u(x)| \to 1$ *as* $|x| \to \infty$,
  (i) $\deg(u, \infty) = 1$.

Observe that the sextic P_8 can be written by virtue of the Cartan cubic C_5 as
$$P_8(x,y) := C_5(\pi(x,y)),$$
where
$$x, y \in \mathbb{H} = \mathbb{R}^4, \quad \pi(x,y) := (|x|^2 - |y|^2, 2x\bar{y}) \in \mathbb{R}^5 = \mathbb{R} \oplus \mathbb{R}^4 = \mathbb{R} \oplus \mathbb{H}.$$
Both solutions (5.5.5) and (5.5.6) are not of the form (5.5.4) modulo translations and isometries.

CHAPTER 6

Cubic Minimal Cones

The main goal of this chapter is to demonstrate another aspect of applications of nonassociative algebras to elliptic PDEs, more precisely, to the minimal surface equation. This very classical equation appears as the Euler-Lagrange equation for hypersurfaces in a Euclidean space minimizing the area. Minimal hypercones constitute an important subclass of singular minimal hypersurfaces strongly related to minimal submanifolds of the Euclidean sphere, playing a crucial role in the investigation of the global structure of general minimal hypersurfaces. The study and classification of minimal hypercones remains a challenging problem in differential geometry; see for instance [**277**]. All the examples of minimal hypercones known thus far are algebraic, coming essentially from two classical algebraic structures: the Jordan and the Clifford algebras and their representations. This makes it natural to look for an adequate algebraic setup explaining the appearance of the concrete algebraic structures in the context of minimal cones. It turns out that a good deal of such a program can be fulfilled in the first nontrivial case of *cubic* minimal hypercones, or actually in the most important particular case of radial eigencubics.

6.1. Brief overview of the main results

For further convenience, let us briefly outline our approach. The main idea of the method is to associate a (nonassociative) algebra to a given minimal cubic cone (actually to its defining polynomial $u(x)$) in such a way that the differential-analytical structure of u becomes transparent from the algebraic side, and vice versa. The main tool for doing this is the so-called Freudenthal multiplication defined in Section 6.3. It associates to any fixed cubic form u on a vector space V with a symmetric nondegenerate bilinear form Q the multiplication $(x, y) \to xy$ by virtue of $\partial_x \partial_y u|_z = Q(xy; z)$. The algebra $V(u)$ thus defined is obviously commutative and we call it the Freudenthal-Springer algebra of the cubic form u. It is nonassociative in general but still has the nice property of being metrized in the sense of [**29**]; i.e., the multiplication on $V(u)$ is associative with respect to the bilinear form Q: $Q(xy, z) = Q(x; yz)$ for any $x, y, z \in V$. Furthermore, Euler's homogeneous function theorem implies that the gradient of u at the point x is just $Du(x) = \frac{1}{2}x^2$. In this algebraic setup, the cubic form $u(x)$ is interpreted as the *norm* of the element x recovered by virtue of $u(x) = \frac{1}{6}Q(x^2; x)$. Now any relevant analytic or geometric properties of the norm can be translated to the algebraic language, and vice versa. For instance, according to the definitions introduced, the Hessian operator of u is the multiplication operator L_x by x which, in particular, implies that a differential relation for $u(x)$ becomes a defining relation in the algebra $V(u)$.

As is often the case with algebraic problems, the knowledge of the idempotent and nilpotent elements of an algebra provides effective information for a classification. Fortunately, these notions possess very natural analytical interpretations in the present setup. Namely, translated into algebraic language, any critical point c of the restriction of $u(x)$ onto the unit sphere in V, being suitably normalized, becomes an idempotent in $V(u)$; in other words $c^2 = c$. In the most important case of a positive definite inner product Q this interpretation implies the existence of nonzero idempotents in $V(u)$. On the other hand, the points $w \in V$ where the gradient of u vanishes correspond exactly to the square zero elements satisfying $w^2 = 0$.

Next, in Section 6.4 we specify the cubic form u by assuming that the zero-locus $u^{-1}(0)$ is a cubic minimal hypercone in $\mathbb{R}^n$, i.e., the mean curvature of $u^{-1}(0)$ vanishes at all its regular points. Hsiang [**110**] pointed out that all known (irreducible) cubic minimal hypercones are actually obtained as solutions of the following elliptic PDE:

$$(6.1.1) \qquad \Delta_1 u = |Du|^2 \Delta u - \tfrac{1}{2} Du \cdot D|Du|^2 = \lambda |x|^2 u, \quad x \in \mathbb{R}^n,$$

where $\lambda \in \mathbb{R}$ and Δ_1 is defined as in (1.1.4). A cubic polynomial solution of (6.1.1) is said to be a *radial eigencubic*. According to the above definitions, any radial eigencubic u gives rise to the corresponding Freudenthal-Springer algebra $V(u)$ on $V = \mathbb{R}^n$. Spelling out the differential equation (6.1.1) in algebraic terms one arrives at a fifth-order defining identity on $V(u)$

$$(6.1.2) \qquad \langle x^2, x^2 \rangle \operatorname{tr} L_x - \langle x^2, x^3 \rangle = \tfrac{2}{3} \lambda \langle x, x \rangle \langle x^2, x \rangle,$$

which becomes the starting point of our study. Any commutative metrized algebra satisfying (6.1.2) for some real constant λ is said to be a radial eigencubic algebra, or just a *REC algebra*. It follows from the definitions that if V is a REC algebra, then a cubic form $u(x) = \langle x, x^2 \rangle$ satisfies (6.1.1) in the inner metric on V induced by $\langle\,,\,\rangle$, and, thus defining a radial eigencubic. It naturally identifies two REC algebras which give rise to congruent minimal cones. Such algebras are said to be *similar*. Thus, the classification of radial eigencubics becomes equivalent to that of REC algebras.

There exist two principal classes of REC algebras, namely those coming from Clifford and Jordan structures, respectively. The first class includes (up to a similarity) any nonassociative metrized algebra possessing an orthogonal decomposition $V = V_0 \oplus V_1$ satisfying $V_1 V_1 \subset V_0$, $V_0 V_0 = \{0\}$, and such that unit vectors of V_0 act by multiplication on V_1 as isometric involutions. Such a REC algebra is said to be *polar*. We consider polar algebras in Section 6.5 and show that they are in one-to-one correspondence with (the equivalence classes of) symmetric Clifford systems. In particular, this readily yields the following existence result: a polar algebra on a Euclidean vector space V exists if and only if the inequalities

$$0 \leq \dim V - 2k - 1 \leq \rho(k)$$

hold for some positive integer k, in which case $\dim V_0 = \dim V - 2k$ and $\dim V_1 = 2k$. Here ρ denotes the Hurwitz-Radon function (6.5.18). A typical example of a polar REC algebra is the Freudenthal-Springer algebra on $V_0 \oplus V_1 = \mathbb{R}^2 \oplus \mathbb{R}^2$ associated with the Lawson minimal cubic cone $2x_1 x_2 x_3 + (x_1^2 - x_2^2) x_4 = 0$.

The second class includes all REC algebras which are nonsimilar to any polar algebra; these will be called *exceptional* REC algebras. A typical example is the

Freudenthal-Springer algebra based on the Cartan isoparametric cubic (3.1.1). The latter is also equivalently defined as the trace free subspace V of the simple Jordan algebra $J = \mathrm{Herm}_3(\mathbb{F}_d)$ (for any $d = 1, 2, 4, 8$) turned into a metrized algebra by virtue of the generic norm on J. The Freudenthal multiplication on V is essentially the orthogonal projection of the corresponding product in J. In Subsection 6.4.5 we provide some further examples of exceptional REC algebras by subtraction of a subalgebra from a simple formally real rank 3 Jordan algebra.

The question naturally arises then of classifying in an intrinsic way the above two classes of REC algebras. As the first step of the classification we establish in Section 6.6 by virtue of standard methods of nonassociative algebra theory (in particular, the Peirce decomposition) that any REC algebra is harmonic, i.e., satisfies the zero trace condition $\mathrm{tr}\, L_x = 0$ for any x. (Translated into PDE language, this property simply states that $u(x)$ is a harmonic function.) The zero trace condition yields that any REC algebra satisfies a nice polynomial identity

$$x^2 x^2 + 4xx^3 - 4|x|^2 x^2 = -\tfrac{4\lambda}{3}\langle x^2, x\rangle x, \quad x \in V.$$

This in particular shows that REC algebras can be thought of as a natural generalization of similar algebraic structures possessing cubic and quartic identities considered in a pure algebraic context in [**159**], [**212**], [**266**], [**81**]. On the other hand, despite a certain similarity between the case of REC algebras and that of Jordan-like structures, the former case is somewhat more complicated as is shown, for instance, by the Peirce decomposition of a REC algebra V associated to a nonzero idempotent c,

$$V = V_c(1) \oplus V_c(-1) \oplus V_c(-\tfrac{1}{2}) \oplus V_c(\tfrac{1}{2}), \qquad L_c = t_i \text{ on } V_c(t_i),$$

consisting generically of four summands (recall that in the Jordan algebra case there are three summands corresponding to the eigenvalues 0, $\tfrac{1}{2}$, and 1). One can show that $\dim V_c(1) = 1$ and by virtue of the harmonicity of V the dimensions $n_1(c) = \dim V_c(-1)$, $n_2(c) = \dim V_c(-\tfrac{1}{2})$, and $n_3(c) = \dim V_c(\tfrac{1}{2})$ are subject to the following condition: $n_3(c) = 2n_1(c) + n_2(c) - 2$. Thus defined, the numbers $n_i(c)$ a priori depend on a particular choice of Peirce decomposition, but we are able to show that the choice of an idempotent c does not affect the dimensions of the invariant subspaces. Moreover, it can be shown that (n_1, n_2) is a similarity invariant of V; we call these the characteristic dimensions of V. In fact, besides the trivial constraint $\dim V = 3n_1 + 2n_2 - 1$, one can show that the characteristic dimensions satisfy a more delicate condition, namely,

(6.1.3) $$n_1 - 1 \leq \rho(n_2 + n_1 - 1),$$

where ρ is the Hurwitz-Radon function (6.5.18), which considerably reduces the number of admissible pairs (n_1, n_2). The latter inequality is a corollary of the existence a hidden Clifford algebra structure inside any REC algebra, which is established in Section 6.8.

The second part of the classification is to establish the existence of an invariant Jordan algebra structure hidden inside any REC algebra. More precisely, in Section 6.9 we show that the invariant subspace $V_c(1) \oplus V_c(-\tfrac{1}{2})$ can naturally be turned into a formally real cubic Jordan algebra, denoted by $V_c^\bullet$. Then the main result of the chapter, Theorem 6.10.1, states that $V_c^\bullet$ completely determines the class of the corresponding REC algebra: V is a polar (resp. exceptional) REC algebra if and only if $V_c^\bullet$ is a reducible (resp. simple) Jordan algebra. Furthermore, the quadratic trace form $x \to \operatorname{tr} L_x^2$ carries essential information: Theorem 6.11.1 shows that in the case of an exceptional REC algebra one has $\operatorname{tr} L_x^2 = \sigma |x|^2$ for some constant σ, while in the case of polar algebras the quadratic trace form $\operatorname{tr} L_x^2$ has generically two different eigenvalues, with the only exception for four (polar) REC algebras having the property $n_2 = 2$. The latter tetrad is represented exactly by the quotients of $\operatorname{Herm}_3(\mathbb{F}_d)$ over its centers, of dimensions $3, 6, 12$, and 24, respectively, and occupies an intermediate place in the classification by sharing certain characteristic properties of both polar and exceptional REC algebras.

Since the rank of $V_c^\bullet$ is at most 3, the Jordan-von Neumann-Wigner classification implies the only possible dimensions of $V_c^\bullet$, and thus those of $V_c(-\frac{1}{2})$: $n_2 = 3d + 2$, $d \in \{1, 2, 4, 8\}$, i.e., exactly the dimensions of Cartan isoparametric cubics! Combining these observations with (6.1.3) readily yields by virtue of the logarithmic character of ρ the finiteness of possible characteristic dimensions (n_1, n_2) for exceptional REC algebras. In fact, in contrast to the case of polar algebras, only finitely many (similarity classes of) exceptional REC algebras do exist and one can prove that the only possible characteristic dimensions of exceptional REC algebras are those displayed in Table 6.1.1.

TABLE 6.1.1. The possible characteristic dimensions (n_1, n_2) of exceptional algebras.

Ex. #	1	2	3	4	5	6	7	8	9	10	11	12	13	14	15	16	17	18	19	20	21	22	23	24
dim V	2	5	8	14	26	9	12	15	21	15	18	21	24	30	42	27	30	33	36	51	54	57	60	72
n_1	1	2	3	5	9	0	1	2	4	0	1	2	3	5	9	0	1	2	3	0	1	2	3	7
n_2	0	0	0	0	0	5	5	5	5	8	8	8	8	8	8	14	14	14	14	26	26	26	26	26

Some further comments are in order. As a byproduct, we obtain the cubic trace identity

$$\operatorname{tr} L_x^3 = (1 - n_1)\langle x, x^2 \rangle$$

established for any REC algebra in Section 6.11. This identity is crucial in the proof of the independence of the characteristic dimensions (n_1, n_2) on a particular choice of an idempotent c. Both the cubic trace identity and the harmonicity of V, which reads $\operatorname{tr} L_x = 0$, can be thought of as some form of "conservation laws on REC algebras". In particular, translated into analytic language, it reveals that any cubic polynomial solution of equation (6.1.1) is *a priori* a harmonic solution of the Hessian type equation

$$\operatorname{tr}(D^2 u(x))^3 = Cu(x), \qquad C \in \mathbb{R}.$$

It is, however, unclear if it is possible to derive the latter equation or the harmonicity of u directly from (6.1.1).

Another interesting feature of general REC algebras covered here only superficially is the existence of a special orthogonal decomposition:

$$V = J \oplus J' \oplus \mathcal{C}_0 \oplus \mathcal{C}_1 \oplus \mathcal{C}_2.$$

Here J and J' are two copies of the Jordan algebra $V_c^\bullet$, and the three subspaces $\mathcal{C}_i$, $i = 0, 1, 2$, possess the Clifford type identity $x_i^2 = -\frac{4}{3}|x_i|^2 c_i$, where three distinct idempotents c_i form a triality system, i.e., $c_1 + c_2 + c_3 = 0$. This above decomposition is somewhat reminiscent of the Kantor-Koecher-Tits construction and triality descriptions of exceptional Lie algebras [247] (see also [158], [19]).

Finally, we mention the papers [**159**], [**212**], and [**266**], where some related algebraic structures, namely general (pseudo-composition) commutative algebras satisfying $x^3 = a(x)x^2 + b(x)x$, were considered.

6.2. Algebraic minimal hypercones in $\mathbb{R}^n$ and radial eigencubics

6.2.1. Minimal hypercones.
Here we explain how the radial eigencubic equation (6.1.1) enters the discussion. Recall that a hypersurface $\mathcal{M}$ of a Euclidean space is called minimal if its mean curvature vanishes in all regular points of $\mathcal{M}$. If $\mathcal{M}$ is a graph of a function $u(x)$, $x \in \mathbb{R}^{n-1}$, then the minimality condition is equivalent to that $u(x)$ satisfying the minimal surface equation

$$(6.2.1) \qquad \sum_{i,j=1}^{n-1} ((1 + |Du|^2)\delta_{ij} - u_{x_i}u_{x_j})u_{x_i x_j} = 0.$$

The famous result of S. N. Bernstein [**25**] states that any entire solution of (6.2.1) is an affine function for $n = 3$. The Bernstein property holds true in all dimensions $n \leq 8$, [**89**], [**73**], [**11**], [**228**], but for any $n \geq 9$ there exists a nonaffine entire solution of (6.2.1) [**28**]. Although many nonaffine examples of entire solutions of the minimal surface equation were shown to exist for $n \geq 9$ (see, for instance, [**226**], [**227**]), no explicit examples are known. Even a simpler question, whether or not there exists a solution of (6.2.1) which is actually a polynomial in x_i, is still unanswered [**225**], [**164**].

An important class of minimal surfaces which naturally appears in the context of the Bernstein property is that of minimal hypercones (and stable minimal hypercones especially). It is well known that the cone over an immersed minimal submanifold E of the standard sphere $\mathbb{S}^{n-1} \subset \mathbb{R}^n$ is itself a (singular) minimal hypersurface in $\mathbb{R}^n$ [**61**]. For instance, if E is the equator of S^{n-1}, then the corresponding cone is a hyperplane; such a minimal cone is said to be trivial. The Fleming-De Giorgi approach [**89**], [**73**] to establishing the Bernstein property was based on a crucial observation that a nonaffine entire solution of (6.2.1) produces a nontrivial minimizing[1] hypercone in $\mathbb{R}^n$ by means of (a subsequence of) contractions $u_\varepsilon(x) = \varepsilon^{-1} u(\varepsilon x)$. For example, the minimal cone

$$x_1^2 + \cdots + x_m^2 - x_{m+1}^2 - \cdots - x_{2m}^2 = 0$$

is minimizing if and only if $m \geq 4$ [**228**], [**28**]. The particular case $m = 4$ was exploited in [**28**] for constructing the first counterexample to the Bernstein property in $\mathbb{R}^9$.

So far all the available examples of minimal hypecones are algebraic, i.e., homogeneous polynomial varieties. W. Y. Hsiang [**110**] also showed that any homogeneous minimal submanifold of the Euclidean sphere is always algebraic and proposed the general problem of a classification of irreducible minimal hypercones of arbitrary higher degree. It is well known (see, for example, [**110**]) that any algebraic minimal hypercone with vertex at the origin can be equivalently defined as the zero-locus of a homogeneous polynomial $u(x) : \mathbb{R}^n \to \mathbb{R}$ satisfying the similarity

$$(6.2.2) \qquad \Delta_1 u = |Du|^2 \Delta u - \tfrac{1}{2}\langle Du, D|Du|^2\rangle \equiv 0 \bmod u,$$

---

[1] A minimal cone is called minimizing if the second variation of the (reduced) volume functional is nonnegative.

where Δ_1 denotes the 1-Laplace operator and (6.2.2) is understood in the usual sense that $\Delta_1 u$ is divisible[2] by u in the polynomial ring $\mathbb{R}[x_1, \ldots, x_n]$. In fact, (6.2.2) is equivalent to the vanishing of the mean curvature of the cone $u^{-1}(0)$ at all its regular points.

REMARK 6.2.1. A related problem on the existence of a (nonidentically zero) homogeneous polynomial solution to the general p-Laplace equation

$$\Delta_p u = |Du|^2 \Delta u - \tfrac{p-1}{2} Du \cdot D|Du|^2 = 0 \quad \text{in } \mathbb{R}^n, \quad p > 1,$$

was considered and studied in the early 1980s by J. L. Lewis [147] (observe that in that case a solution is understood in the usual sense with the right-hand side being identically zero). Lewis also proved that there are no polynomial solutions for $n = 2$, but the general case of $n \geq 3$ is still unsettled.

A homogeneous polynomial solution of (6.2.2) is said to be an *eigenfunction* of Δ_1, and it is called a *regular* eigenfunction if its gradient is nonzero outside the origin. The following lemma shows that the class of eigenfunctions is closed under multiplications by homogeneous polynomials, which makes it natural to study irreducible eigenfunctions only.

LEMMA 6.2.1. *If u is an eigenfunction of Δ_1 and v is a homogeneous polynomial, then uv is also an eigenfunction.*

PROOF. Indeed, let $\equiv$ denote the similarity modulus u. Then one easily verifies that $|D(uv)|^2 \equiv v^2|Du|^2$ and $\Delta uv \equiv v\Delta u + 2\langle Du, Dv \rangle$, and furthermore,

$$\langle D(uv), D|D(uv)|^2 \rangle \equiv v^3 \langle Du, D|Du|^2 \rangle + 4v^2 |Du|^2 \langle Du, Dv \rangle.$$

Combining the relations obtained yields $\Delta_1 uv \equiv v^3 \Delta_1 u \equiv 0$, hence implying the desired conclusion. □

In what follows, we consider entirely the case of eigenfunctions of Δ_1 of degree 3. In fact, as was already mentioned in the introduction, we shall consider a particular case when a cubic minimal hypercone is defined by a cubic polynomial solution u of (6.1.1), in which case u is said to be a *radial eigencubic*.

For geometric reasons, we make no distinction between two general cubic homogeneous polynomials u_1 and u_2 whose cones $u_1^{-1}(0)$ and $u_2^{-1}(0)$ (obviously nonempty) are congruent as subsets of $\mathbb{R}^n$, i.e., agree upon an isometry. Such polynomials will also be called *similar*. It readily follows by the real Nullstellensatz [162] (see also [251, Proposition 2.4]) that two irreducible homogeneous cubic polynomials u_1 and u_2 are similar in the above sense if and only if they agree up to the action of the group of $\mathbb{R}^n$ consisting of similarities, i.e., compositions of isometries and dilations: there exists an orthogonal endomorphism $F \in O(\mathbb{R}^n)$ and a constant $c \in \mathbb{R}$, $c \neq 0$, such that $u_1(x) = cu_2(Fx)$.

6.2.2. Examples of radial eigenfunctions. Any linear form on $\mathbb{R}^n$ is a trivial example of an eigenfunction (the associated minimal hypercone is a hyperplane). Hsiang [110] proved that the quadratic forms

$$u(x) = (q-1)(x_1^2 + \cdots + x_p^2) - (p-1)(x_{p+1}^2 + \cdots + x_{p+q}^2), \qquad x \in \mathbb{R}^{p+q},\ p, q \geq 2,$$

---

[2] It is easy to see that $\Delta_1 u$ is a homogeneous polynomial of degree $3 \deg u - 4$.

are the only (upon isometries and dilations of $\mathbb{R}^n$) eigenfunctions of degree 2. Indeed, it is straightforward to see that the quadric $u(x)$ above satisfies

$$\Delta_1 u = -8(p-1)(q-1)u.$$

Examples of *regular* eigenfunctions are very rare, and essentially the only known examples are those associated with the cones over isoparametric minimal hypersurfaces of the Euclidean sphere [230], [231] and the minimal hypercones over homogeneous minimal submanifolds in the Euclidean spheres, in particular [109], [111], [90], [117]. The situation in the lower-dimensional case of $n \leq 4$ is somewhat better understood. An infinite family of (nonhomogeneous) immersed algebraic minimal submanifolds of the sphere $\mathbb{S}^3 \subset \mathbb{R}^4$ of arbitrarily high degree was constructed and studied by H. B. Lawson in [142]. The defining polynomial of a Lawson cone of degree $p+q \geq 2$ is given explicitly by

$$(6.2.3) \qquad u(x) = \mathrm{Re}(x_1 + x_2\sqrt{-1})^p(x_3 + x_4\sqrt{-1})^q$$

for any unordered pair of coprime positive integers p, q. Despite the fact that the defining polynomials are not regular eigencubics for $p+q \geq 3$, it can be verified that the associated cones are regular immersions of a torus or the Klein bottle. It follows from a recent result of O. Perdomo [198] that any regular degree 3 cubic minimal cone in $\mathbb{R}^4$ is that over Lawson's minimal tori (6.2.3) for $(p,q) = (1,2)$ (see also Example 6.5.1 below). It is also worthy to mention that according to a recent result of S. Brendle [31] any embedded minimal torus in the unit sphere S^3 is congruent to the Clifford torus.

Lawson also proposed a higher-dimensional generalization of the above family on $\mathbb{R}^{2m} = \mathbb{C}^m$ with the defining polynomial given explicitly as

$$u(x) = \mathrm{Re}\prod_{i=1}^{m}(x_k + x_{i+m}\sqrt{-1})^{p_i}, \quad \gcd(p_1,\ldots,p_m) = 1,$$

although in the latter case the cones associated with the latter eigenfunctions are not always immersions of nonsingular varieties. Some further examples of minimal hypercones of arbitrary higher degree were recently provided in [251] where, for instance, it was shown that the determinant form on the vector space of square matrices of size n is an eigenfunction of degree n in $\mathbb{R}^{n^2}$.

6.3. Metrized and Freudenthal-Springer algebras

6.3.1. Metrized algebras. Let us consider a quadratic vector space V over $\mathbb{F} = \mathbb{R}$ or $\mathbb{C}$ endowed with a nonsingular symmetric bilinear form $Q(x,y) : V \times V \to \mathbb{F}$ (see also Subsection 3.1.2). An algebra on $V = (V, Q)$ is called *metrized* (cf. [29], [128, p. 453]) if the bilinear form Q is *associative*, i.e.,

$$Q(xy, z) = Q(x, yz), \qquad \forall x, y, z \in V.$$

Abusing terminology slightly, we say that V is a positive definite metrized algebra if the bilinear form $Q(x,y)$ is positive definite. For geometrical reasons it is convenient to think of the bilinear form Q as an inner product on V.

An example of a metrized algebra is a finite-dimensional full matrix algebra with the trace form or any finite-dimensional real or complex semisimple Lie algebra with its Killing form. In the present context another relevant example of a metrized algebra is obtained on an arbitrary quadratic vector space with a distinguished cubic form on it. We consider this case more fully below.

6.3.2. Freudenthal-Springer algebras. Let $u: V \to \mathbb{F}$ be a cubic form on a quadratic vector space (V, Q). Using the notation of Subsection 3.2.7, consider the full polarization

(6.3.1) $\quad u(x; y; z) = \frac{1}{6}(u(x+y+z) - u(x+z) - u(y+z) - u(x+z) + u(x) + u(y) + u(z))$

which is a symmetric trilinear form. Then one can define a composition law $(x, y) \to xy$ on V by virtue of the dual relation

(6.3.2) $\qquad\qquad\qquad 6u(x; y; z) = Q(xy, z).$

This product is obviously commutative: $xy = yx$, and by virtue of the symmetry of the trilinear form (6.3.1) the corresponding algebra on V is metrized: $Q(xy, z) = Q(x, yz)$.

DEFINITION 6.3.1. The composition law defined by (6.3.2) is called Freudenthal multiplication and the corresponding metrized algebra, denoted by (V, Q, u), or just $V(u)$, is said to be Freudenthal-Springer algebra of the cubic form u. The cubic form $u(x) = \frac{1}{6}\langle x, xx \rangle$ is called the norm of $V(u)$.

Notice that though for a generic cubic form u the algebra $V(u)$ is neither associative no power associative, the small powers $x^2 = xx$ and $x^3 = xx^2$ are well-defined (while $x^2x^2 \ne xx^3$ in general).

REMARK 6.3.1. The above definition is essentially that considered earlier by H. Freudenthal [**91**] and T. A. Springer [**232**] in the context of cubic Jordan algebras (and the exceptional Albert algebras, in particular). More explicitly, applying the above definitions to a formally real cubic Jordan algebra V with a cubic Jordan form $N: V \to \mathbb{F}$ and the trace bilinear form $T(x; y)$ (see Subsection 3.2.7), one readily finds that the above product xy on $V(N)$ is essentially the Freudenthal cross product #. Indeed, it follows from (3.2.24) and (3.2.18) that $N(x; y; z) = \frac{1}{2}T((x+y)^\# - x^\# - y^\#; z) = \frac{1}{2}T(x\#y; z)$, which yields $xy = 3x\#y$, and therefore $xx = x^2 = 6x^\#$. The latter yields by the sharp identity (3.2.25) that $x^2x^2 = 216N(x)x$. The latter relation is exactly (up to a constant factor) the basic equation (1) in [**232**].

6.3.3. Similar metrized algebras. Our further aim is to associate a metrized algebra to a radial eigencubic; therefore it is natural not to distinguish similar eigencubics, i.e., those which agree up to the action of the group of similarities consisting of isometries and dilatations of the ambient space. This motivates the following definition.

DEFINITION 6.3.2. Two metrized algebras V and V' are called similar if there exists a nonzero scalar $k \in \mathbb{F}^\times$ and an isometry $F: V \to V'$ such that

(6.3.3) $\qquad\qquad k\, F(xy) = F(x)F(y), \qquad \forall x, y \in V.$

It is straightforward to see that two metrized algebras V and V' are similar if and only if the corresponding cubic forms $u(x) = Q(x, x^2)$ and $u'(x') = Q'(x'^2, x')$ agree up to an isometry and dilatation.

6.3.4. Similarity invariants of metrized algebras. Showing the similarity of two metrized algebras is typically a difficult task and would simplify greatly if one could construct an invariant with respect to the similarity group. A function defined on a subclass of metrized algebras is said to be a *similarity invariant* if it takes a constant value on each algebra in the subclass and this value is equal for similar algebras.

The following proposition is useful for the construction of similarity invariants.

PROPOSITION 6.3.1. *Let V and V' satisfy Definition 6.3.2. Then*

$$\operatorname{tr} L_z^i = k^i \operatorname{tr} L_x^i,$$

$$Q'(L_z^i z, L_z^j z) = k^{i+j} Q(L_x^i x, L_x^j x), \qquad i, j \geq 0,$$

where $z = F(x)$.

PROOF. We have from (6.3.3) that $L_{F(x)} F(y) = k\, F(L_x y)$ for any $x, y \in V$. Hence $L_z^i z = k^i L_x^i x$, which yields the first identity in the proposition. Furthermore, if $\{e_i\}$ is an orthonormal basis of V, then the vectors $e'_i = F(e_i)$ form an orthonormal basis of V'. This yields $\operatorname{tr} L_z^i = k^i \operatorname{tr} L_x^i$, as required. □

6.3.5. Metrized subalgebras. Given a vector subspace $V' \subset V$ of an algebra V it of course need not be a subalgebra with respect to the algebra multiplication on V'. On the other hand, the situation with metrized algebra is somewhat different in the sense that one can turn any vector subspace of a metrized algebra into a new metrized algebra simply by projecting the ambient multiplication onto the subspace.

More precisely, if (V, Q) is a metrized algebra and $V' \subset V$ is a vector subspace, we define the new multiplication $\diamond$ on V' by virtue of

$$x \diamond y = \pi_{V'}(xy).$$

Then for any $x, y, z \in V'$: $Q(x \diamond y, z) = Q(xy, z) = Q(x, yz) = Q(x, y \diamond z)$; hence the new algebra is also metrized with respect to the restriction of Q on V'.

DEFINITION 6.3.3. A vector subspace $V' \subset V$ of a metrized algebra V endowed with the multiplication $\diamond$ is said to be a *metrized subalgebra* of V if $Q(x; x^{\diamond 2}) = Q(x; x^2)$ for all $x \in V'$.

Observe that it readily follows from the definition that a usual subalgebra of a metrized algebra is also a metrized subalgebra.

REMARK 6.3.2. The above definition of a metrized subalgebra is very close to that of the covariant derivative on immersed Riemannian submanifolds (see for instance [**129**]) which is defined as the orthogonal projection of the ambient covariant derivative onto the submanifold's tangent space.

6.4. Radial eigencubic algebras

With the above definitions at hand, we introduce below the principal object of our study, cubic REC algebras. We restrict ourselves here to examining some important examples of REC algebras coming from the Cartan isoparametric cubic and formally real Jordan algebras, and then we proceed with further examples in the next section.

6.4.1. Radial eigencubics and metrized algebras.

Let $V = \mathbb{R}^n$ be endowed with the standard inner product $Q(x;y) = \langle x,y\rangle$. Then a cubic homogeneous polynomial solution $u: V \to \mathbb{R}$ of (6.1.1) gives rise to the Freudenthal multiplication, which will be denoted by xy. We have by (6.3.2)

$$u(x) = \tfrac{1}{6}\langle x, x^2\rangle, \tag{6.4.1}$$

and the associativity of $\langle\,,\,\rangle$ reads as follows:

$$6u(x;y;z) = \langle xy, z\rangle = \langle x, yz\rangle. \tag{6.4.2}$$

In order to rewrite (6.1.1) in the new notation we notice that by (6.3.2)

$$\partial_y u|_x = 3u(x;y) = 3u(x;x;y) = \tfrac{1}{2}\langle x^2, y\rangle,$$

which leads to the explicit expression for the gradient of u

$$Du(x) = \tfrac{1}{2} x^2. \tag{6.4.3}$$

Moreover, spelling out the definition of the Hessian operator, one obtains by virtue of (6.4.2) that $\langle y, D^2 u(x) z\rangle = 6u(x;y;z) = \langle L_x y, z\rangle$, where $L_x y = xy : V \to V$ denotes the multiplication operator by x. This shows that

$$D^2 u(x) = L_x. \tag{6.4.4}$$

An important corollary of the characteristic property of a metrized algebra (6.4.2) is that the multiplication operator L_x is selfadjoint with respect to the inner product $\langle\,,\,\rangle$.

PROPOSITION 6.4.1. *A cubic form $u: V = \mathbb{R}^n \to \mathbb{R}$ satisfies (6.1.1) if and only if the Freudenthal-Springer algebra $V(u)$ satisfies the defining identity*

$$\langle b,x\rangle \langle x^2, x^2\rangle - \langle x^2, x^3\rangle = \tfrac{2}{3}\lambda\langle x,x\rangle \langle x^2, x\rangle, \tag{6.4.5}$$

where the vector $b \in V$ is implicit by virtue of $\operatorname{tr} L_x = \langle b, x\rangle$, *or equivalently*

$$b = b(V) := \sum_{i=1}^{n} e_i^2, \tag{6.4.6}$$

where $e_1, \ldots, e_n$ is an arbitrary orthonormal basis of $\mathbb{R}^n$.

PROOF. Let u satisfy (6.1.1). Using (6.4.3) and (6.4.4), one obtains

$$\langle Du, D^2 u|_x Du\rangle = \tfrac{1}{4}\langle L_x x^2, x^2\rangle = \tfrac{1}{4}\langle x^3, x^2\rangle.$$

One also has for the Laplacian,

$$\Delta u(x) = \operatorname{tr} D^2 u|_x = \operatorname{tr} L_x = \sum_{i=1}^{n} \langle L_x e_i, e_i\rangle = \sum_{i=1}^{n} \langle e_i^2, x\rangle = \langle b, x\rangle, \tag{6.4.7}$$

where b is defined by (6.4.6). Using (6.4.3) and inserting the resulting relations into (6.1.1) yields by virtue of (6.4.1) the defining identity (6.4.5). In the converse direction, spelling out (6.4.5) by virtue of the Euler homogeneous function theorem and (6.4.2) yields (6.1.1). □

6.4.2. The main definition. Equation (6.4.5) is a starting point for further analysis. It is convenient to single out this as a defining relation in an abstract metrized algebra, which motivates the following definition.

DEFINITION 6.4.1. A (nonassociative) commutative algebra[3] V with an associative positive definite symmetric bilinear form $\langle\,,\,\rangle$ is called a *radial eigencubic algebra* (or just a *REC algebra*) if the identity

(6.4.8) $$\langle x^2, x^2\rangle \operatorname{tr} L_x - \langle x^2, x^3\rangle = \tfrac{2}{3}\lambda\langle x, x\rangle\langle x^2, x\rangle$$

holds for some real scalar $\lambda = \lambda(V) \in \mathbb{R}$, referred to as the *structural constant* of the REC algebra V.

A REC algebra is said to be *harmonic* if $\operatorname{tr} L_x = 0$.

It follows from Proposition 6.4.1 that any REC algebra gives rise to a radial eigencubic $u(x) = \langle x, x^2\rangle$, and vice versa.

Observe that a metrized subalgebra of a REC algebra is not necessarily a REC algebra itself, but a weaker property holds.

PROPOSITION 6.4.2. *Any algebra similar to a REC algebra is also a REC algebra. More precisely, if V and V' satisfy Definition 6.3.2 and V satisfies (6.4.8), then V' is a REC algebra with $\lambda(V') = k^2\lambda(V)$.*

PROOF. It follows from (6.3.3) that $L_{F(x)}F(y) = k\,F \circ L_x y$ for any $x, y \in V$. This yields in particular that $F(x)^{\diamond 2} = kF(x^2)$ and $F(x)^{\diamond 3} = k^2 F(x^3)$. Furthermore, since an isometry takes an orthonormal basis to another one, one also has $\operatorname{tr} L_{F(x)} = k \operatorname{tr} L_x$. Thus, (6.4.8) yields

$$Q'(z^{\diamond 2}, z^{\diamond 2})\operatorname{tr} L_z - Q'(z^{\diamond 2}, z^{\diamond 2}) = \tfrac{2}{3}\lambda k^2 Q'(z,z)Q'(z^{\diamond 2}, z)$$

where $z = F(x)$, which yields the desired conclusion. □

6.4.3. The defining identities in REC algebras. We denote by $\{A, B\} = AB + BA$ the anticommutator of the operators A and B and also denote by $\{x \otimes y\}$ the symmetrized tensor product associated to vectors $x, y \in V$ acting by the rule $\{x \otimes y\} = x\langle y, z\rangle + y\langle x, z\rangle$ for any $z \in V$. Then a further polarization of the scalar structure equation (6.4.5) yields a polynomial and operator identities which determine the algebraic structure on a REC algebra V more fully.

PROPOSITION 6.4.3. *A commutative positive definite metrized algebra $(V, \langle\,,\,\rangle)$ is a REC algebra if and only if any of the following equivalent identities is valid:*

(6.4.9) $$4xx^3 + x^2 x^2 - 4\langle b, x\rangle x^3 + 2\lambda |x|^2 x^2 + \tfrac{4}{3}\lambda\langle x^2, x\rangle x = \langle x^2, x^2\rangle b,$$

and

(6.4.10) $$\begin{aligned} L_{x^3} + 2L_x^3 + \{L_x, L_{x^2}\} + \lambda\{x^2 \otimes x\} + \lambda|x|^2 L_x + \tfrac{1}{3}\lambda\langle x^2, x\rangle \\ = \{x^3 \otimes b\} + \langle x, b\rangle(L_{x^2} + 2L_x^2), \end{aligned}$$

where the vector $b \in V$ is implicitly defined by $\langle b, x\rangle = \operatorname{tr} L_x$.

---

[3]To avoid the trivial case corresponding to $u \equiv 0$ we shall always tacitly assume that there exists a nonzero element $x \in V$ such that $\langle x^2, x\rangle \neq 0$.

PROOF. Let V be a REC algebra. The associativity of the inner product gives for any $y \in V$, $\partial_y \langle x^2, x^2 \rangle = 4 \langle xy, x^2 \rangle = 4 \langle y, x^3 \rangle$ and similarly $\partial_y \langle x^2, x^3 \rangle = \langle y, 4xx^3 + x^2x^2 \rangle$; hence differentiating (6.4.8) with respect to y yields

$$\langle x^2, x^2 \rangle \langle b, y \rangle + 4 \langle y, x^3 \rangle \langle b, x \rangle - \langle y, 4xx^3 + x^2x^2 \rangle = \tfrac{4}{3}\lambda \langle x, y \rangle \langle x^2, x \rangle + 2\lambda |x|^2 \langle x^2, y \rangle,$$

which by virtue of the arbitrariness of y yields (6.4.9). Furthermore,

$$\partial_y xx^3 = yx^3 + 2x(x(xy)) + x(x^2 y) = (L_{x^3} + 2L_x^3 + L_x L_{x^2})y,$$

and similarly one finds $\partial_y x^2 x^2 = 4 L_{x^2} L_x y$, $\partial_y \langle x^2, x^2 \rangle b = (4b \otimes x^3)y$, and

$$\tfrac{1}{2} \partial_y \langle b, x \rangle x^3 = (x^3 \otimes b + \langle x, b \rangle L_{x^2} + 2 \langle x, b \rangle L_x^2)y,$$

$$\tfrac{1}{2} \partial_y |x|^2 x^2 = (2x^2 \otimes x + 2|x|^2 L_x)y,$$

$$\tfrac{1}{2} \partial_y \langle x^2, x \rangle x = 3 \langle x^2, y \rangle x + \langle x^2, x \rangle y = (3x \otimes x^2 + \langle x^2, x \rangle)y,$$

which yields by polarization of (6.4.9) the operator identity (6.4.10).

In the converse direction, any of the identities (6.4.9) and (6.4.10) yields by the associativity of the inner product the defining identity (6.4.8), thus showing that V satisfies the axioms of REC algebra. □

6.4.4. An example: The isoparametric Cartan cubics revisited.

Let $u(x)$ denote the Cartan cubic (3.1.1) and let $V(u)$ be the corresponding Freudenthal-Springer algebra $V = \mathbb{R}^n$. Using the Cartan theorem [55], u is equivalently defined as an irreducible cubic homogeneous solution of the differential system (3.1.2). Hence spelling out the first equation of (3.1.2) in the present algebraic notation yields by virtue of (6.4.3) that

(6.4.11) $$\langle x^2, x^2 \rangle = 36 |x|^4,$$

while the second equation of (3.1.2) yields by (6.4.7) that $\operatorname{tr} L_x = 0$ for any $x \in V$. Further polarization of (6.4.11) gives $\langle x^2, xy \rangle = 36 |x|^2 \langle x, y \rangle$ for any $y \in V$, implying the vector identity

(6.4.12) $$x^3 = 36 |x|^2 x.$$

Applying the inner product to the latter identity yields $\langle x^3, x^2 \rangle = 36 |x|^2 \langle x, x^2 \rangle$, thereby proving that $V(u)$ is a harmonic REC algebra with the structural constant $\lambda(V) = -54$.

The Freudenthal-Springer algebra $V(u)$ of the cubic form (3.1.1) is said to be the *special Cartan REC algebra*. Notice that by Proposition 6.4.2 any commutative positive definite metrized algebra which is harmonic ($\operatorname{tr} L_x = 0$) and satisfies

(6.4.13) $$\langle x^2, x^2 \rangle = C |x|^4$$

for some real $C > 0$ is a REC algebra, and it will be called a *Cartan REC algebra*.

A further polarization of (6.4.12) yields the operator equation

(6.4.14) $$L_{x^2} + 2 L_x^2 = 36 |x|^2 + 72 x \otimes x,$$

which yields by the harmonicity of V the following trace identity:

(6.4.15) $$\operatorname{tr} L_x^2 = 18(2 + \dim V) |x|^2.$$

We point out that by the Cartan theorem the dimension $\dim V$ of a Cartan REC algebra can take one of the four values: $n = 3d+2$, where $d = 1, 2, 4, 8$ (cf. with [**250**]).

REMARK 6.4.1. A useful corollary of (6.4.14) is the commutativity of L_{x^2} and L_x^2. Indeed, since $L_a(b \otimes c) = L_a b \otimes c$, it follows from (6.4.14) and (6.4.12) that

$$\tfrac{1}{72}[L_{x^2}, L_x^2] = [x \otimes x, L_x^2] = (x \otimes x^3 - x^3 \otimes x) = 36|x|^2(x \otimes x - x \otimes x) = 0.$$

This property makes the algebraic structure $V(u)$ "nearly Jordan" (recall that a stronger commutativity property $[L_{x^2}, L_x] = 0$ is equivalent to the definition of a Jordan algebra (3.2.4)). In fact, $V(u)$ can be realized as the trace free subspace of a simple Jordan algebra of rank 3 (see also Proposition 6.4.4 below).

6.4.5. Radial eigencubics via Jordan algebras. The above example admits a further generalization in the context of formally real Jordan algebras.

PROPOSITION 6.4.4. *Let $(X, N, \cdot)$ be a simple formally real cubic Jordan algebra with the generic norm N and let $Y \subset X$ be a Jordan subalgebra of X containing the unit. Then the Freudenthal-Springer algebra of the pullback of the normalized generic form $\tfrac{1}{6}N$ on $X \ominus Y = Y^\perp$ is a harmonic REC algebra with the structural constant $\lambda = -\tfrac{3}{4}$.*

PROOF. Let $x \bullet y$ and xy denote the Jordan product on X and the Freudenthal multiplication on $Y^\perp$, respectively. Then

(6.4.16) $$\langle xy, z \rangle = N(x; y; z) = \tfrac{1}{2}T(x\#y; z), \qquad \forall x, y, z \in Y^\perp,$$

where T is the bilinear trace form on X. This implies $\langle x^2, x \rangle = N(x)$ and also the following relation between the Freudenthal multiplications on $Y^\perp$ and X:

(6.4.17) $$xy = \tfrac{1}{2}(x\#y)^\perp, \qquad \forall x, y \in Y^\perp,$$

where $\perp$ and $\top$ denote the orthogonal projections onto $Y^\perp$ and Y, respectively. Applying (3.2.26) to the latter identity and using $\operatorname{Tr}(x) = \operatorname{Tr}(y) = 0$ one readily finds

(6.4.18) $$xy = (x \bullet y)^\perp, \qquad \forall x, y \in Y^\perp,$$

which shows that the Freudental multiplication on $Y^\perp$ is essentially the projection of the Jordan multiplication onto this subspace.

Next, let $\{e_i\}_{1 \le i \le m}$, $m = \dim X$, be an orthonormal basis of X chosen such that the first $k = \dim Y$ vectors form an orthogonal basis of Y, and moreover $e_1 = e$ is the unit of X. Then since Y is a Jordan subalgebra of X, we have $\sum_{i=1}^k e_i^{\bullet 2} \in Y$. Furthermore, since X is simple, we have $\sum_{i=1}^m e_i^{\bullet 2} = e$ (see [**85**, 6b, p. 59]); hence

(6.4.19) $$b = \sum_{j=k+1}^m e_j^2 = \left(\sum_{j=k+1}^m e_j^{\bullet 2}\right)^\perp = 0,$$

and hence V is harmonic.

Next, by (6.4.18) we have for any $x, y \in Y^\perp$: $\langle x, y^2 \rangle = \langle x, y^{\bullet 2} \rangle = \langle x \bullet y, y \rangle$. Choosing $y = x^2 = x^{\bullet 2} - (x^{\bullet 2})^\top$, we obtain

$$\langle x^2 x^2, x \rangle = \langle x^{\bullet 2} - (x^{\bullet 2})^\top, x \bullet (x^{\bullet 2} - (x^{\bullet 2})^\top) \rangle$$
$$= \langle x^{\bullet 2} - (x^{\bullet 2})^\top, x^{\bullet 3} - x \bullet (x^{\bullet 2})^\top \rangle.$$

Since Y is a Jordan subalgebra, we find $\langle (x^{\bullet 2})^\top, x\bullet(x^{\bullet 2})^\top\rangle = \langle (x^{\bullet 2})^\top \bullet (x^{\bullet 2})^\top, x\rangle = 0$, which yields

(6.4.20) $$\tfrac{1}{2}\langle x^2 x^2, x\rangle = \langle x^{\bullet 3}, x^{\bullet 2} - 2(x^{\bullet 2})^\top\rangle.$$

Next, notice that $\operatorname{Tr} x = 3\langle x, e\rangle = 0$, $\operatorname{Tr}(x^{\bullet 2}) = 3|x|^2$, and since X is a cubic Jordan algebra, we have by (3.2.27) and (3.2.30) that $S(x) = \tfrac{1}{2}(\operatorname{Tr}(x)^2 - \operatorname{Tr}(x^{\bullet 2})) = -\tfrac{3}{2}|x|^2$ and $x^{\bullet 3} = \tfrac{3}{2}|x|^2 x + N(x)e$. Inserting the latter identity into (6.4.20) yields by virtue of the orthogonality of x and Y that

$$\langle x^2 x^2, x\rangle = \tfrac{3}{2}|x|^2\langle x, x^{\bullet 2}\rangle - N(x)|x|^2.$$

On the other hand, by (6.4.16) and (6.4.18), $N(x) = \langle x, x^2\rangle = \langle x, x^{\bullet 2}\rangle$; hence $\langle x^2, x^3\rangle = \tfrac{1}{2}\langle x, x^2\rangle |x|^2$, which by (6.4.19) and (6.4.8) finishes the proof. □

We mention below some explicit examples of the above construction. Notice that the Jordan-von Neumann-Wigner classification yields that the only simple rank 3 Euclidean Jordan algebras are the Hermitian matrix algebras $\operatorname{Herm}_3(\mathbb{F}_d)$, $d = 1, 2, 4, 8$, endowed with the commutative Jordan multiplication $A \cdot B = \tfrac{1}{2}(AB+BA)$. Observe also that, for $1 \leq r \leq d$, there exists a natural embedding of $\operatorname{Herm}_3(\mathbb{F}_r)$ into $\operatorname{Herm}_3(\mathbb{F}_d)$ as a Jordan subalgebra. Furthermore, given a primitive Jordan frame $e_1 + e_2 + e_3 = e$ in $\operatorname{Herm}_3(\mathbb{F}_d)$ one also has three "diagonal" (commutative and associative) unital Jordan subalgebras obtained as the linear spans of e, $\{e, e_1\}$, and $\{e, e_1, e_2\}$; we denote these by $\mathcal{D}_1$, $\mathcal{D}_2$, and $\mathcal{D}_3$, respectively.

TABLE 6.4.1. The dimensions of the subtracted Jordan algebras.

	$\operatorname{Herm}_3(\mathbb{F}_1)$	$\operatorname{Herm}_3(\mathbb{F}_2)$	$\operatorname{Herm}_3(\mathbb{F}_4)$	$\operatorname{Herm}_3(\mathbb{F}_8)$
$\mathcal{D}_1$	$\mathbb{R}^5$	$\mathbb{R}^8$	$\mathbb{R}^{14}$	$\mathbb{R}^{26}$
$\mathcal{D}_2$	$\mathbb{R}^4$	$\mathbb{R}^7$	$\mathbb{R}^{13}$	$\mathbb{R}^{25}$
$\mathcal{D}_3$	$\mathbb{R}^3$	$\mathbb{R}^6$	$\mathbb{R}^{12}$	$\mathbb{R}^{24}$
$\operatorname{Herm}_3(\mathbb{F}_1)$	–	–	$\mathbb{R}^9$	$\mathbb{R}^{21}$
$\operatorname{Herm}_3(\mathbb{F}_2)$	–	–	–	$\mathbb{R}^{18}$
$\operatorname{Herm}_3(\mathbb{F}_4)$	–	–	–	–

In the described notation, the construction given in Proposition 6.4.4 yields the following examples of REC algebras (also summarized in Table 6.4.1):

(a) Four (exceptional)[4] Cartan REC algebras $\operatorname{Herm}_3(\mathbb{F}_d) \ominus \mathcal{D}_1$ in $\mathbb{R}^5$, $\mathbb{R}^8$, $\mathbb{R}^{14}$, and $\mathbb{R}^{26}$ (see Subsection 6.4.4).

(b) Four reducible REC algebras $\operatorname{Herm}_3(\mathbb{F}_d) \ominus \mathcal{D}_2$ in $\mathbb{R}^4$, $\mathbb{R}^7$, $\mathbb{R}^{13}$, and $\mathbb{R}^{25}$.

(c) Four reducible REC algebras $\operatorname{Herm}_3(\mathbb{F}_d) \ominus \mathcal{D}_3$ in $\mathbb{R}^3$, $\mathbb{R}^6$, $\mathbb{R}^{12}$, and $\mathbb{R}^{24}$, the so-called mutants[5].

---

[4] For the definition and discussion of reducible and exceptional REC algebras see Section 6.10.
[5] See Subsection 6.11.5.

(d) Three (exceptional) REC algebras in $\mathbb{R}^9$, $\mathbb{R}^{18}$, and $\mathbb{R}^{21}$ corresponding to REC algebras on $\mathrm{Herm}_3(\mathbb{H}) \ominus \mathrm{Herm}_3(\mathbb{C})$, $\mathrm{Herm}_3(\mathbb{O}) \ominus \mathrm{Herm}_3(\mathbb{H})$, and $\mathrm{Herm}_3(\mathbb{O}) \ominus \mathrm{Herm}_3(\mathbb{C})$, respectively (one can show that in the remaining cases, the induced subspace REC algebra is trivial). We also mention that the REC algebra $\mathrm{Herm}_3(\mathbb{H}) \ominus \mathrm{Herm}_3(\mathbb{C})$ corresponds exactly to an example of a nonhomogeneous minimal cone in $\mathbb{R}^9$ considered by Hsiang in [**110**].

6.5. Polar and Clifford REC algebras

Here we define and study another, infinite family of REC algebras which is closely related to symmetric representations of Clifford algebras. A typical example of such an algebra is the Freudenthal-Springer algebra associated with the Lawson cubic cone.

6.5.1. Polar REC algebras.

DEFINITION 6.5.1. A (nonassociative) commutative metrized algebra $(V, \langle\,,\,\rangle)$ is called *polar* if there exists a proper orthogonal decomposition $V = V_0 \oplus V_1$ such that the following conditions are satisfied:

(6.5.1) $$V_0 V_0 = \{0\},$$
(6.5.2) $$V_1 V_1 \subset V_0,$$
(6.5.3) $$L_x^2 = |x|^2 \pi_{V_1}, \qquad \forall x \in V_0,$$

where $\pi_X : V \to X$ is the orthogonal projection on X. We assume additionally that $\mathrm{tr}\, L_x = 0$ for any $x \in V_0$ whenever $\dim V_0 = 1$.

Given $x \in V$, the orthogonal decomposition $x = x_0 + x_1$, $x_i \in V_i$, is said to be the *polar decomposition* of x.

LEMMA 6.5.1. *Let V be a polar algebra. Then*

(6.5.4) $$V_0 V_1 \subset V_1,$$
(6.5.5) $$L_y^2|_{V_0} = |y|^2 \mathbf{1}_{V_0}, \qquad \forall y \in V_1,$$
(6.5.6) $$\{L_x, L_y\} = 2\langle x, y\rangle \pi_{V_1}, \qquad x, y \in V_0,$$
(6.5.7) $$\{L_x, L_y\}|_{V_0} = 2\langle x, y\rangle \mathbf{1}_{V_0}, \qquad x, y \in V_1.$$

PROOF. By (6.5.1) and the associativity of the inner product $0 = \langle xx', y\rangle = \langle x', yx\rangle$ for any $x, x' \in V_0$ and $y \in V_1$, which by the arbitrariness of x' proves that $yx \in V_1$; hence (6.5.4) follows. Furthermore, under the same assumptions, one has by (6.5.3)
$$|x|^2 |y|^2 = \langle L_x^2 y, y\rangle = \langle x(xy), y\rangle = \langle xy, xy\rangle = \langle L_y^2 x, x\rangle,$$
which implies (6.5.5) by virtue of the selfadjointness of L_y^2. Finally, polarizing (6.5.3) and (6.5.5) yields (6.5.6) and (6.5.7), respectively. $\square$

PROPOSITION 6.5.1. *Any polar algebra V is a harmonic REC algebra with the structural constant $\lambda(V) = -2$.*

PROOF. Let $x = x_0 + x_1$ be the polar decomposition. Then by (6.5.1),
(6.5.8) $$x^2 = x_1^2 + 2x_0 x_1,$$

where $x_1^2 \in V_0$ by (6.5.2). Then by virtue of the orthogonality of V_0 and V_1,

(6.5.9) $$\langle x, x^2 \rangle = \langle x_0, x_1^2 \rangle + 2\langle x_1, x_1 x_0 \rangle = 3\langle x_0, x_1^2 \rangle.$$

Similarly, $x^3 = x_1^2 x_0 + 2x_1(x_0 x_1) + 2x_0(x_0 x_1) + x_1^3$, where $x_1^2 x_0 = 0$ by virtue of (6.5.1); therefore applying (6.5.3) and (6.5.5), one finds

$$x^3 = 2|x_1|^2 x_0 + 2|x_0|^2 x_1 + x_1^3.$$

Arguing similarly and using (6.5.9), one obtains

(6.5.10)
$$\begin{aligned}\langle x^2, x^3 \rangle &= \langle x_1^2, 2|x_1|^2 x_0 \rangle + \langle 2x_0 x_1, 2|x_0|^2 x_1 + x_1^3 \rangle \\ &= 2|x_1|^2 \langle x_1^2, x_0 \rangle + 4|x_0|^2 \langle x_0, x_1^2 \rangle + 2\langle x_1 x_0, x_1{}^3 \rangle \\ &= 2|x_1|^2 \langle x_1^2, x_0 \rangle + 4|x_0|^2 \langle x_0, x_1^2 \rangle + 2|x_1|^2 \langle x_1 x_0, x_1 \rangle \\ &= \tfrac{4}{3}|x|^2 \langle x, x^2 \rangle.\end{aligned}$$

In order to verify (6.4.8), by Definition 6.5.1 it remains to show that $\operatorname{tr} L_x = 0$ for $\dim V_0 \geq 2$. We have $L_x = L_{x_0} + L_{x_1}$, where $\operatorname{tr} L_{x_1} = 0$ because L_{x_1} interchanges V_0 and V_1 by (6.5.4) and (6.5.2). Moreover, $L_{x_0}|_{V_1} = 0$ by (6.5.1); hence $\operatorname{tr} L_x = \operatorname{tr} L_{x_0}|_{V^-}$. Since $\dim V_0 \geq 2$, there exists $0 \neq x_0' \in V_0$ such that $\langle x_0, x_0' \rangle = 0$. Then (6.5.6) yields $\{L_{x_0}, L_{x_0'}\} = 0$; hence $L_{x_0'}^2 L_{x_0} = -L_{x_0'} L_{x_0} L_{x_0'}$. This yields by (6.5.3) that $|x_0'|^2 L_{x_0} = -L_{x_0'} L_{x_0} L_{x_0'}$; hence applying the trace to the latter identity and using the fact that the trace is invariant under cyclic permutations, we get

$$|x_0'|^2 \operatorname{tr} L_{x_0} = -\operatorname{tr} L_{x_0'} L_{x_0} L_{x_0'} = -\operatorname{tr} L_{x_0} L_{x_0'}^2 = -|x_0'|^2 \operatorname{tr} L_{x_0},$$

proving that $\operatorname{tr} L_x = \operatorname{tr} L_{x_0}|_{V_1} = 0$. Thus, it follows from (6.5.10) that V satisfies (6.4.5) with $\lambda = -2$, and the proposition follows. $\square$

LEMMA 6.5.2. *If V is a polar algebra, then*

(6.5.11) $$\operatorname{tr} L_x^2 = |x_0|^2 \dim V_1 + 2|x_1|^2 \dim V_0,$$
(6.5.12) $$\operatorname{tr} L_x^3 = (2 - \dim V_0)\langle x, x^2 \rangle.$$

Moreover, the dimension $\dim V_1$ is even.

PROOF. Let $x_i \in V_i$ and let f_{ij} stand for the restriction of L_{x_i} onto V_j, where $i, j = 0, 1$. Then by the axioms of polar algebra $f_{00} = 0$, $f_{01} : V_1 \to V_1$ with $f_{01}^2 = |x_0|^2 \mathbf{1}_{V_1}$, which yields $\operatorname{tr} L_{x_0}^2 = |x_0|^2 \dim V_1$. Similarly, $f_{10} : V_0 \to V_1$ and $f_{11} : V_1 \to V_0$, and by (6.5.5), $f_{11} f_{10} = |x_1|^2 \mathbf{1}_{V_0}$. This yields $\operatorname{tr} L_{x_1}^2 = \operatorname{tr} f_{10} f_{11} + \operatorname{tr} f_{11} f_{10} = 2|x_1|^2 \dim V_0$. Furthermore, $\operatorname{tr} L_{x_1} L_{x_0} = 0$ because $L_{x_1} L_{x_0}$ interchanges V_0 and V_1. Hence, writing the polar decomposition $x = x_0 + x_1$ for an arbitrary $x \in V$ we arrive at the desired identity:

$$\operatorname{tr} L_x^2 = \operatorname{tr} L_{x_0}^2 + \operatorname{tr} L_{x_1}^2 + 2 \operatorname{tr} L_{x_0} L_{x_1} = |x_0|^2 \dim V_1 + 2|x_1|^2 \dim V_0.$$

Next, by Proposition 6.5.1, $b(V) = 0$ and $\lambda(V) = 0$; hence applying the trace to (4.4.10) one finds

(6.5.13) $$\operatorname{tr} L_x^3 + \operatorname{tr} L_x L_{x^2} - \langle x^2, x \rangle(\tfrac{1}{3} \dim V + 2) = 0.$$

On the other hand, polarizing (6.5.11) we get

$$\operatorname{tr} L_x L_y = \langle x_0, y_0 \rangle \dim V_1 + 2\langle x_1, y_1 \rangle \dim V_0;$$

hence by virtue of $x^2 = x_1^2 + 2x_0 x_1$ and $\langle x, x^2 \rangle = 3\langle x_0, x_1^2 \rangle$ we have

$$\operatorname{tr} L_x L_{x^2} = \langle x_0, x_1^2 \rangle \dim V_1 + 4 \langle x_1, x_0 x_1 \rangle \dim V_0 = \tfrac{1}{3}(4 \dim V_0 + \dim V_1) \langle x, x^2 \rangle.$$

Combining the latter equation with (6.5.13) and observing that $\dim V = \dim V_0 + \dim V_1$ we get (6.5.12).

Finally, notice that it follows from the axioms of polar algebra that for any $x \in V_0$ the operator L_x is selfadjoint with the properties $V_0 = \ker L_x$, $L_x : V_1 \to V_1$, and $L_x^2 = |x|^2$ on V_1; thus the only nonzero eigenvalues of L_x are $\pm |x_0|$ of equal multiplicities (by virtue of $\operatorname{tr} L_x = 0$), implying that $\dim V_1$ is even. □

An important corollary of the cubic trace identity (6.5.12) is the invariance of the polar dimensions.

COROLLARY 6.5.1. *If V is polar algebra, then*

$$(6.5.14) \qquad \tau(x) := \frac{4|x|^2 \operatorname{tr} L_x^3}{3\langle x^2, x^3 \rangle} = 2 - \dim V_0 \qquad \forall x \in V.$$

In particular, if V' is a polar algebra similar to V, then $\dim V_0 = \dim V_0'$, $\dim V_1 = \dim V_1'$.

PROOF. The formula (6.5.14) follows immediately from (6.5.12) and Proposition 6.5.1. Suppose V' be a polar algebra similar to V and let $F: V \to V'$ be an isometry according to Definition 6.3.2. Then it readily follows by Proposition 6.3.1 that $\tau(F(x)) = \tau(x)$ for all $x \in V$, which yields $\dim V_0 = \dim V_0'$, and by virtue of $\dim V = \dim V'$ it also shows the remaining identity. □

Let us also give another important corollary of the definition of a polar algebra and Lemma 6.5.2. Recall that a homomorphism σ of a commutative algebra V over a field $\mathbb{F}$ is called involution if $\sigma^2 = \operatorname{id}_V$. We refer the reader to [**128**] for more detailed information on involutions. Then the following is easily verified.

LEMMA 6.5.3. *Let V be a polar algebra. The map $x \to \bar{x} = x_0 - x_1$, where $x = x_0 + x_1$ is the polar decomposition, is an involution of V.*

Let $\sigma(x) = \bar{x}$ be defined according to Lemma 6.5.3. Then the subspaces V_0 and V_1 can be thought of, respectively, as symmetric and skew-symmetric elements with respect to σ.

6.5.2. Polar algebras and symmetric Clifford systems. The existence of polar algebras can be seen as follows. Let X and Y be Euclidean vector spaces and let $\operatorname{End}_s(Y)$ denote the real vector space of symmetric endomorphisms of Y.

A triple (X, Y, A) is called a *symmetric Clifford system*, denoted for short by $A \in \operatorname{Cliff}(X, Y)$, if the linear map $A: X \to \operatorname{End}_s(Y)$ and

$$(6.5.15) \qquad A(x)^2 = |x|^2 \mathbf{1}_Y, \quad \forall x \in X.$$

If $\dim X = 1$, one requires additionally that $\operatorname{tr} A(x) = 0$.

REMARK 6.5.1. Observe that in the case $\dim X \geq 2$ the trace free condition follows from the definition. Indeed, the polarization of (6.5.15) yields

$$(6.5.16) \qquad \{A(x_1), A(x_2)\} = 2\langle x_1, x_2 \rangle \mathbf{1}_Y, \quad \forall x_1, x_2 \in X.$$

If $x_1 \in X$, then by virtue of $\dim X \geq 2$ there exists $x_2 \neq 0$ and $\langle x_1, x_2 \rangle = 0$ which yields by (6.5.16) and the invariance of the trace under the cyclic permutations

$$0 = \operatorname{tr}\{A(x_1), A(x_2)\}A(x_2) = 2\operatorname{tr} A(x_1)A(x_2)^2 = |x_2|^2 \operatorname{tr} A(x_1),$$

implying $\operatorname{tr} A(x_1) = 0$.

It follows from (6.5.15) that the only possible eigenvalues of $A(x)$ are $\pm|x|$, and the trace free condition implies that these have equal multiplicities. In particular, this shows that $\dim Y$ is even. A more delicate constraint on the dimensions of X and Y due to Hurwitz [112] and Radon [207] states that the existence of a symmetric Clifford system (X, Y, A) implies

(6.5.17) $$\dim X - 1 \leq \rho(\tfrac{1}{2} \dim Y),$$

where the Hurwitz-Radon function ρ is defined by

(6.5.18) $$\rho(m) = 8a + 2^b \quad \text{if } m = 2^{4a+b} \cdot \text{odd}, \ 0 \leq b \leq 3.$$

Now we are ready to establish a correspondence between polar algebras and symmetric Clifford systems. Then we establish in the next section that this correspondence is actually a bijection in a natural case.

PROPOSITION 6.5.2. *If (X, Y, A) is a symmetric Clifford system, then the Freudenthal algebra on $V = X \times Y$ generated by the cubic form $u(z) = \frac{1}{2}\langle y, A(x)y \rangle$, $z = (x, y) \in X \times Y$, is a polar REC algebra. In the converse direction, if $V = V_0 \oplus V_1$ is a polar algebra, then the triple (V_0, V_1, A) with $A(x) = L_x|_{V_0}$ is a symmetric Clifford system.*

PROOF. Setting $V_0 = X \times \{0\}$ and $V_1 = \{0\} \times Y$ and polarizing the expression for u, we find the explicit expression for the Freudenthal multiplication

(6.5.19) $$z_1 z_2 = (\hat{A}(y_1), \ A(x_1)y_2 + A(x_2)y_1)$$

where $z_1 = (x_1, y_1)$, $z_2 = (x_2, y_2)$, and the linear map $\hat{A}(y_1) : Y \to X$ is well-defined by the dual relation

(6.5.20) $$\langle \hat{A}(y_1)y_2, x_1 \rangle = \langle y_1, A(x_1)y_2 \rangle$$

for any $x_1 \in X$ and $y_2 \in Y$. Then (6.5.19) shows that $z_1 z_2 = 0$ if $z_1, z_2 \in V_0$; hence (6.5.1) holds. Moreover, for any $z_1, z_2 \in V_1$ one has $\pi_{V_1}(z_1 z_2) = 0$; therefore $z_1 z_2 \in V_0$, and hence (6.5.2) follows. Finally, (6.5.19) yields $z_1 z_2 = (0, A(x)y') \in V_1$ for any $z_1 = (x, 0) \in V_0$ and $z_2 = (0, y') \in V_1$; hence using (6.5.15)

$$L_{z_1}^2 z_2 = (0, A(x)^2 y') = |x|^2 (0, y') = |z_1|^2 z_2,$$

thus implying (6.5.3) and thereby proving that V is polar. The converse statement of the proposition follows by the definition of a polar algebra. □

EXAMPLE 6.5.1 (Lawson minimal cubic). In [142], Lawson constructed an infinite family of algebraic minimal cones in $V = \mathbb{R}^4$ which are either immersions of the torus or the Klein bottle. The cubic member of the Lawson family is given by

$$u(x) = x_1 x_3 x_4 + \frac{1}{2}(x_3^2 - x_4^2)x_2$$

and by a result of Perdomo [198], $u(z) = 0$ is the only minimal eigencubic in $\mathbb{R}^4$ with $|Du(x)| \neq 0$ for all $x \neq 0$. The Freudenthal-Springer algebra of u is a typical example of a polar algebra. Namely, identifying V with $\mathbb{C} \times \mathbb{C}$ in an obvious way, we have for the inner product $\langle x, y \rangle = \operatorname{Re}(z_1\bar{v}_1 + z_2\bar{v}_2)$, where $x = (z_1, z_2)$ and $y = (v_1, v_2)$, which easily yields $u(x) = \frac{1}{2}\operatorname{Re} z_1\bar{z}_2^2$ and the explicit form for the Freudenthal multiplication:

$$xy = (z_2 v_2, \bar{z}_2 v_1 + \bar{v}_2 z_1).$$

The latter readily yields that V is a polar algebra with respect to the polar decomposition $V_0 = \mathbb{C} \times \{0\}$ and $V_1 = \{0\} \times \mathbb{C}$. Furthermore, it is easily verified that $u = \langle y, A(x)y \rangle$, where $x = (x_1, x_2)$, $y = (x_3, x_4)$, and

$$A(x) = \begin{pmatrix} x_2 & x_1 \\ x_1 & -x_2 \end{pmatrix} \in \operatorname{Cliff}(\mathbb{R}^2, \mathbb{R}^2).$$

6.5.3. The classification of polar REC algebras. First we recall some facts and definitions following [**58**] (see also [**113**]). Given two symmetric Clifford systems (X, Y_1, A_1) and (X, Y_2, A_2) it is straightforward to see that $(X, Y_1 \oplus Y_2, A_1 \oplus A_2)$ is a symmetric Clifford system again. A symmetric Clifford system (X, Y, A) is called *reducible* if there is a nontrivial orthogonal decompositions $Y = Y_1 \oplus Y_2$ such that each Y_i is an invariant subspace of $A(x)$ for any $x \in X$; otherwise it is called *irreducible*.

Two symmetric Clifford systems (X, Y, A) and (X', Y', A') are called *geometrically equivalent* if there exist isometries $f : X \to X'$ and $g : Y \to Y'$ such that $A(x) = g^{-1} A'(f(x)) g$. Then the following properties are well known.

- Any symmetric Clifford system $\mathcal{A}$ is geometrically equivalent to a direct sum of irreducible ones.
- A symmetric Clifford system (X, Y, A) is irreducible if and only if $\dim Y = 2\delta(\dim X - 1)$, where $\delta(q)$ is defined by Table 5.4.1 on page 144.
- If $\dim X \not\equiv 1 \pmod 4$, then there exists exactly one class of geometrically equivalent irreducible symmetric Clifford systems on (X, Y).
- If $\dim X \equiv 1 \pmod 4$, then there exists exactly $\lfloor \frac{\dim Y}{4\delta(\dim X - 1)} \rfloor + 1$ distinct geometrically equivalent irreducible symmetric Clifford systems on (X, Y).

The following proposition reduces the classification of polar REC algebras to that of symmetric Clifford systems.

THEOREM 6.5.1. *The correspondence defined in Proposition* 6.5.2 *is natural in the sense that it associates similar polar algebras to geometrically equivalent symmetric Clifford systems, and vice versa.*

PROOF. First we assume that two symmetric Clifford systems (X, Y, A) and (X', Y', A') are geometrically equivalent and let $f : X \to X'$ and $g : Y \to Y'$ be the corresponding isometries satisfying $gA(x) = A'(f(x))g$. Then $F = f \times g$ is an isometry of $V = X \times Y$ onto $V' = X' \times Y'$ and it is easily verified that $\hat{A}(y) = f^{-1}\hat{A}'(y')g$, where $y' = g(y)$ and $\hat{A}$ is defined by (6.5.20). Then using (6.5.19) we find for any $z = (x, y) \in V = X \times Y$

$$F(z^2) = F(\hat{A}(y)y,\ 2A(x)y) = (\hat{A}'(y')y',\ 2gA(x)y)$$
$$= (\hat{A}'(y')y',\ 2A'(f(x))y') = F(z)^2.$$

Polarizing the latter identity yields that F is an algebra homomorphism of V and V', which shows by virtue of the isometry of F that the algebras are similar (in fact isomorphic).

Now, let V and V' be similar polar algebras; i.e., there exists an isometry $F: V \to V'$ and $k \in \mathbb{R}^\times$ such that

(6.5.21) $$kF(xy) = F(x)F(y), \qquad \forall x, y \in V.$$

By Corollary 6.5.1

(6.5.22) $$\dim V_0 = \dim V_0', \quad \dim V_1 = \dim V_1'.$$

We claim that either $F: V_0 \to V_0'$ or $F: V_0 \to V_1'$. Indeed, (6.5.1) yields $F(x)^2 = 0$ for any $x \in V_0$. Therefore, if $F(x) = x_0' + x_1' \in V_0' \oplus V_1'$ is the polar decomposition, one easily finds by (6.5.2) and (6.5.4) that $2x_0' x_1' = (x_1')^2 = 0$; hence by (6.5.5), $0 = x_1'(x_1' x_0') = |x_1'|^2 x_0'$, implying that either $x_0' = 0$ or $x_1' = 0$. This shows that $F: V_0 \to V_0' \cup V_1'$ and yields the claim by virtue of the linearity of F. Therefore we have two cases.

CASE I. If $F: V_0 \to V_0'$, then since F is an isometry, we have by (6.5.22), $F: V_1 \to V_1'$; hence (6.5.3) applied subsequently to V and V' yields for any nonzero $x \in V_0$ and $y \in V_1$

(6.5.23) $$\begin{aligned} k^2 |x|^2 F(y) &= k^2 F(x(xy)) = kF(x)F(xy) \\ &= F(x)(F(x)F(y)) = |F(x)|^2 F(y) = |x|^2 F(y) \end{aligned}$$

which shows that $k^2 = 1$. Let $f = kF|_{V_0}$ and $g = F|_{V_1}$. Then $x \in V_0$ and $y \in V_1$ yields $xy \in V_1$; hence

$$g(xy) = F(L_x y) = kF(x)F(y) = L_{f(x)} g(y),$$

which yields $L_x = g^{-1} L_{f(x)} g$ on V_1, thus implying that (V_0, V_1, Λ) is geometrically equivalent to (V_0', V_1', Λ').

CASE II. Now suppose $F: V_0 \to V_1'$ and denote $W_0 = V_0$ and $W_0' = V_0'$. It follows that $F^{-1} W_0' \subset V_1$, readily implying by the isometry of F the orthogonal decompositions $V_1 = W_1 \oplus W_2$ and $V_1' = W_1' \oplus W_2'$ such that

(6.5.24) $$F: \begin{array}{l} W_0 \to W_1', \\ W_1 \to W_0', \\ W_2 \to W_2'. \end{array}$$

Then (6.5.21) readily implies $W_1 W_1 = W_1' W_1' = \{0\}$. In particular, $0 = \langle W_1 W_1, W_0 \rangle = \langle W_1, W_0 W_1 \rangle$; hence by virtue of $W_0 W_1 \subset W_0 V_1 \subset V_1$ it yields $W_0 W_1 \subset W_2$. Similarly one verifies $W_0' W_1' \subset W_2'$.

Furthermore, the inclusion $W_2 W_2 \subset V_1 V_1 \subset V_0 = W_0$ yields by (6.5.21) and (6.5.24) for any $x \in W_2$ that $kF(x)^2 = F(x^2) \in W_1' \subset V_1'$. On the other hand, by (6.5.24), $F(x) \in W_2' \subset V_1'$; hence $F(x)^2 \in W_0'$. Combining the inclusions obtained implies $F(x)^2 = x^2 = 0$; thus, $W_2 W_2 = W_2' W_2' = \{0\}$. In particular, using the associativity of the inner product $0 = \langle W_2 W_2, W_0 \rangle = \langle W_2, W_0 W_2 \rangle$ and $W_0 W_2 \subset V_1$ yields $W_2 W_0 \subset W_1$. Similarly one verifies $W_2' W_0' \subset W_1'$. Therefore,

$$W_i W_j \subset W_k \text{ for any pairwise distinct } i, j, k, \text{ and } W_i W_i \subset \{0\}.$$

Clearly, similar relations are also valid for V'.

Now, observe that for any nonzero $x \in W_0$ and $y \in W_1$, $F(x) \in W_1' \subset V_1'$ and $F(y) \in W_0'$; hence using (6.5.3), (6.5.5), and the isometry of F and arguing as in (6.5.23), we find $k^2 = 1$. We can assume that $k = 1$ (replace F by kF, if needed); hence F is an isometric algebra homomorphism of V onto V'. Furthermore, setting

$x \in W_1$, $|x| = 1$, and $y \in W_2$ we find by (6.5.24) that $F(x) \in W_0'$ and $F(y) \in W_2' \subset V_1'$; therefore (6.5.3) yields $F(x(xy)) = F(y)$, implying by the nonsingularity of F that $L_x^2 = \mathbf{1}_{W_2}$. This implies in particular that $\dim W_1 = \dim W_2 = \dim W_0$ (where the latter equality is by (6.5.22) and (6.5.24)). This obviously implies $\dim W_i = \dim W_j'$ for any $0 \leq i, j \leq 2$. In summary, for any $x_i \in W_i$, L_{x_i} interchanges W_j and W_k, and moreover $L_{x_i}^2 = |x_i|^2$ on $W_j \oplus W_k$ and $L_{x_i}^2 = 0$ on W_i. In particular,

$$(6.5.25) \qquad \{L_{x_i}, L_{y_i}\}|_{W_j \oplus W_k} = 2\langle x_i, y_i \rangle, \qquad \forall x_i, y_i \in W_i.$$

Then using the latter identity, we obtain for arbitrary $x_j \in W_j$, $x_k \in W_k$, and $x_i, y_i \in W_i$

$$\langle (L_{x_i} x_j)(L_{x_i} x_k), y_i \rangle = \langle L_{x_i} x_j, L_{y_i}(L_{x_i} x_k) \rangle$$
$$= 2\langle x_j x_k, x_i \rangle \langle x_i, y_j \rangle - \langle L_{x_i} x_j, L_{x_i}(L_{y_i} x_k) \rangle$$
$$= 2\langle x_j x_k, x_i \rangle \langle x_i, y_j \rangle - |x_i|^2 \langle x_j x_k, y_j \rangle,$$

implying by the arbitrariness of y_i

$$(6.5.26) \qquad (L_{x_i} x_j)(L_{x_i} x_k) = 2\langle x_j x_k, x_i \rangle x_i - |x_i|^2 x_j x_k.$$

Now we are ready to present the geometric equivalence. Let $e_2 \in W_2$ be a fixed unit vector and let

$$\sigma(x_2) := 2\langle x_2, e_2 \rangle e_2 - x_2 : W_2 \to W_2.$$

It is straightforward to see that σ is an isometric involution. Let us also define the following linear maps: $f = F \circ L_{e_2} : W_0 \to W_0'$ and $g : V_1 \to V_1'$ by virtue of

$$g = \begin{cases} F \circ L_{e_2}, & \text{on } W_1, \\ F \circ \sigma, & \text{on } W_2. \end{cases}$$

Then it follows from the isometry of F and L_{e_2} that f and g are also isometries. We have by (6.5.26) for any $x_i \in W_i$,

$$(L_{e_2} x_0)(L_{e_2} x_1) = \sigma(x_0 x_1);$$

hence by virtue of $x_0 x_1 \in W_2$

$$g(x_1) f(x_0) = F(L_{e_2} x_1) F(L_{e_2} x_0) = F((L_{e_2} x_1)(L_{e_2} x_0)) = F(\sigma(x_1 x_0)) = g(x_1 x_0).$$

On the other hand, by the definition of σ and (6.5.25)

$$L_{\sigma(x_2)} L_{e_2} - L_{e_2} L_{x_2} = 2\langle x_2, e_2 \rangle L_{e_2}^2 - \{L_{e_2}, L_{x_2}\} = 0;$$

hence

$$g(x_2) f(x_0) - g(x_2 x_0) = F(L_{\sigma(x_2)} L_{e_2} x_0 - L_{e_2} x_2 x_0) = 0.$$

This yields $L_x = g^{-1} \circ L'_{f(x)} \circ g$, thus implying that $f \times g : V_0 \times V_1 \to V_0' \times V_1'$ is the desired geometric equivalence between (V_0, V_1, L) and (V_0', V_1', L'), where L' denotes the multiplication operator on V'. $\square$

REMARK 6.5.2. The only nontrivial case in the proof of Theorem 6.5.1 (Case II) is interesting due to its close relation to composition algebras and trialities, which can be seen as follows. By the properties of the operator L established in the course of the proof of Case II, for any fixed pair of unit vectors $e_j \in W_j$ and $e_k \in W_k$, the vector space W_i endowed with the new (noncommutative and nonassociative, in general) multiplication $x_i \star y_i = (e_j x_i)(y_i e_k)$ permits a composition, i.e., satisfies $|x_i \star y_i|^2 = |x_i|^2 |y_i|^2$. Since $|x|^2$ is a nonsingular quadratic form, the celebrated Hurwitz theorem implies $\dim W_i \in \{1, 2, 4, 8\}$. The interested reader is referred to [**233**] for further information about composition algebras. Furthermore, the isometry $f \times g$ above is an example of a triality [**233**, p. 42]. Another relevant observation in the present context is that the four REC algebras occurring in Case II correspond exactly to the so-called "mutants", i.e., REC algebras having the property $n_2 = 2$; see also Subsection 6.11.5.

6.6. The harmonicity of radial eigencubics

In this section we shall establish the following principal result.

THEOREM 6.6.1. *Any REC algebra of* $\dim V \geq 2$ *is harmonic. Equivalently, any radial eigencubic in* $\mathbb{R}^n$, $n \geq 2$, *is a harmonic function. In particular, any REC algebra V contains a nonzero idempotent.*

6.6.1. Idempotents in a REC algebra.
Recall that an element c of an algebra V is called an idempotent if $c^2 = c$. We shall denote by $I(V)$ the set of all nonzero idempotents of V.

LEMMA 6.6.1. *Let V be a metrized algebra with a positive definite bilinear form Q and let there exist $x \in V$ such that $Q(x^2, x) \neq 0$. Then $I(V) \neq \emptyset$.*

PROOF. Set $S = \{x \in V : Q(x) = 1\}$ for the unit hypersphere S in V. Then S is compact in the standard Euclidean topology on V. Since the cubic form $u(x) = Q(x^2; x)$ is continuous as a function on S, it attains its maximum value there, say at some point $y \in V$, $Q(y) = 1$. Since u is an odd function, the maximum value $u(y)$ is strictly positive. We have the stationary equation $0 = \partial_x u|_y$ whenever $x \in V$ satisfies the tangential condition $Q(x; y) = 0$. Thus, using (6.3.2) we obtain
$$0 = \partial_x u|_y = 3u(y; x) = \tfrac{1}{2} Q(y^2; x),$$
which implies immediately that y and y^2 are parallel, i.e., $y^2 = ky$, for some $k \in \mathbb{R}^\times$ (observe also that $k > 0$ by virtue of $0 < u(y) = Q(y^2; y) = kQ(y; y)$). Scaling y appropriately, namely setting $c = y/k$, yields $c^2 = c$. □

Let V be a REC algebra and let $c \in I(V)$. Setting $x = c$ in (6.4.5) yields
$$(6.6.1) \qquad \langle b, c \rangle = 1 + \tfrac{2}{3}\lambda |c|^2 = 1 + 2\alpha, \quad \alpha = \tfrac{1}{3}\lambda |c|^2,$$
and by (6.4.9)
$$(6.6.2) \qquad (1 + 2\alpha)c = |c|^2 b.$$

REMARK 6.6.1. It follows that $b(V)$ is uniquely determined by (6.6.2). In particular, if V permits two distinct nonzero idempotents, then (6.6.2) immediately implies $b(V) = 0$.

Setting $x = c$ in the operator identity (6.4.10) yields
$$(6.6.3) \qquad 2L_c^3 - 4\alpha L_c^2 + \alpha L_c + \alpha = \frac{2(1-\alpha)}{|c|^2} c \otimes c.$$

Notice that L_c acts on $\mathbb{R}c$ as an identity operator and we have
$$2L_c^3 - 4\alpha L_c^2 + \alpha L_c + \alpha = 0$$
on the orthogonal complement $c^\perp = \{x \in V : \langle x, c \rangle = 0\}$. Let us denote by $\tau(c)$ the spectrum of $L_c|_{c^\perp}$. Then obviously the polynomial
$$(6.6.4) \qquad \chi(t) := t^3 - 2\alpha t^2 + \tfrac{1}{2}\alpha t + \tfrac{1}{2}\alpha$$

vanishes on $\tau(c)$. We denote by t_i the roots of (6.6.4) counted according to the multiplicity. Since L_c is selfadjoint, one has

$$c^\perp = \bigoplus_{t \in \tau(c)} V_c(t),$$

where

$$V_c(t) = \ker(L_c - t) \cap c^\perp, \quad t \in \tau(c),$$

and one easily obtain

(6.6.5) $$\sum_{t \in \tau(c)} \dim V_c(t) = \dim c^\perp = \dim V - 1 = n - 1.$$

On the other hand, it follows by (6.4.6) that $\langle b, c \rangle = \operatorname{tr} L_c$; hence by (6.6.1)

(6.6.6) $$\sum_{t \in \tau(c)} t \dim V_c(t) = \operatorname{tr} L_c|_{c^\perp} = \langle b, c \rangle - 1 = 2\alpha.$$

EXAMPLE 6.6.1. It is easy to characterize nonzero idempotents in a polar algebra. Suppose that V is such an algebra and let $c = z_0 + z \in I(V)$, where $z_0 \in V_0$ and $z \in V_1$. By (6.5.8), $z^2 + 2zz_0 = z_0 + z$. Equating the even and odd parts yields the equations $z^2 = z_0$, $2zz_0 = z$, and hence

(6.6.7) $$2z^3 = z.$$

Conversely, for any element $z \in V_1$ satisfying (6.6.7) we have $c = z^2 + z \in I(V)$. Indeed, using (6.5.1), $c^2 = z^2 z^2 + 2z^3 + z^2 = 2z^3 + z^2 = z + z^2 = c$.

We have some corollaries of the existence of an idempotent on REC algebras.

PROPOSITION 6.6.1. *If* $\dim V \geq 2$, *then* $\lambda(V) \neq 0$; *equivalently*, $\alpha(c) \neq 0$ *for any* $c \in I(V)$. *In particular, if* $\dim V \geq 2$, *then* $0 \notin \tau(c)$ *for any* $c \in I(V)$.

PROOF. Indeed, if $\lambda = 0$, then by (6.6.1), $\alpha = 0$ for any choice of an idempotent c of V. Then (6.6.3) yields $L_c^3 = \frac{1}{|c|^2} c \otimes c$; hence $L_c|_{c^\perp} \equiv 0$. We claim that $c^\perp$ is a trivial algebra (i.e., $xy = 0$ for all $x, y \in c^\perp$). Indeed, we have for any $x, y \in c^\perp$, $0 = \langle xc, y \rangle = \langle c, xy \rangle$, which yields $xy \in c^\perp$, i.e., $c^\perp$ is a subalgebra. It follows, that $L_x e = L_{x^2} e = L_{x^3} e = 0$; hence, applying (4.4.10) to c, we obtain by the assumption $\lambda = 0$ that $x^3 \langle b, c \rangle = 0$. Using (6.6.1), we have $x^3 = 0$, and taking the scalar product with x we obtain $0 = \langle x^3, x \rangle = \langle x^2, x^2 \rangle$, which implies $x^2 = 0$ for all $x \in c^\perp$. Polarizing the latter equation we find $xy = 0$ for all $x, y \in c^\perp$, as claimed. Therefore V is similar to $\mathbb{R}c$ and thus is a 1-dimensional algebra.

The last conclusion is a corollary of $\alpha(c) \neq 0$ and the characteristic equation (6.6.4). $\square$

6.6.2. Auxiliary lemmas. In what follows we shall assume that $\dim V \geq 2$. We follow a standard convention of using vector space notation in a formula if the formula is valid for any choice of the vector space represented. For instance, $V_1 V_2 \subset V_3$ means that the product $xy \in V_3$ for any $x \in V_1$ and $y \in V_2$ etc.

LEMMA 6.6.2. *Let V be a REC algebra, $c \in I(V)$, and suppose $V_c(s_1)$, $V_c(s_2)$, and $V_c(s_3)$ are nontrivial eigenspaces of L_c, $s_i \in \tau(c)$ (repeated eigenvalues are allowed). If $\langle V_c(s_1) V_c(s_2), V_c(s_3) \rangle \neq 0$, then*

(6.6.8) $$R(s_i, s_j, s_k) := 2\sigma_1^2 - 2\sigma_2 - (1 + 4\alpha)\sigma_1 + 3\alpha = 0,$$

where $\sigma_1 = \sum_i s_i$, $\sigma_2 = \sum_{i<j} s_i s_j$.

PROOF. If $x \in c^\perp$, then by (6.6.2), $\langle x, b \rangle = 0$; hence applying (6.4.10) to c followed by the scalar product by c one readily finds

(6.6.9) $\qquad 2\langle L_x^3 c, c \rangle + 2\langle xc, x^2 c \rangle + \alpha \langle x^2, x \rangle = (4\alpha + 1)\langle x^3, c \rangle.$

Now suppose that $x_i \in V_c(s_i)$, $i = 1, 2, 3$. Then

$$\partial_{123}\langle x^3, c \rangle = 2(\langle x_1 x_2, c x_3 \rangle + \langle x_1 x_2, c x_3 \rangle + \langle x_1 x_2, c x_3 \rangle)$$
$$= 2\langle x_1 x_2, x_3 \rangle \sum_i s_i = 2\langle x_1 x_2, x_3 \rangle \sigma_1,$$

where $\partial_{123} = \partial_{x_1} \partial_{x_2} \partial_{x_3}$, and similarly one also finds $\partial_{123}\langle x^2, x \rangle = 6\langle x_1 x_2, x_3 \rangle$, $\partial_{123}\langle L_x^3 c, c \rangle = 2\langle x_1 x_2, x_3 \rangle \sigma_2$, and $\partial_{123}\langle x^3, c \rangle = 2\langle x_1 x_2, x_3 \rangle \sigma_1$. Moreover, we have

$$\partial_{123}\langle xc, x^2 c \rangle = 2\langle x_1 x_2, x_3 \rangle \sum_i s_i^2 = 2\langle x_1 x_2, x_3 \rangle (\sigma_1^2 - 2\sigma_2).$$

Using (6.6.9) one obtains $(4\sigma_1^2 - 4\sigma_2 - 2(1 + 4\alpha)\sigma_1 + 6\alpha)\langle x_1 x_2, x_3 \rangle = 0$, which yields the desired conclusion. $\square$

LEMMA 6.6.3. *If $\alpha \neq t \in \tau(c)$, then $\langle V_c(t) V_c(t), V_c(t) \rangle \equiv 0$.*

PROOF. Assume by contradiction $\langle V_c(t) V_c(t), V_c(t) \rangle \equiv 0$; hence by Lemma 6.6.2, $R(t, t, t) = 4t^2 - (1 + 4\alpha)t + \alpha = (4t - 1)(t - \alpha) = 0$. Hence $t = \frac{1}{4}$ and (6.6.4) yields $\alpha = -\frac{1}{32}$, in which case equation (6.6.4) has a unique real root. The latter implies that the only possible eigenvalue of L_c on $(\mathbb{R}c)^\perp$ is $\frac{1}{4}$; hence by (6.6.6), $\dim V_c(t) = 2\alpha/\lambda = -\frac{1}{4}$, and a contradiction follows. $\square$

LEMMA 6.6.4. *If $(2\alpha + 1)(68\alpha^2 - 38\alpha - 27) \neq 0$, then $V_c(t) V_c(t) \subset \mathbb{R}c \oplus V_c(t)$ for any $t \in \tau(c)$.*

PROOF. Observe that the statement is trivial when $\tau(c)$ contains a single point. Therefore we suppose that there are at least two distinct eigenvalues of L_c and notice that it suffices to show that $\langle V_c(t) V_c(t), V_c(s) \rangle \equiv 0$ for any $s \neq t$, $s \in \tau(c)$. Observe that by the assumptions of the lemma and $\alpha \neq 0$ the desired property follows immediately from Lemma 6.6.2 and the identity

(6.6.10) $\qquad \prod_{i \neq j} R(t_i, t_i, t_j) = \frac{1}{2}\alpha^2(2\alpha + 1)(68\alpha^2 - 38\alpha - 27),$

where the latter product is taken over all pairs of (maybe, equal) roots t_i and t_j of (6.6.4). In order to establish the latter identity we denote by t_1, t_2, t_3 the roots of (6.6.4) and observe that the following relations follow easily by (6.6.4) and (6.6.8):

$$R(t_1, t_1, t_2) = 2\chi'(t_1) + (2t_2 - 1)(2t_1 + t_2 - 2\alpha) = 2\chi'(t_1) + (2t_2 - 1)(t_1 - t_3),$$

where the last equality is by Viète's formula $2\alpha = t_1 + t_2 + t_3$ for (6.6.4). Note also that $\chi'(t_1) = (t_1 - t_2)(t_1 - t_3)$; hence $R(t_1, t_1, t_2) = (2t_1 - 1)(t_1 - t_3)$, and by virtue of symmetry one also has

(6.6.11) $\qquad R(t_i, t_i, t_j) = (2t_i - 1)(t_i - t_j), \qquad \forall i \neq j.$

This implies $R(t_1, t_1, t_2) R(t_1, t_1, t_3) = (2t_1 - 1)^2 \chi'(t_1)$; thus

$$\prod_{i \neq j} R(t_i, t_i, t_j) = \prod_{i=1}^3 (2t_i - 1)^2 \chi'(t_i) = 2^6 \chi^2(\tfrac{1}{2}) \cdot D(\chi') = 2(2\alpha + 1) D(\chi'),$$

where $D(\chi') = \frac{1}{4}\alpha^2(68\alpha^2 - 38\alpha - 27)$ is the discriminant of χ', proving (6.6.10). The lemma follows. $\square$

PROPOSITION 6.6.2. *Let $V_c(t)V_c(t) \subset \mathbb{R}c$ for some $t \in \tau(c)$. Then $\alpha = -\frac{1}{2}$, $b(V) = 0$, $t = -1$, and*

$$(6.6.12) \qquad x^2 = -\frac{1}{|c|^2}|x|^2 c, \quad \forall x \in V_c(t).$$

PROOF. The assumption implies the existence of a bilinear form

$$q(x,y) : V_c(t) \times V_c(t) \to \mathbb{R}, \qquad xy = q(x,y)c.$$

We have $x^2 = q(x,x)c$, $x^3 = tq(x,x)x$, $xx^3 = tq^2(x,x)c$, and $x^2 x^2 = q^2(x,x)c$. Furthermore, $V_c(t)V_c(t) \subset \mathbb{R}c$ yields $\langle x^2, x \rangle = 0$. Applying the relations obtained to (6.4.9) we find by virtue of $\langle b, x \rangle = 0$ and (6.6.2) that

$$(6.6.13) \qquad ((2t-\alpha)q(x,x) + \lambda|x|^2)q(x,x) = 0.$$

Observe that $q(x,x) \not\equiv 0$ because otherwise $V_c(t)V_c(t) = \{0\}$; hence

$$0 \equiv \langle x^2, c \rangle = \langle x, L_c x \rangle = t|x|^2 \quad \forall x \in V_c(t),$$

implying $t = 0$, and thus $\alpha = 0$ by (6.6.4); a contradiction to Proposition 6.6.1 follows. Then the standard argument shows that (6.6.13) implies $(\alpha - 2t)q(x,x) \equiv \lambda|x|^2$ on $V_c(t)$. By Proposition 6.6.1, $\lambda \neq 0$, which implies $\alpha - 2t \neq 0$, and therefore $q(x,x) = \lambda|x|^2/(\alpha - 2t)$. Thus

$$(6.6.14) \qquad x^2 = \frac{\lambda|x|^2}{\alpha - 2t}c, \quad \forall x \in V_c(t).$$

On the other hand, $\langle x^2, c \rangle = \langle x, L_c x \rangle = t|x|^2$ for any $x \in V_c(t)$; hence (6.6.14) yields by virtue of (6.6.1) that $2t^2 - t\alpha + 3\alpha = 0$. Thus, the latter quadratic polynomial and the polynomial in (6.6.4) have a common root t, which implies the vanishing of their resultant with respect to t. The latter is found explicitly to be $8\alpha^2(2\alpha+1)(12\alpha+1)$ (see, for instance, [**265**]). Thus eliminating by Proposition 6.6.1 the case $\alpha = 0$, one has either $\alpha = -\frac{1}{2}$ or $\alpha = -\frac{1}{12}$. The latter case is impossible because for $\alpha = -\frac{1}{12}$ equation (6.6.4) has a single real root $t_1 = \frac{1}{3}$ and (6.6.6) yields $\dim V_c(t) = \frac{2\alpha}{t} = -\frac{1}{2}$; a contradiction follows. Thus, $\alpha = -\frac{1}{2}$ and $b(V) = 0$ by (6.6.2). Furthermore, since t is a common root of (6.6.4) and $2t^2 - t\alpha + 3\alpha = 2t^2 + \frac{1}{2}t - \frac{3}{2} = 0$, we obtain $t = -1$, and the relation (6.6.12) follows by (6.6.14). $\square$

6.6.3. Proof of Theorem 6.6.1. We argue by contradiction and suppose that $b(V) \neq 0$, which is equivalent by (6.6.2) to $\alpha \neq -\frac{1}{2}$. Then Proposition 6.6.2 yields that there exist $t, s \in \tau(c)$ such that $\langle V_c(t)V_c(t), V_c(s) \rangle \neq 0$. We consider two cases.

CASE 1. $t = s$. Then by Lemma 6.6.3, $t = \alpha$; therefore (6.6.4) yields

$$\alpha(2\alpha + 1)(\alpha - 1) = 0,$$

and by Proposition 6.6.1 and the assumption we have $\alpha = 1$. The characteristic equation (6.6.4) has three distinct real roots $t_1 = \alpha = 1$ and $t_{2,3} = \frac{1 \pm \sqrt{3}}{2}$. A simple analysis of (6.6.6) shows that $\dim V_c(t_2) = \dim V_c(t_3)$ and furthermore $\dim V_c(t_1) + \dim V_c(t_2) = 2$. By the assumption, $t = 1 \in \tau(c)$; hence $\dim V_c(1) \geq 1$, implying that $(\dim V_c(1), \dim V_c(t_2))$ is either $(1,1)$ or $(2,0)$.

Consider $(\dim V_c(1), \dim V_c(t_2)) = (1,1)$. Note that $t_2 \neq \alpha$ implies by Lemma 6.6.3 that $V_c(t_2)V_c(t_2) \subset V_c(t_2)^\perp$. On the other hand, $\alpha = 1$ yields that $R(t_2, t_2, t_j) \neq 0$ for $j = 1,3$; thus by Lemma 6.6.2, $\langle V_c(t_2)V_c(t_2), V_c(t_j)\rangle \equiv 0$, for $j = 1,3$. In summary, $V_c(t_2)V_c(t_2) \subset \mathbb{R}c$; hence by Proposition 6.6.2, $\alpha = -\frac{1}{2}$, a contradiction.

Now suppose $(\dim V_c(1), \dim V_c(t_2)) = (2,0)$, in other words, $V = \mathbb{R}c \oplus V_c(1)$, and choose arbitrarily an orthonormal basis e_1, e_2 of $V_c(1)$. Then by virtue of (6.4.6) and (6.6.2),
$$e_1^2 + e_2^2 = \frac{2\alpha}{|c|^2}c = \frac{2}{|c|^2}c.$$

On the other hand, $L_c e_k = t e_k = e_k$; hence $\langle e_k^2, c\rangle = \langle e_1, e_1 c\rangle = |e_1|^2 = 1$ for $k = 1,2$. Therefore one readily finds
$$e_1^2 = a_1 e_1 + a_2 e_2 + \frac{1}{|c|^2}c, \qquad e_2^2 = -a_1 e_1 - a_2 e_2 + \frac{1}{|c|^2}c.$$

By the associativity of the scalar product, $\langle e_1 e_2, e_1\rangle = \langle e_2, e_1^2\rangle = a_2$, and similarly $\langle e_1 e_2, e_2\rangle = -a_1$. Since $\langle e_1 e_2, c\rangle = \langle e_1, e_2\rangle = 0$, we get $e_1 e_2 = a_2 e_1 - a_1 e_2$. This yields
$$e_1^3 = a_1 e_1^2 + a_2 e_2 e_1 + \frac{1}{|c|^2}e_1 = \left(a_1^2 + a_2^2 + \frac{1}{|c|^2}\right)e_1 + \frac{a_1}{|c|^2}c.$$

Applying $x = e_1$ in (6.4.5) we find by virtue of $\lambda = \frac{3}{|c|^2}$ that $-\langle e_1^2, e_1^3\rangle = \frac{2}{|c|^2}\langle e_1^2, e_1\rangle$, and using the above identities yields $-(a_1^2 + a_2^2 + \frac{2}{|c|^2})a_1 = \frac{2}{|c|^2}a_1$; hence $a_1 = 0$. A similar argument yields $a_2 = 0$; thus $e_1^2 = e_2^2 = \frac{1}{|c|^2}c$ and $e_1 e_2 = 0$. This immediately yields $V_c(1)V_c(1) \subset V_c(1)$; hence $\alpha = -\frac{1}{2}$ by Proposition 6.6.2. A contradiction follows.

CASE 2. $s \neq t$. Then Lemma 6.6.4 obviously yields $68\alpha^2 - 38\alpha - 27 = 0$ (because $2\alpha + 1 \neq 0$ by the assumption). Since the analysis of the roots of the latter equation is similar, it suffices to consider the case $\alpha = \frac{19 + 13\sqrt{13}}{68}$. Then equation (6.6.4) has two distinct roots: $t_1 = \frac{1-2\sqrt{13}}{17}$ and $t_2 = \frac{1+\sqrt{13}}{4}$. Applying this to (6.6.6) one easily finds the only possible integer solution $(\dim V_c(t_1), \dim V_c(t_2)) = (2,1)$; hence $\tau(c) = \{t_1, t_2\}$. On the other hand, $\alpha \notin \tau(c)$ yields by Lemma 6.6.3, $\langle V_c(t_1)V_c(t_1), V_c(t_1)\rangle = 0$. Furthermore, one easily verifies that $R(t_1, t_1, t_2) \neq 0$, implying by Lemma 6.6.2 that $\langle V_c(t_1)V_c(t_1), V_c(t_2)\rangle = 0$. Thus, $V_c(t_1)V_c(t_1) \subset \mathbb{R}c$; hence $\alpha = -\frac{1}{2}$ by Proposition 6.6.2, a contradiction. The theorem follows.

6.7. The Peirce decomposition of a REC algebra

6.7.1. Normalized REC algebras.
By Theorem 6.6.1, $b(V) = 0$ holds in any REC algebra; thus $\alpha = -\frac{1}{2}$, which yields by (6.6.1)

(6.7.1) $$\lambda = -\frac{3}{2|c|^2}.$$

Observe that λ is negative; hence any REC algebra is similar to a REC algebra satisfying the condition

(6.7.2) $$\lambda = -2,$$

which is equivalent by virtue of (6.7.1) to the requirement that

(6.7.3) $$|c|^2 = \tfrac{3}{4}, \qquad \forall c \in I(V).$$

It is convenient to work with some fixed normalization like (6.7.2) and we shall refer to a REC algebra satisfying any of the equivalent conditions (6.7.2) and (6.7.3) as a *normalized* REC algebra.

REMARK 6.7.1. We wish to point out that any particular choice of the normalization affects the defining equation (6.4.8) and, as a corollary, its specialization for harmonic algebras (see (6.7.5) below), as well as its further polarizations and the exact form of the multiplication on V. But, of course, it does not affect the algebraic properties of V defined up to a similarity, i.e., those properties of REC algebras which are of the most interest for us. Our concrete choice of the λ in (6.7.2) is due to some computational reasons and does not affect the generality of our main results because we make no distinction between similar algebras.

6.7.2. The structural identities. In what follows, unless otherwise stated, we tacitly assume that V stands for a normalized REC algebra. Applying $b(V) = 0$ and (6.7.2) to (6.4.6), (6.4.5), (6.4.9), and (6.4.10), one obtains, respectively,

$$\sum_{i=1}^{\dim V} e_i^2 = 0, \tag{6.7.4}$$

$$\langle x^2, x^3 \rangle = \tfrac{4}{3} \langle x, x \rangle \langle x, x^2 \rangle, \tag{6.7.5}$$

$$4xx^3 + x^2 x^2 - 4|x|^2 x^2 - \tfrac{8}{3} \langle x^2, x \rangle x = 0, \tag{6.7.6}$$

$$L_{x^3} + 2L_x^3 + \{L_x, L_{x^2}\} - 2\{x^2 \otimes x\} - 2|x|^2 L_x - \tfrac{2}{3}\langle x^2, x \rangle = 0. \tag{6.7.7}$$

We shall also use a polarization of the latter equation obtained by differentiating it with respect to $y \in V$; in other words,

$$L_{M_x y} + \{M_x, L_y\} + L_x L_y L_x + \{L_x, L_{xy}\} - \{x^2 \otimes y\} - |x|^2 L_y - \langle x^2, y \rangle = 0, \tag{6.7.8}$$

where

$$M_x = \tfrac{1}{2} L_{x^2} + L_x^2 - 2x \otimes x. \tag{6.7.9}$$

Applying (6.7.8) to y we find

$$\begin{aligned}L_x^2 y^2 + L_y^2 x^2 + 2\{L_x, L_y\}xy - 4\langle x, y\rangle xy + \tfrac{1}{2}x^2 y^2 + (xy)^2 \\ = x^2|y|^2 + y^2|x|^2 + 2\langle x^2, y\rangle y + 2\langle y^2, x\rangle x.\end{aligned} \tag{6.7.10}$$

6.7.3. The characteristic dimensions. It follows from (6.6.4) that for any $c \in I(V)$ the characteristic polynomial $\chi(t)$ takes the form

$$\chi(t) = t^3 + t^2 - \tfrac{1}{4}t - \tfrac{1}{4} = (t+1)(t+\tfrac{1}{2})(t-\tfrac{1}{2});$$

hence L_c has three eigenvalues, $t_1 = -1$, $t_2 = -\tfrac{1}{2}$, and $t_3 = \tfrac{1}{2}$. Let

$$V = V_c(1) \oplus V_c(-1) \oplus V_c(-\tfrac{1}{2}) \oplus V_c(\tfrac{1}{2}) \tag{6.7.11}$$

be the Peirce decomposition of V with respect to the idempotent c, and we agree on the following notation for the corresponding dimensions:

$$n_1(c) = \dim V_c(-1), \quad n_2(c) = \dim V_c(-\tfrac{1}{2}), \quad n_3(c) = \dim V_c(\tfrac{1}{2}).$$

Then (6.6.5) and (6.6.6) yield, respectively,

(6.7.12) $$n_3(c) = 2n_1(c) + n_2(c) - 2$$

and

(6.7.13) $$n = 3n_1(c) + 2n_2(c) - 1.$$

It follows from the latter relation that $\min\{n_1(c), n_2(c)\} \geq 1$.

DEFINITION 6.7.1. We refer to
$$(n_1(c), n_2(c)) = (\dim V_c(-1), \dim V_c(-\tfrac{1}{2}))$$
as to the *characteristic dimensions* of the idempotent c.

Observe that if one knows one of the characteristic dimensions, the other is uniquely determined by virtue of the formulas (6.7.13) and (6.7.12).

It is convenient to summarize the above results in the proposition below.

PROPOSITION 6.7.1. *Any normalized REC algebra V satisfies the defining equation (6.7.5) subject to condition (6.7.4). Moreover, the following conditions are satisfied:*

- *The set of nonzero idempotents $I(V)$ is nonempty and for any $c \in I(V)$, (6.7.3) holds.*
- *The set of eigenvalues of the multiplication operator L_c is a subset of $\{-1, -\tfrac{1}{2}, \tfrac{1}{2}\}$.*
- *The characteristic dimensions $(n_1(c), n_2(c))$ satisfy conditions (6.7.13) and (6.7.12).*

6.7.4. Some explicit formulas for the characteristic dimensions.

COROLLARY 6.7.1. *Let V be a Clifford REC algebra. Then for any $c \in I(V)$ and any polar algebra $V_0 \oplus V_1$ similar to V*
$$(n_1(c), n_2(c)) = (\dim V_0 - 1, \; \tfrac{1}{2}\dim V_1 - \dim V_0 + 2).$$

PROOF. By Corollary 6.5.1, $\dim V_0$ is a similarity invariant. Applying (6.5.12) to $x = c \in I(V)$ we find by virtue of $\langle c, c^2 \rangle = \tfrac{3}{4}$ that $\operatorname{tr} L_c^3 = \tfrac{3}{4}(2 - \dim V_0)$. On the other hand, the Peirce decomposition (6.7.11) yields $\operatorname{tr} L_c^3 = 1 - n_1(c) + \tfrac{1}{8}(n_3(c) - n_2(c))$, which by virtue of (6.7.13) and (6.7.12) yields the desired formula. □

COROLLARY 6.7.2. *Let V be a Cartan REC algebra. Then for any $c \in I(V)$,*
$$(n_1(c), n_2(c)) = (\tfrac{1}{3}(1 + \dim V), \; 0)$$
for any choice of an idempotent $c \in I(V)$.

PROOF. Without loss of generality we may assume that V is the special Cartan algebra; see Subsection 6.4.4. Then for any $c \in I(V)$ we have by (6.4.11), $|c|^2 = 36$, and setting $x = c$ in (6.4.14) we also obtain $2L_c^2 + L_c - 1 = 72 c \otimes c$, which shows that the only possible eigenvalues of L_c on $V_c(1)^\perp$ are -1 and $\tfrac{1}{2}$, implying $n_2(c) = \dim V_c(-\tfrac{1}{2}) = 0$. Using (6.7.13) and (6.7.12) yields the desired formula. □

6.7.5. The multiplication table of $V_c(t_i)$.

Let us again assume that V is a normalized REC algebra and $c \in I(V)$. Applying the explicit values of $t_1 = -1$, $t_2 = \frac{1}{2}$, and $t_3 = \frac{1}{2}$ to (6.6.11), one finds by Proposition 6.6.2 that $\langle V_c(t_i)V_c(t_j), V_c(t_k)\rangle \equiv 0$, for all ordered triples (i,j,k) except maybe for the following eight:

$$(i,j,k) \in \{(0,0,0),(0,1,1),(0,2,2),(0,3,3),(1,2,3),(2,2,2),(1,3,3),(2,3,3)\}.$$

This yields the corresponding multiplication table for the eigenspaces presented by Table 6.7.1.

TABLE 6.7.1. The multiplication table of $V_i = V_i(c)$, $c \in I(V)$.

	$V_c(1)$	$V_c(-1)$	$V_c(-\frac{1}{2})$	$V_c(\frac{1}{2})$
$V_c(1)$	$V_c(1)$	$V_c(-1)$	$V_c(-\frac{1}{2})$	$V_c(\frac{1}{2})$
$V_c(-1)$	$V_c(-1)$	$V_c(1)$	$V_c(\frac{1}{2})$	$V_c(-\frac{1}{2}) \oplus V_c(\frac{1}{2})$
$V_c(-\frac{1}{2})$	$V_c(-\frac{1}{2})$	$V_c(\frac{1}{2})$	$V_c(1) \oplus V_c(-\frac{1}{2})$	$V_c(-1) \oplus V_c(-\frac{1}{2})$
$V_c(\frac{1}{2})$	$V_c(\frac{1}{2})$	$V_c(-\frac{1}{2}) \oplus V_c(\frac{1}{2})$	$V_c(-1) \oplus V_c(-\frac{1}{2})$	$V_c(1) \oplus V_c(-1) \oplus V_c(-\frac{1}{2})$

It follows from the multiplication table that $V_c(1,-1)$ and $V_c(1,-\frac{1}{2})$ are subalgebras[6] of V:

$$(6.7.14) \quad V_c(1,-1)V_c(1,-1) \subset V_c(1,-1), \quad V_c(1,-\tfrac{1}{2})V_c(1,-\tfrac{1}{2}) \subset V_c(1,-\tfrac{1}{2}).$$

Here and in what follows we adopt a convention to write

$$V_c(a,b,\ldots) = V_c(a) \oplus V_c(b) \oplus \cdots.$$

6.7.6. Spectral properties of nilpotent elements.

Besides the idempotents we also make use of degree 2 nilpotent elements of V. More precisely, let us denote by

$$\mathcal{N}(V) = \{w \in V : |w| = 1 \text{ and } w^2 = 0\}$$

the set of normalized square zero elements.

The following lemma is basic to many of the results in the next sections.

LEMMA 6.7.1. *If $w \in \mathcal{N}(V)$, then $L_w : V \to V$ has the three eigenvalues 0 and ± 1, and $\dim V_w(-1) = \dim V_w(1) \geq 1$, $\dim V_w(0) \geq 1$, where*

$$V = V_w(0) \oplus V_w(-1) \oplus V_w(1)$$

is the corresponding eigen-decomposition. Moreover, if $z \in \mathcal{N}(V)$ is another nilpotent element, then $\dim V_w(\varepsilon) = \dim V_z(\varepsilon)$ for any $\varepsilon \in \{0, \pm 1\}$. Furthermore, the

---

[6]Observe, however, that these algebras need not be *REC subalgebras* because the zero trace condition (the harmonicity condition) is not satisfied in general for the subalgebras. We discuss these subalgebras more fully below.

following relations hold:

(6.7.15) $$xy = \lambda \langle x, y \rangle w, \forall x, y \in V_w(\varepsilon), \quad \varepsilon^2 = 1,$$

(6.7.16) $$xy \in V'_w(0), \forall x \in V_w(-1), \ y \in V_w(1),$$

(6.7.17) $$xy \in V_w(1, -1), \forall x, y \in V'_w(0), |$$

(6.7.18) $$xy \in V_w(-\varepsilon) \oplus V'_w(0), \forall x \in V'_w(0), \ y \in V_w(\varepsilon), \quad \varepsilon^2 = 1,$$

where $V'_w(0) := V_w(0) \ominus \mathbb{R}w$. See also Table 6.7.2.

TABLE 6.7.2. The multiplication table of $V_w(\varepsilon)$, $w \in \mathcal{N}(V)$.

	$\mathbb{R}w$	$V'_w(0)$	$V_w(-1)$	$V_w(1)$
$\mathbb{R}w$	0	0	$V_w(-1)$	$V_w(1)$
$V'_w(0)$	0	$V_w(-1, 1)$	$V_w(1) \oplus V'_w(0)$	$V_w(-1) \oplus V'_w(0)$
$V_w(-1)$	$V_w(-1)$	$V_w(1) \oplus V'_w(0)$	$\mathbb{R}w$	$V'_w(0)$
$V_w(1)$	$V_w(1)$	$V_w(-1) \oplus V'_w(0)$	$V'_w(0)$	$\mathbb{R}w$

PROOF. Setting $x = w$, (6.7.7) yields $L_w^3 = L_w$, implying that the only possible eigenvalues of L_w are 0 and ± 1. Since $L_w w = 0$, we have $\dim V_w(0) \geq 1$, and since any REC algebra is harmonic, we also have $\operatorname{tr} L_w = 0$; hence $\dim V_w(1) = \dim V_w(-1)$. On the other hand, since $I(V)$ is nonempty, $L_c w \neq 0$ for any $c \in I(V)$ (otherwise L_c would have zero eigenvalue, which contradicts Proposition 6.7.1); hence $V \neq V_w(0)$, which yields $\dim V_w(1) \geq 1$. Thus, all three numbers 0 and ± 1 are eigenvalues of L_w of nonzero multiplicities.

Applying (6.7.8) to $x = w$ and $y = z$, where $w, z \in \mathcal{N}(V)$, yields
$$L_w^2 L_z + L_z L_w^2 + L_w L_z L_w - 2\langle z, w \rangle L_w = L_z;$$
hence
$$2 \operatorname{tr} L_w^2 L_z^2 + \operatorname{tr} L_z L_w L_z L_w - 2\langle z, w \rangle L_w L_z = \operatorname{tr} L_z^2 = \dim V - \dim \ker L_z = 2 \dim V_z(1).$$
Observe that by virtue of the invariance of the trace under cyclic permutations the left-hand side of the latter identity holds and does not change if we interchange the role of z and w, hence implying
$$\dim V_z(1) = \dim V_z(-1) = \dim V_w(1) = \dim V_w(-1)$$
and, thus, also $\dim V_w(0) = \dim V_z(0)$.

To proceed we need an auxiliary identity. Notice that (6.7.10) applied to $y = w$ yields

(6.7.19) $$L_w^2 x^2 + 2\{L_x, L_w\}wx - \langle w, x \rangle wx + (wx)^2 = x^2 + 2\langle x^2, w \rangle w, \quad \forall x \in V,$$

and that further polarization yields for any $x, y \in V$

(6.7.20)
$$L_w^2 xy + \{L_x, L_w\}wy + \{L_y, L_w\}wx - \tfrac{1}{2}(\langle w, y \rangle wx + \langle w, x \rangle wy) + (wx)(wy)$$
$$= xy + 2\langle xy, w \rangle w.$$

Now, setting $x \in V_w(\varepsilon)$ in (6.7.19) we find $(L_w^2 + 2\varepsilon L_w + 2)x^2 = 2\varepsilon|x|^2 w$. Applying L_w to the latter identity and using $L_w^3 = L_w$ we obtain $(2\varepsilon L_w^2 + 3L_w)x^2 = 0$, which easily yields $x^2 \subset V_w(0)$ by virtue of the spectral properties of L_w. Hence the former identity yields $x^2 = \varepsilon|x|^2 w$, and (6.7.15) follows by polarization. Next, applying (6.7.20) to $x \in V_w(-1)$, $y \in V_w(1)$ we get $L_w^2 xy = 0$; thus $xy \in V_w(0)$. Moreover, we have $\langle xy, w\rangle = -\langle x, y\rangle = 0$; hence (6.7.16) follows. Similarly, setting $x \in V_w(0)$ in (6.7.19) one finds $L_w^2 x^2 = x^2$; i.e., $x^2 \in V_w(-1) \oplus V_w(1) = V_w(1,-1)$. Polarizing the latter relation yields (6.7.17). Finally, applying (6.7.20) to $x \in V_w'(0)$ and $y \in V_w(\varepsilon)$ with $\varepsilon^2 = 1$ we find $L_w(L_w + \varepsilon)xy = 0$; hence $xy \in V_w(0) \oplus V_w(-\varepsilon)$. But $\langle xy, w\rangle = \langle y, xw\rangle = 0$; hence (6.7.18) follows. □

COROLLARY 6.7.3. *If $\mathcal{N}(V) \neq \emptyset$, then for any $w \in \mathcal{N}(V)$ and $w' \in V_w(\varepsilon)$, $\varepsilon^2 = 1$, normalized by $|w'| = 2$,*

$$c := \tfrac{1}{2}(\varepsilon w + \mu w') \in I(V), \qquad \mu^2 = 1.$$

Moreover, $V_w(\varepsilon) \ominus \mathbb{R}w' \subset V_c(\tfrac{1}{2})$ and $\mathrm{span}(w, w')$ is an invariant subspace of L_c with two eigenvalues 1 and $-\tfrac{1}{2}$. In particular, $n_2(c) \geq 1$.

PROOF. Notice by Lemma 6.7.1 that $V_w(\varepsilon) \neq \{0\}$; hence w' does exists, and $c \neq 0$ because w and w' are linearly independent. Moreover, by (6.7.15), $c^2 = \tfrac{1}{4}(2\mu w' + 2w) = c$; hence $c \in I(V)$. We also have $L_c w = \tfrac{1}{2}\mu ww' = \tfrac{1}{2}\mu\varepsilon w'$ and $L_c w' = \tfrac{1}{2}w' + \varepsilon\mu w$; hence $\mathrm{span}(w, w')$ is an invariant subspace of L_c. One easily finds that L_c has two eigenvalues 1 and $-\tfrac{1}{2}$ on $\mathrm{span}(w, w')$; therefore $n_2(c) \geq 1$. Moreover, for any $x \in V_w(\varepsilon) \ominus \mathbb{R}w'$ we have $\langle x, w'\rangle = 0$; hence (6.7.15) yields $xw' = 0$. Therefore $L_c x = \tfrac{1}{2}\varepsilon wx = \tfrac{1}{2}x$, and the result follows. □

6.8. Triality systems and a hidden Clifford algebra structure

6.8.1. Preliminary remarks. It follows from Table 6.7.1 that $V_c(1,-1)$ is a subalgebra of V if $n_1(c) = \dim V_c(-1) \geq 1$ for some $c \in I(v)$. Observe, however, that in general $V_c(1,-1)$ itself is not a REC algebra except for the case when $n_1(c) = 1$. More precisely, one has the following description of $V_c(1,-1)$.

PROPOSITION 6.8.1. *If $c \in I(V)$ and $n_1(c) \geq 1$, then*

(6.8.1) $$y^2 = -\tfrac{4}{3}|y|^2 c, \quad \forall y \in V_c(-1)$$

and

(6.8.2) $$y'y'' = -\tfrac{4}{3}\langle y', y''\rangle c, \quad \forall y', y'' \in V_c(-1).$$

In particular, $V_c(1,-1)$ is an algebra with an involution L_c and the minimum polynomial

(6.8.3) $$x^2 + 2tx + (\tfrac{4}{3}|y|^2 - 3t^2)c = 0$$

and

$$\sum e_i^2 = \tfrac{4}{3}(n_1(c) - 1)c$$

for any orthonormal basis $\{e_i\}_{1 \leq i \leq n_1(c)+1}$ of $V_c(1,-1)$.

PROOF. Table 6.7.1 yields $V_c(-1)V_c(-1) \subset V_c(1)$; hence (6.8.1) follows readily from Proposition 6.6.2 and (6.7.3), and then (6.8.2) follows from (6.8.1) by polarization. Applying (6.8.1) to an arbitrary $x = tc + y \in V_c(1,-1)$, where $t \in \mathbb{R}$ and $y \in V_c(-1)$, we obtain

$$x^2 = t^2 c - 2ty - \tfrac{4}{3}|y|^2 c = -2tx + (3t^2 - \tfrac{4}{3}|y|^2)c,$$

which proves (6.8.3). Since y and c are linearly independent, the algebra $V_c(1,-1)$ has rank 2. Moreover, we have $L_c x = ct - y$; hence $L_c^2 = 1$. Similarly one finds that $L_c(xx') = L_c x \, L_c x'$ for all $x, x' \in V_{01(c)}$; i.e., L_c is an involution on $V_c(1,-1)$.

Finally, we notice that the sum $\sum e_i^2$ does not depend on a particular choice of an orthonormal basis of $V_c(1,-1)$; thus, choosing the basis such that $e_1 = \frac{2}{\sqrt{3}} c$ and the remaining vectors form an orthonormal basis of $V_1(c)$, the desired formula for the sum easily follows from (6.8.1). $\qquad\square$

6.8.2. The main definition.

DEFINITION 6.8.1. *Any system of idempotents $c_i \in I(V)$, $i = 0, 1, 2$, satisfying*

$$c_0 + c_1 + c_2 = 0$$

is said to be a triality system.

If the idempotents $c_i \in I(V)$, $i = 0, 1, 2$, form a triality system, then for any permutation of indices $0, 1, 2$ one has $c_k = c_k^2 = (c_i + c_j)^2 = c_i + 2c_i c_j + c_j = 2c_i c_j - c_k$, implying

(6.8.4) $$c_i c_j = c_k.$$

PROPOSITION 6.8.2. *Let V be a normalized REC algebra, $c_0 \in I(V)$, and $n_1(c_0) \geq 1$. Then for any $\xi \in V_{c_0}(-1)$, $|\xi| = \frac{3}{4}$, the vectors*

(6.8.5) $$c_0, \quad c_1 = -\tfrac{1}{2} c_0 + \xi, \quad c_2 = -\tfrac{1}{2} c_0 - \xi$$

form a triality system. Conversely, any triality system c_0, c_1, c_2 gives rise to $\xi = \frac{1}{2}(c_1 - c_2) \in V_{c_0}(-1)$ and $|\xi| = \frac{3}{4}$ such that (6.8.5) holds.

PROOF. The fact that c_0, c_1, c_2 form a triality system is true by (6.8.1). In the converse direction, if $\{i, j, k\} = \{1, 2, 3\}$ is a permutation, then (6.8.4) yields $c_k c_{ij} = \frac{1}{2}(c_k c_i - c_k c_j) = -c_{ij}$ by the definition of a triality system; hence $c_{ij} \in V_{c_k}(-1)$. Furthermore, (6.7.3) yields $0 = \langle c_k, c_i \rangle + \langle c_k, c_j \rangle + \langle c_k, c_k \rangle$; hence $\langle c_i, c_j \rangle = -\frac{3}{8}$ for all $i \neq j$, and thus $4 \langle c_{ij}, c_{ij} \rangle = \frac{9}{16}$. This proves that $\xi = \frac{1}{2}(c_1 - c_2) \in V_{c_0}$, $|\xi| = \frac{3}{4}$, and (6.8.5) holds. $\qquad\square$

COROLLARY 6.8.1. *In a normalized REC algebra V, if c_0, c_1, c_2 is a triality system, then*

$$c_{ij} = \tfrac{1}{2}(c_i - c_j) \in V_{c_k}(-1), \quad |c_{ij}| = \tfrac{3}{4},$$

for any permutation $\{i, j, k\} = \{0, 1, 2\}$.

If c_0, c_1, c_2 is a triality system in V, then the 2-dimensional subspace $\Pi = \mathrm{span}(c_0, c_1, c_2)$ is easily seen to be invariant under action of any L_{c_i}, $i = 0, 1, 2$. Indeed, one has $L_{c_i} c_j = -c_i - c_j$ and $L_{c_i} c_k = c_j$, showing that L_{c_i} has the eigenvalues ± 1 (each of multiplicity one) on Π. Let us denote

$$\mathcal{C}_i = V_{c_i}(-1) \cap \Pi^\perp.$$

The following proposition shows that any triality system gives rise to a special orthogonal eigen-decomposition of V.

6.8. TRIALITY SYSTEMS AND A HIDDEN CLIFFORD ALGEBRA STRUCTURE

PROPOSITION 6.8.3. *If c_0, c_1, c_2 form a triality system and $\{i, j, k\}$ is a permutation of $\{0, 1, 2\}$, then*

$$(6.8.6) \qquad V = \Pi \oplus \mathcal{C}_0 \oplus \mathcal{C}_1 \oplus \mathcal{C}_2 \oplus V_{c_k}(-\tfrac{1}{2}) \oplus c_{ij} V_{c_k}(-\tfrac{1}{2}),$$

$$(6.8.7) \qquad V_{c_k}(\tfrac{1}{2}) = \mathcal{C}_i \oplus \mathcal{C}_j \oplus c_{ij} V_{c_k}(-\tfrac{1}{2});$$

see also Table 6.8.1. Moreover,

$$(6.8.8) \qquad \{L_{c_i}, L_{c_j}\} = -\tfrac{1}{4} \quad \text{on } V_{c_k}(-\tfrac{1}{2}) \oplus c_{ij} V_{c_k}(-\tfrac{1}{2}),$$

and

$$(6.8.9) \qquad L_{c_{ij}}^2 = \tfrac{3}{16} \quad \text{on } V_{c_k}(-\tfrac{1}{2}) \oplus c_{ij} V_{c_k}(-\tfrac{1}{2}).$$

TABLE 6.8.1. The spectral properties of a triple system c_0, c_1, c_2.

	Π	$\mathcal{C}_0$	$\mathcal{C}_1$	$\mathcal{C}_2$
L_{c_0}	± 1	-1	$\tfrac{1}{2}$	$\tfrac{1}{2}$
L_{c_1}	± 1	$\tfrac{1}{2}$	-1	$\tfrac{1}{2}$
L_{c_2}	± 1	$\tfrac{1}{2}$	$\tfrac{1}{2}$	-1

PROOF. First notice that

$$(6.8.10) \qquad \mathcal{C}_i \subset V_{c_j}(\tfrac{1}{2}), \quad \forall i \neq j.$$

Indeed, given $x \in \mathcal{C}_i$ we have $c_i x = -x$. By Proposition 6.8.2, $c_{jk} \in \mathcal{C}_i$, and by virtue of $x \in \Pi^\perp$, we also have $\langle x, c_{jk} \rangle = 0$; hence by (6.8.2), $c_{jk} x = 0$. On the other hand, $c_0 + c_1 + c_2 = 0$ yields $c_{jk} = c_j + \tfrac{1}{2} c_i$; hence $c_j x = -\tfrac{1}{2} c_i x = \tfrac{1}{2} x$, implying $x \in V_{c_j}(\tfrac{1}{2})$. Hence (6.8.10) follows. Furthermore, interchanging i and j in (6.8.10) we have $\mathcal{C}_j \subset V_{c_i}(\tfrac{1}{2})$, which implies $\langle \mathcal{C}_i, \mathcal{C}_j \rangle = 0$ for all $i \neq j$. Thus (6.8.10) yields $\mathcal{C}_i \oplus \mathcal{C}_j \subset V_{c_k}(\tfrac{1}{2})$, implying the existence of a subspace $H_k \subset V_{c_k}(\tfrac{1}{2})$ such that

$$(6.8.11) \qquad V_{c_k}(\tfrac{1}{2}) = \mathcal{C}_i \oplus \mathcal{C}_j \oplus H_k.$$

Now, (6.8.6) and (6.8.7) will be established if we verify that $H_k = c_{ij} V_{c_k}(-\tfrac{1}{2})$. To this end, let us notice that Π is an invariant subspace of $L_{c_{ij}}$ and also using Table 6.8.1 one finds that

$$(6.8.12) \qquad L_{c_{ij}} = \begin{cases} -\tfrac{3}{4}, & x \in \mathcal{C}_i, \\ \tfrac{3}{4}, & x \in \mathcal{C}_j, \\ 0, & x \in \mathcal{C}_k, \end{cases}$$

which implies that $V_{c_k}(-\tfrac{1}{2}) \oplus H_k$ is an invariant subspace of $L_{c_{ij}}$. On the other hand, by Proposition 6.8.2, $c_{ij} \in V_{c_k}(-1)$; hence (see Table 6.7.1) $L_{c_{ij}} : V_{c_k}(-\tfrac{1}{2}) \to V_{c_k}(\tfrac{1}{2})$, thus proving that $L_{c_{ij}}(V_{c_k}(-\tfrac{1}{2})) \subset H_k$. In order to prove the converse inclusion, we notice that $L_{c_{ij}}$ is an injection on $V_{c_k}(-\tfrac{1}{2})$. Indeed, if $c_{ij} x = 0$ holds for some $x \in V_{c_k}(-\tfrac{1}{2})$, then $c_i x = c_j x$, implying by the definition of a triality system $-\frac{1}{2}x = c_k x = -c_i x - c_j x = -2c_i x$, and therefore $c_i x = \frac{1}{4}x$. But $c_i \in I(V)$; hence by Proposition 6.7.1, L_{c_i} has no eigenvalue $\frac{1}{4}$. Therefore $x = 0$ and the injectivity follows. In particular, $\dim L_{c_{ij}}(V_{c_k}(-\frac{1}{2})) = \dim V_{c_k}(-\frac{1}{2}) = n_2(c_k)$; i.e., $\dim H_k \geq n_2(c_k)$. Hence using (6.8.11) we obtain by (6.7.12) that

$$(6.8.13) \quad \begin{aligned} n_2(c_k) &\leq \dim H_k = n_3(c_k) - n_1(c_i) - n_1(c_j) + 2 \\ &= n_2(c_k) + (2n_1(c_k) - n_1(c_i) - n_1(c_j)), \end{aligned}$$

which shows that $2n_1(c_k) \geq n_1(c_i) + n_1(c_j)$. Since the latter inequality holds true for any permutations of indices, we have $n_1(c_i) = n_1(c_j) = n_1(c_k)$; therefore (6.8.13) implies $n_2(c_k) = H_k$. Thus, $H_k = c_{ij} V_{c_k}(-\frac{1}{2})$.

Note by (6.8.6) that $H := V_{c_k}(-\frac{1}{2}) \oplus c_{ij} V_{c_k}(-\frac{1}{2})$ does not depend of the choice of $k \in \{0, 1, 2\}$, and by virtue of $H \subset V_{c_k}(-\frac{1}{2}, \frac{1}{2})$, H is a common invariant subspace of all L_{c_k}. Moreover, the latter inclusion also yields $L_{c_k}^2 = \frac{1}{4}$ on H for any $k \in \{0, 1, 2\}$; hence

$$\tfrac{1}{4} = L_{c_k}^2 = (L_{c_i} + L_{c_j})^2 = \tfrac{1}{2} + \{L_{c_i}, L_{c_j}\} \quad \text{on } H,$$

which proves (6.8.8). This also yields

$$4L_{c_{ij}}^2 = (L_{c_i} - L_{c_j})^2 = \tfrac{1}{2} - \{L_{c_i}, L_{c_j}\} = \tfrac{3}{4} \quad \text{on } H,$$

hence establishing (6.8.9) and finishing the proof. $\square$

6.8.3. The hidden Clifford algebra structure.

PROPOSITION 6.8.4. *For any $x \in V_c(-1)$,*

$$(6.8.14) \qquad A(x)^2 = |x|^2 \quad \text{on } V_c(-\tfrac{1}{2}, \tfrac{1}{2}),$$

where $A(x) = \sqrt{3} L_x - (1 + \sqrt{3})\{L_x, L_c\}$. In particular,

$$(6.8.15) \qquad L_x^2 = \tfrac{1}{3}|x|^2 \quad \text{on } V_c(-\tfrac{1}{2})$$

and the triple $(V_c(-1), V_c(-\frac{1}{2}, \frac{1}{2}), A)$ with A defined by (6.8.14) is a symmetric Clifford system.

PROOF. Setting $\xi = \frac{3}{4} x/|x|$, we have in the notation of Proposition 6.8.2, $c_{12} = \xi$; hence (6.8.9) yields $L_\xi^2 = \frac{3}{4}$ on $V_c(-\frac{1}{2})$, which yields (6.8.15). Furthermore, by Proposition 6.8.2, $c_0 = c$, $c_1 = -\frac{1}{2}c + \xi$, and $c_2 = -\frac{1}{2}c - \xi$ form a triality system; hence (6.8.7) with $(i, j, k) = (1, 2, 0)$ yields

$$V_c(\tfrac{1}{2}) = \mathcal{C}_1 \oplus \mathcal{C}_2 \oplus \xi V_c(-\tfrac{1}{2}),$$

while (6.8.12) and (6.8.9) yield, respectively,

$$(6.8.16) \qquad L_\xi x = \begin{cases} -\tfrac{3}{4}x, & x \in \mathcal{C}_1, \\ \tfrac{3}{4}x, & x \in \mathcal{C}_2 \end{cases}$$

and

$$(6.8.17) \qquad L_\xi^2 = \tfrac{3}{16} \quad \text{on } H = V_c(-\tfrac{1}{2}) \oplus \xi V_c(-\tfrac{1}{2}).$$

Using (6.8.8),

$$(6.8.18) \qquad \{L_\xi, L_c\} = \{L_{c_{12}}, L_{c_0}\} = \{L_{c_1}, L_{c_0}\} - \{L_{c_2}, L_{c_0}\} = 0 \quad \text{on } H.$$

Now suppose $x \in \mathcal{C}_i$. Then (6.8.10) yields $L_{c_{ij}}x = \frac{1}{2}(L_{c_i} - L_{c_j})x = -\frac{3}{4}x$ and $L_{c_k}x = \frac{1}{2}x$; therefore
$$\{L_{c_{ij}}, L_{c_k}\}x = L_{c_{ij}}L_{c_k}x + L_{c_k}L_{c_{ij}}x = -\frac{3}{4}x,$$
implying $\{L_\xi, L_c\} = -\frac{3}{4}$ on $\mathcal{C}_1$. Similarly one finds $\{L_\xi, L_c\} = \frac{3}{4}$ on $\mathcal{C}_2$; hence comparing with (6.8.16) and (6.8.18) yields

(6.8.19) $$\{L_\xi, L_c\} = \begin{cases} L_\xi, & x \in \mathcal{C}_1 \oplus \mathcal{C}_2, \\ 0, & x \in H. \end{cases}$$

This shows by virtue of (6.8.17) that
$$(L_\xi - a\{L_\xi, L_c\})^2 = \begin{cases} \frac{9}{16}(1-a)^2, & x \in \mathcal{C}_1 \oplus \mathcal{C}_2, \\ \frac{3}{16}, & x \in H; \end{cases}$$

hence $(L_\xi + a\{L_\xi, L_c\})^2 = \frac{3}{16}$ on $H \oplus \mathcal{C}_1 \oplus \mathcal{C}_2 \equiv V_c(-\frac{1}{2}, \frac{1}{2})$ for $(1-a)^2 = \frac{1}{3}$, which by our choice of ξ yields (6.8.14) for any $x \in V_c(-1)$. This immediately implies that the triple $(V_c(-1), V_c(-\frac{1}{2}, \frac{1}{2}), A)$ is a symmetric Clifford system.

Finally, notice that by (6.8.19), $\{L_\xi, L_c\} = 0$ on $V_c(-\frac{1}{2})$, which proves by virtue of (6.8.14) the identity (6.8.15). $\square$

As an immediate corollary of the existence of a symmetric Clifford system we have the following restriction on possible values of the characteristic dimensions of a REC algebra.

COROLLARY 6.8.2. *If V is a REC algebra and $c \in I(V)$, then*
$$n_1(c) - 1 \leq \rho(n_1(c) + n_2(c) - 1),$$
where ρ is the Hurwitz-Radon function ρ of (6.5.18).

PROOF. Indeed, $\frac{2}{3}A_\xi$ is obviously a selfadjoint endomorphism of $Y = V_c(-\frac{1}{2}, \frac{1}{2})$ for any $\xi \in X = V_c(-1)$. It follows from (6.8.14) that $\frac{2}{3}A : X \to \text{End}_s(Y)$ is a Clifford symmetric system. Since $\dim X = \dim V_c(-1) = n_1(c)$ and by (6.7.12) $\dim Y = n_2(c) + n_3(c) = 2(n_1(c) + n_2(c) - 1)$, we arrive at the desired conclusion by virtue of (6.5.17). $\square$

6.9. Jordan algebra $V_c^\bullet$

In this section we establish the existence of a hidden Jordan algebra structure inside any REC algebra and associated with the invariant subspace $V_c(1, -\frac{1}{2})$.

THEOREM 6.9.1. *Let V be a REC algebra and let $n_2(c) \geq 1$ for some $c \in I(V)$. Then the cubic form $N(x) = \frac{1}{6}\langle x, x^2 \rangle$ on $V_c(1, -\frac{1}{2})$ with a basepoint $e = 2c$ is Jordan. Let $V_c^\bullet$ denote the Jordan algebra on the invariant subspace $V_c(1, -\frac{1}{2})$ obtained by virtue of the Springer construction. Then the Jordan multiplication on $V_c^\bullet$ is given by*

(6.9.1) $\qquad x \bullet y = \frac{1}{2}xy + \langle x, c \rangle y + \langle y, c \rangle x - 2\langle xy, c \rangle c, \qquad x, y \in V_c(1, -\frac{1}{2}).$

Moreover, the bilinear trace form $T(x; y) = \frac{1}{3}\langle x, y \rangle$ is associative with respect to $\bullet$.

We shall see that the resulting Jordan algebra $V_c^\bullet$ plays a crucial role in the classification of REC algebras. In particular, the main result of the next section, Theorem 6.10.1, claims that the property of a REC algebra V being Clifford is equivalent to the reducibility of the associated Jordan algebra $V_c^\bullet$.

6.9.1. An auxiliary lemma. We begin below with some general observations concerning the invariant subspaces $V_c(-\frac{1}{2})$ and $V_c(1,-\frac{1}{2})$ and finish the proof of Theorem 6.9.1 at the end of Subsection 6.9.2.

Let V be a REC algebra and let $c \in I(V)$. It follows from Table 6.7.1 that the vector space $V_c(1,-\frac{1}{2}) = V_c(1) \oplus V_c(-\frac{1}{2})$ is closed under the Freudental multiplication which means that $V_c(1,-\frac{1}{2})$ is a subalgebra of V. It is easy to see that the defining equation (6.7.5) is still valid for any $x \in V_c(1,-\frac{1}{2})$. Observe, however, that the subalgebra in not necessarily a REC algebra itself because condition (6.7.4) may not be satisfied. We begin with the following simple observation:

$$(6.9.2) \qquad 2L_c + 1 - 4c \otimes c = 0 \quad \text{on} \quad V_c(1,-\tfrac{1}{2}).$$

Indeed, the left-hand side vanishes for c, and thus for any vector in $V_c(1)$. Moreover, since $L_c x = -\frac{1}{2}x$ on $V_c(-\frac{1}{2})$, the formula (6.9.2) is also valid for any $x \in V_c(-\frac{1}{2})$.

LEMMA 6.9.1. *For any* $x \in V_c(1,-\frac{1}{2})$,

$$(6.9.3) \quad x^3 + 2\langle x,c\rangle x^2 + \big(4\langle x,c\rangle^2 - |x|^2\big)x + \big(\tfrac{2}{3}\langle x^2, x\rangle - 16\langle x,c\rangle^3 + 4|x|^2\langle x,c\rangle\big)c = 0,$$

$$(6.9.4) \qquad x^2 x^2 = \tfrac{4}{3}\langle x^2, x\rangle x,$$

$$(6.9.5) \qquad xx^3 - |x|^2 x^2 - \tfrac{1}{3}\langle x^2, x\rangle x = 0.$$

PROOF. By (6.9.2),

$$(6.9.6) \qquad yc = 2\langle y,c\rangle c - \tfrac{1}{2}y, \quad \forall y \in V_c(1,-\tfrac{1}{2}).$$

Using (6.9.6) and setting $a_k = \langle x^k, c\rangle$, we obtain for any $x \in V_c(1,-\frac{1}{2})$

$$(6.9.7) \qquad a_2 = \langle x, xc\rangle = 2a_1^2 - \tfrac{1}{2}|x|^2,$$

$$(6.9.8) \qquad a_3 = \langle x^2, xc\rangle = 2a_1 a_2 - \tfrac{1}{2}\langle x^2, x\rangle,$$

Similarly one finds that $L_x c = 2a_1 c - \frac{x}{2}$, $L_{x^3} c = 2a_3 c - \frac{1}{2}x^3$, $L_x^3 c = 8a_1^3 c - 2a_1^2 x - a_1 x^2 - \frac{1}{2}x^3$, and $\{L_x, L_x^2\}c = 8a_1 a_2 c - a_2 x - a_2 x^2 - x^3$. Hence, applying (6.7.7) to e and using the relations obtained above, one readily obtains on eliminating a_3 by virtue of (6.9.8) that

$$(6.9.9) \quad x^3 + 2a_1 x^2 + 2a_2 x + \big(\tfrac{2}{3}\langle x^2, x\rangle - 8a_1 a_2\big)c = 0, \quad \forall x \in V_c(1,-\tfrac{1}{2}),$$

and (6.9.3) follows. Furthermore, multiplying the latter relation by x and eliminating x^3 by virtue of (6.9.9) readily yields $xx^3 = (4a_1^2 - 2a_2)x^2 + \frac{1}{3}\langle x^2, x\rangle x$, which proves (6.9.5) by virtue of (6.9.7). Then (6.9.4) follows from (6.9.5) and (6.7.6). □

COROLLARY 6.9.1. *If* $c \in I(V)$ *and* $n_2(c) \geq 1$, *then*

$$\langle x^2, x^2\rangle = |x|^4, \qquad x \in V_c(-\tfrac{1}{2}).$$

PROOF. Let $x \in V_c(-\frac{1}{2})$. Then $\langle c, x\rangle = 0$; hence one finds from (6.9.5) that $\langle xx^3, c\rangle = |x|^2 \langle x^2, c\rangle$, and hence $\langle x^2, (xc)x\rangle = |x|^2 \langle x, xc\rangle$, which yields the desired relation by virtue of $xc = -\frac{1}{2}x$. □

6.9.2. Proof of Theorem 6.9.1.
We have by (3.2.17) that $N(x;y) = \frac{1}{6}\langle x^2, y\rangle$; hence by virtue of $e^2 = 2e$ we find

(6.9.10) $$\operatorname{Tr}(x) = 3N(e;x) = \langle e, x\rangle$$

and $S(x) = 3N(x;e) = \frac{1}{2}\langle x^2, e\rangle$. Hence $S(x;y) = \langle xy, e\rangle$, which yields the expression for the bilinear trace form:

$$3T(x;y) = \operatorname{Tr}(x)\operatorname{Tr}(y) - S(x;y) = \langle e, x\rangle\langle e, y\rangle - \langle e, xy\rangle.$$

It also follows from (6.9.2) that $\langle e, x\rangle\langle e, y\rangle - \langle e, xy\rangle = \langle y, e\langle e, x\rangle - ex\rangle = \langle y, x\rangle$; hence $3T(x;y) = \langle x, y\rangle$. Then the definition (3.2.24) of the adjoint element yields $\langle x^\#, y\rangle = \frac{1}{2}\langle x^2, y\rangle$; hence by virtue of the nonsingularity of $\langle\,,\,\rangle$ we have $x^\# = \frac{1}{2}x^2$. In order to justify the adjoint identity (3.2.25) we remark that it follows by expressing $x^2 x^2$ by virtue of (6.7.6) and using (6.9.5):

$$x^{\#\#} = \tfrac{1}{8}x^2 x^2 = -\tfrac{1}{2}xx^3 + \tfrac{1}{2}|x|^2 x^2 + 2N(x)x = N(x)x.$$

Now, when it is proved that $N(x)$ is a Jordan form with a basepoint $e = 2c$, the Jordan structure on $V_c^\bullet$ is obtained by applying the Springer construction to the cubic form $\frac{1}{6}\langle x, x^2\rangle$ (as was described in Subsection 3.2.7) and observing that (3.2.30) follows from (6.9.1).

6.9.3. The multiplication operator on $V_c^\bullet$.
We will not make much use of the Jordan product $\bullet$; instead, we will be working with the multiplication operator $R_x y = x \bullet y$ on $V_c^\bullet$ expressed by virtue of (6.9.1) as

(6.9.11) $$R_x = \tfrac{1}{2}L_x + \langle x, c\rangle + x \otimes c - 2c \otimes cx, \qquad x \in V_c(1, -\tfrac{1}{2}).$$

It is convenient for further applications to rewrite (6.9.11) in a more symmetric manner. To this end, we notice that (6.9.2) yields $2cx = 4\langle c, x\rangle c - x$ for any $x \in V_c(1, -\tfrac{1}{2})$; hence eliminating cx in (6.9.11) yields

(6.9.12) $$R_x = \tfrac{1}{2}L_x + \langle x, c\rangle - 4\langle x, c\rangle c \otimes c + \{c \otimes x\},$$

which also yields that R_x is a selfadjoint operator.

Proposition 6.9.1. *In the described notation,*

(6.9.13) $$\tfrac{1}{2}\sum_{i=1}^{n_2(c)+1} e_i^2 = \tfrac{2-n_2(c)}{3}c + \beta(c), \qquad \beta(c) \in V_c(-\tfrac{1}{2}),$$

and

(6.9.14) $$\operatorname{tr} R_x = \langle \beta(c), x\rangle + \tfrac{2(n_2(c)+1)}{3}\langle x, c\rangle,$$

where $\{e_i\}$ is an arbitrary orthonormal basis of $V_c(1, -\tfrac{1}{2})$.

Proof. We have from (6.9.11)

(6.9.15) $$\operatorname{tr} R_x = \langle y, x\rangle + n_2(c)\langle x, c\rangle,$$

where $y = \frac{1}{2}\sum_{i=1}^{n_2(c)+1} e_i^2 \in V_c(1, -\tfrac{1}{2})$ and $\{e_i\}$ is an arbitrary orthonormal basis of $V_c(1, -\tfrac{1}{2})$. We have $y = \tfrac{2}{3}\langle y, c\rangle c + \beta(c)$. Choosing an orthonormal basis such that $e_1 = c/|c|$ with the remaining vectors being an orthonormal basis of $V_c(-\tfrac{1}{2})$, we

find
$$\langle y,c\rangle = 1 + \sum_{i=1}^{n_2(c)}\langle e_i, ce_i\rangle = 1 - \frac{1}{2}\sum_{i=1}^{n_2(c)}\langle e_i, e_i\rangle = \frac{2-n_2(c)}{2},$$
which readily yields the claimed formulas by virtue of (6.9.15). □

6.9.4. An auxiliary lemma.

LEMMA 6.9.2. *Let $x \in V_c(-\frac{1}{2})$, $x \ne 0$, satisfy $x^2 = px + qc$ for some $p, q \in \mathbb{R}$. Then $-q = p^2 = \frac{2}{3}|x|^2$ and $\langle x, x^2\rangle = \frac{3}{2}p^3$. Moreover $\mathrm{span}(c,x)$ is an invariant subspace of L_x, and if additionally $n_2(c) \ge 2$ holds, then the only possible eigenvalues of L_x on $V_c(-\frac{1}{2}) \ominus \mathbb{R}x$ are $-p$, $-\frac{p}{2}$, and $\frac{p}{2}$.*

PROOF. We have $\langle x^2, x\rangle = p|x|^2$ and also $-\frac{1}{2}|x|^2 = \langle x^2, c\rangle = \frac{3}{4}q$; hence $|x|^2 = -\frac{3}{2}q$. This shows, in particular, that $q \ne 0$. Further, using (6.9.4), we find
$$\tfrac{4}{3}p|x|^2 x = x^2 x^2 = (p^2 - q)px + q(p^2 + q)c,$$
which by virtue of $q \ne 0$ yields $q = -p^2$, and therefore $p \ne 0$ and $\frac{4}{3}|x|^2 = 2p^2$. Next, it follows from $L_x x = px + qc$ and $L_x c = -\frac{1}{2}x$ that the linear span $\mathrm{span}(c,x)$ is an invariant subspace of L_x. Thus, $V_c(-\frac{1}{2}) \ominus \mathbb{R}x$ is also an eigenspace of L_x and the following relations are valid there: $L_c = -\frac{1}{2}$ and $c \otimes c = v \otimes c = c \times v = 0$; hence (6.7.7) readily yields
$$4L_v^3 + 4pL_v^2 - p^2 L_v - p^3 = 0.$$
The latter identity implies that the only possible eigenvalues of L_v on $V_c'(-\frac{1}{2})$ are $-p$, $-\frac{p}{2}$, and $\frac{p}{2}$. □

COROLLARY 6.9.2. *In the above notation, the following are true:*
- *If $n_2(c) = 0$, then $V_c^\bullet = \mathbb{R}c$ is a simple Jordan algebra of rank 1.*
- *If $n_2(c) = 1$, then $V_c^\bullet$ is a reducible Jordan algebra of rank 2.*
- *If $n_2(c) \ge 2$, then $V_c^\bullet$ is a Jordan algebra of rank 3.*

PROOF. The case $n_2(c) = 0$ is trivial. In the case $n_2(c) = 1$ we have $\dim V_c^\bullet = 2$ and it is easily seen that rank $V_c^\bullet = 2$. The Jordan-von Neumann-Wigner classification [**124**] states that the only simple Jordan algebras of rank 2 are the spin factors (see Example 3.2.4). But these are at least 3-dimensional; thus $V_c^\bullet$ must be reducible.

Now suppose $n_2(c) \ge 2$. By the Springer construction, rank $V_c^\bullet \le 3$, and since $n_2(c) \ge 2$, we also see that rank $V_c^\bullet \ge 2$. Thus it suffices to show that rank $V_c^\bullet \ne 2$. Arguing by contradiction, assume that rank $V_c^\bullet = 2$; hence any element of $V_c^\bullet$ must satisfy a quadratic relation $x^{\bullet 2} + t(x)x + n(x)e = 0$, where $e = 2c$ and $t: V_c^\bullet \to \mathbb{R}$ is a linear form. Applying (6.9.1) to the latter identity yields for any $x \in V_c(-\frac{1}{2}) \subset V_c^\bullet$ that $x^2 + t_1(x)x + n_1(x)c = 0$, where $t_1(x)$ is a new linear form. Applying Lemma 6.9.2 we obtain $|x|^2 = \frac{3}{2}t_1^2(x)$, which together with $\dim V_c(-\frac{1}{2}) \ge 2$ yields a contradiction to the fact that $t_1(x)$ is a linear form. □

6.9.5. Reducibility of $V_c^\bullet$.
Recall that a vector subspace $\mathcal{I}$ of a Jordan algebra J is called ideal if $\mathcal{I} \bullet J \subset \mathcal{I}$. An ideal is called nontrivial if $\mathcal{I} \ne \{0\}$ and $\mathcal{I} \ne J$. If J is a Euclidean algebra with respect to the associative product $\langle\,,\,\rangle$, then $\mathcal{I}^\perp$ is ideal whenever $\mathcal{I}$ is, and in this case $\mathcal{I} \bullet \mathcal{I}^\perp = \{0\}$.

In this subsection we establish the following criterion.

THEOREM 6.9.2. *The following conditions are equivalent:*
(a) *The Jordan algebra $V_c^\bullet$ is reducible.*
(b) *There exists $w \in V_c(1, -\frac{1}{2}) \cap \mathcal{N}(V)$ such that $L_w = -1$ on*
$$V_{c,w}(-\tfrac{1}{2}) := V_c(1, -\tfrac{1}{2}) \ominus \mathrm{span}(c, w).$$
(c) *At least one of the conditions $n_2(c) = 2$ or $\beta(c) \neq 0$ holds.*

PROOF. The implication (a)$\Rightarrow$(b) is proved as follows. Since $V_c^\bullet$ is a Euclidean Jordan algebra, it is semisimple. Moreover, since the rank of $V_c^\bullet$ is at most 3, it contains a rank 1 ideal, say $\mathcal{I} = \mathbb{R}w$, where w is a nonzero idempotent in $V_c^\bullet$, i.e., $w^{\bullet 2} = w$. Write $w = v + (1-t)c$ for some $v \in V_c(-\frac{1}{2})$, $t \in \mathbb{R}$, and notice that $v \neq 0$ because otherwise by virtue of $e = 2c$ one has $V_c^\bullet = e \bullet V_c^\bullet = \mathcal{I}$, implying that $V_c^\bullet$ is simple. Using $w^{\bullet 2} = w$ and (6.9.1) yields $v^2 = 2tv + (1 - t^2 - 2|v|^2)c$. Applying Lemma 6.9.2 to the obtained relation yields $t = \pm\frac{1}{3}$ and $|v|^2 = \frac{2}{3}$; hence

(6.9.16) $$v^2 = 2tv - \tfrac{4}{9}c.$$

Let us consider first the case $n_2(c) \geq 2$. Then there exists $0 \neq z \in V_c(-\frac{1}{2})$ and $\langle v, z \rangle = 0$; hence $z \in \mathcal{I}^\perp$. It follows that $w \bullet z = 0$; hence expanding the latter equality by (6.9.1) yields $vz = (t-1)z$, and hence $(t-1)$ is an eigenvalue of L_v on $V_c(-\frac{1}{2}) \ominus \mathbb{R}v$. On the other hand, by Lemma 6.9.2 the spectrum of L_v is a subset of $\{-2t, -t, t\}$, which immediately shows that $t = -\frac{1}{3}$ is impossible, thus implying $t = \frac{1}{3}$. It follows that $|v|^2 = \frac{3}{2}p^2 = \frac{2}{3}$ and (6.9.16) readily yields $w^2 = 0$ and $|w| = 1$; therefore $w \in \mathcal{N}(V)$. Moreover, we have $V_{c,w}(-\frac{1}{2}) = V_c(-\frac{1}{2}) \ominus \mathbb{R}v \subset V_c(-\frac{1}{2})$ and the identity $L_w = L_v + \frac{2}{3}L_c = -\frac{2}{3} - \frac{1}{3} = -1$ holds on $V_{c,w}(-\frac{1}{2})$, as desired.

Now suppose $n_2(c) = 1$. By Corollary 6.9.2, $V_c^\bullet$ is a reducible rank 2 Jordan algebra; hence $V_c^\bullet = \mathcal{I}_1 \oplus \mathcal{I}_2$, where $\mathcal{I}_i = \mathbb{R}w_i$, $i = 1, 2$, with w_i being two orthogonal (primitive) idempotents; i.e., $w_i^{\bullet 2} = w_i$ and $e = 2c = w_1 + w_2$. Then arguing as above one readily obtains that $w_1 = v + (1-t)c$ and $w_2 = -v + (1+t)c$ for some v satisfying (6.9.16) and $t = \pm\frac{1}{3}$. Since the substitution $t \to -t$, $v \to -v$ does not change (6.9.16) and interchanges w_1 and w_2, one can assume without loss of generality that $t = \frac{1}{3}$, which yields $w_1^2 = 0$ and $|w_1|^2 = 1$. Thus $w_1 \in w \in V_c(1, -\frac{1}{2}) \cap \mathcal{N}(V)$ and by virtue of $V_c(1, -\frac{1}{2}) = \mathrm{span}(c, w)$, the lemma is proved completely.

In order to prove the implication (b)$\Rightarrow$(a) we suppose that $w \in V_c(1, -\frac{1}{2}) \cap \mathcal{N}(V)$ is such that $L_w = -1$ on $V_{c,w}(-\frac{1}{2})$. Note that the nonemptyness of $\mathcal{N}(V)$ yields $n_1(c) \geq 1$ and we may assume that $n_2(c) \geq 2$ because otherwise by Corollary 6.9.2, $V_c^\bullet$ is reducible. Thus, there exists $0 \neq z \in V_{c,w}(-\frac{1}{2})$ and by the assumption $z \in V_w(-1)$; hence (6.7.15) yields $z^2 = -|z|^2 w$. We have
$$-\tfrac{1}{2}|z|^2 = \langle z, cz \rangle = \langle c, z^2 \rangle = -\langle c, w \rangle |z|^2;$$
hence $2\langle c, w \rangle = 1$. The latter yields by (6.9.1) that $w^{\bullet 2} = 2\langle w, c \rangle w = w$; hence w is a nonzero idempotent of $V_c^\bullet$. Furthermore, we notice by (6.9.12) that
$$R_w = \tfrac{1}{2}L_w + \tfrac{1}{2} - 2c \otimes c + \{c \otimes w\},$$
and $R_w w = w \bullet w = w$ and $R_w c = w \bullet c = \frac{1}{2}w$; hence $\mathrm{span}(c, w)$ is an invariant subspace of R_w with eigenvalues 0 and 1. Thus, $V_{c,w}(-\frac{1}{2})$ is also an invariant subspace of R_w, and by the assumption $R_w = \frac{1}{2}L_w + \frac{1}{2} = 0$ there. In summary,

the only eigenvalues of multiplication operator R_w are 0 and 1, which proves by the well-known property of the Peirce decomposition that the Jordan algebra $V_c^\bullet$ is reducible (see (3.2.16) in Subsection 3.2.6 above or Proposition IV.1.2 in [**85**]).

In order to establish the implication (c)⇒(a) we notice by Corollary 6.9.2 that if $n_2(c) = 1$, then $V_c^\bullet$ is reducible, and if $n_2(c) = 2$, then $V_c^\bullet$ is a 3-dimensional Jordan algebra of rank 3; hence it is also reducible because the least dimension of a simple Jordan algebra of rank 3 by the Jordan-von Neumann-Wigner classification is 6 (the corresponding simple Jordan algebra is $\mathrm{Herm}_3(\mathbb{R})$). Thus, suppose $n_2(c) \geq 3$ and $\beta(c) \neq 0$. Then $V_c^\bullet$ is rank 3 and using (6.9.10) we have $\mathrm{Tr}\, x = \langle x, e \rangle = 2\langle x, c \rangle$; hence (6.9.14) becomes $\mathrm{tr}\, R_x = \frac{\dim J}{\mathrm{rank}\, J} \mathrm{Tr}\, x + \langle \beta(c), x \rangle$. On the other hand, Proposition III.4.2 in [**85**] says that in a simple Jordan algebra $\mathrm{tr}\, R_x = \frac{\dim J}{\mathrm{rank}\, J} \mathrm{Tr}\, x$, which proves by virtue of the assumption on $\beta(c)$ that $V_c^\bullet$ is reducible.

It remains to establish that (b) implies (c). Observe that it is sufficient to consider the case $n_2(c) \neq 2$. Let w satisfy condition (b) and set $w' = 2c - w$. It was established in the course of the proof of the implication (b)⇒(a) that $2\langle w, c \rangle = 1$, which yields $\langle w, w' \rangle = 0$, $\langle w', w' \rangle = 2$ and also by virtue of (6.9.2) that $2cw = w'$. The latter also yields $w'^2 = 2w$. Thus $e_1 = w$, $e_2 = w'/\sqrt{2}$ together with an arbitrary orthonormal basis $e_3, \ldots, e_{n_2+1}$ of $V_{c,w}(-\frac{1}{2})$ form an orthonormal basis of $V_c(1, -\frac{1}{2})$. By the assumption $L_w = -1$ on $V_{c,w}(-\frac{1}{2})$; hence (6.7.15) yields $e_i^2 = -w$ for $3 \leq i \leq n_2(c) + 1$. Therefore $\sum_{i=1}^{n_2(c)+1} e_i^2 = (n_2(c) - 2)w$, which yields $\beta(c) \neq 0$ and, thereby, implies (c). The theorem is proved. $\square$

6.9.6. Triality systems revisited. Combining the results of two preceding sections we arrive at the following observation. Starting from a triality system $c_0 + c_1 + c_2 = 0$, $c_i \in I(V)$, one has the following decomposition:

$$V = J \oplus J' \oplus \mathcal{C}_0 \oplus \mathcal{C}_1 \oplus \mathcal{C}_2,$$

where $J = V_c(-\frac{1}{2}) \oplus \mathbb{R}c$ and $J' = c_{12}J$. According to Theorem 6.9.1, J carries a natural Jordan algebra structure defined by virtue of (6.9.1) as

$$x \bullet y = \tfrac{1}{2}xy + \langle x, c \rangle y + \langle y, c \rangle x - 2\langle xy, c \rangle c, \qquad x, y \in J,$$

while J' is an isotopic copy of J; i.e., the Jordan product on J' is defined by $(c_{12}x) \diamond (c_{12}y) = c_{12}(x \bullet y)$, $x, y \in J$, with the obvious new unit element $2c_{21}$.

On the other hand, it follows from (6.8.1) that the three subspaces $\mathcal{C}_i$, $i = 0, 1, 2$, possess the Clifford type identity $x_i^2 = -\frac{4}{3}|x_i|^2 c_i$. It follows from Proposition 6.8.3 that $\Lambda := \frac{4}{\sqrt{3}} L_{c_{12}}$ is an isometric involution interchanging J and J' and acting on $\mathcal{C}_i$ as follows: $\Lambda = 0$ on $\mathcal{C}_0$, and $\Lambda = (-1)^k \frac{3}{4}$ on $\mathcal{C}_k$, $k = 1, 2$.

6.10. Reducible REC algebras

Now we are ready to establish the central result of this chapter. We begin with the definition.

DEFINITION 6.10.1. A REC algebra V is called *reducible* if the Jordan algebra $V_c^\bullet$ is reducible for some nonzero idempotent c.

THEOREM 6.10.1. *A REC algebra V is Clifford if and only if it is reducible. In particular, any normalized reducible REC algebra V is polar and its polar decomposition $V = V_0 \oplus V_1$ can be chosen such that $V_0 = V_w(0)$ for any $w \in N(V)$.*

Furthermore,

(6.10.1) $$\dim V_0 = 1 + n_1(c), \qquad \forall c \in I(V).$$

We give the proof of the "only if" part separately in Proposition 6.10.1 and consider general reducible REC algebras in Lemma 6.10.1. We finish the proof of the theorem in Subsection 6.10.2.

6.10.1. Auxiliary propositions.

PROPOSITION 6.10.1. *Any Clifford REC algebra V is a reducible REC algebra.*

PROOF. First note that if V is similar to W and V is a reducible REC algebra, then W is also reducible. Therefore, we assume that V is a polar algebra; hence $V = V_0 \oplus V_1$ according to the Definition 6.5.1. Let us also suppose $c \in I(V)$ is chosen arbitrarily. By Example 6.6.1, $c = z^2 + z$ for some $z \in V_1$ satisfying $2z^3 = z$. We claim that $w = 2z^2$ satisfies condition (b) in Theorem 6.9.2. To this end notice that $\frac{3}{4} = |c|^2 = |z|^2 + |z^2|^2$, and by (6.6.7), $2|z^2|^2 = |z|^2$; hence $|z|^2 = \frac{1}{2}$. Further, by $w \in V_0$, $w^2 = 0$ and also
$$|w|^2 = 4\langle z^2, z^2 \rangle = 4\langle z, z^3 \rangle = 2\langle z, z \rangle = 1;$$
hence $w \in \mathcal{N}(V)$. Moreover, $L_c w = c - \frac{1}{2} w$ and $L_c c = c$; hence the linear span $\mathrm{span}(z, z^2) = \mathrm{span}(c, w)$ is an invariant subspace of L_c with eigenvalues 1 and $-\frac{1}{2}$. This yields $w \in \mathcal{N}(V) \cap V_c(1, -\frac{1}{2})$.

It remains to show that $L_w x = -x$ for any $x = x_0 + x_1 \in V_c(1, -\frac{1}{2}) \ominus \mathrm{span}(c, w)$. We have $\langle x_0, z^2 \rangle = \langle x_1, z \rangle = 0$ and $cx = -\frac{1}{2} x$. The latter equation yields $L_w x = 2L_c x - zx = -x - zx$; thus, in order to prove the desired identity it suffices to establish that $zx = 0$. To this end, we notice that

(6.10.2) $$x_0 z + x_1 z^2 = -\tfrac{1}{2} x_1, \qquad x_1 z = -\tfrac{1}{2} x_0.$$

Multiplying the first equation by x_0 yields by virtue of (6.5.3) and (6.5.5) that
$$-\tfrac{1}{2} x_1 x_0 = x_0(x_0 z) + x_0(x_1 z^2) = |x_0|^2 z + 2\langle x_0, z^2 \rangle x_1 - z^2(x_0 x_1) = |x_0|^2 z - z^2(x_0 x_1);$$
hence using (6.6.7) we find $-\tfrac{1}{2}\langle x_1 x_0, z \rangle = |x_0|^2 |z|^2 - \tfrac{1}{2}\langle x_0 x_1, z \rangle$, which yields $|x_0|^2 = 0$ and thereby finishes the proof of the proposition. $\square$

LEMMA 6.10.1. *Let V be reducible with w satisfying condition* (b) *of Theorem 6.9.2. Then*

(6.10.3) $$V_w(1) = \mathbb{R} w' \oplus V_{c,w}(\tfrac{1}{2}),$$

(6.10.4) $$V_w(-1) = V_{c,w}(-\tfrac{1}{2}) \oplus L_w V_c(-1),$$

(6.10.5) $$V_w(0) = \mathbb{R} w \oplus (L_w + 1) V_c(-1),$$

where $w' = 2c - w$, $V_{c,w}(\tfrac{1}{2}) = V_w(1) \ominus \mathbb{R} w' \subset V_c(\tfrac{1}{2})$, and $V_{c,w}(-\tfrac{1}{2}) = V_c(1, -\tfrac{1}{2}) \ominus \mathrm{span}(c, w)$. Furthermore,

(6.10.6) $\dim V_w(-1) = \dim V_w(1) = n_1(c) + n_2(c) - 1, \qquad \dim V_w(0) = n_1(c) + 1.$

PROOF. Applying (6.9.2), we obtain $w'w = w$, which by Corollary 6.7.3 yields that $\mathrm{span}(w,c) = \mathrm{span}(w,w')$ is a 2-dimensional invariant subspace of L_c with the eigenvalues 1 and $-\frac{1}{2}$, each of multiplicity 1, and $V_{c,w}(\frac{1}{2}) := V_w(1) \ominus \mathbb{R}w' \subset V_c(\frac{1}{2})$. In particular, this proves (6.10.3). Note also that $\mathrm{span}(w,c)$ is also an invariant subspace of L_w with two eigenvalues 1 and 0. Denote $V'_c(\frac{1}{2}) = V_c(\frac{1}{2}) \ominus V_{c,w}(\frac{1}{2})$. It is clear that

(6.10.7) $$U := V_c(-1) \oplus V'_c(\tfrac{1}{2}) \subset V_w(0) \oplus V_w(-1);$$

hence

(6.10.8) $$L_w^2 + L_w = L_w(L_w + 1) = 0 \quad \text{on } U.$$

On the other hand, using the spectral properties of L_c we have $(L_c+1)(L_c - \frac{1}{2}) = 0$ on U; hence

(6.10.9) $$L_c^2 = \tfrac{1}{2}(1 - L_c) \quad \text{on } U.$$

Applying (6.7.8) to $x = c$ and $y = w$ yields

(6.10.10) $$L_c L_w L_c + \{L_c^2, L_w\} + 2L_c^2 - \tfrac{3}{4}L_w - \tfrac{1}{2} = 4c \otimes c - \{c \otimes w\}.$$

Since (6.10.10) vanishes on U, we obtain by applying the trace over U and using (6.10.9) that

(6.10.11) $$\tfrac{3}{4}\mathrm{tr}_U L_w - \tfrac{3}{2}\mathrm{tr}_U L_w L_c - \mathrm{tr}_U L_c + \tfrac{1}{2}\dim U = 0.$$

Similarly, applying (6.7.8) to $x = w$ and $y = c$ yields

$$L_w^2 + \tfrac{3}{2}L_w - \{L_w^2 + L_w, L_c\} - L_w L_c L_w = 2w \otimes w - 2\{c \otimes w\};$$

thus, the same argument by virtue of (6.10.8) yields $\mathrm{tr}_U L_w L_c = -\tfrac{1}{2}\mathrm{tr}_U L_w$. Hence combining the latter identity with (6.10.11) we obtain

(6.10.12) $$3\mathrm{tr}_U L_w - 2\mathrm{tr}_U L_c + \dim U = 0.$$

On the other hand, by the assumption, $L_w = -1$ on $V_{c,w}(-\frac{1}{2})$, which yields $\mathrm{tr}_U L_w = -(\dim V_w(-1) - n_2(c) + 1)$, and also $\mathrm{tr}_U L_c = -n_1(c) + \tfrac{1}{2}\dim V'_c(\tfrac{1}{2})$, which yields by virtue of (6.10.12)

(6.10.13) $$\dim V_w(-1) = n_1(c) + n_2(c) - 1.$$

By Lemma 6.7.1, $\dim V_w(1) = \dim V_w(-1)$ and using (6.7.13) and (6.10.13)

$$\dim V_w(0) = n - 2\dim V_w(-1) = n_1(c) + 1;$$

hence (6.10.6) is proved.

To proceed with (6.10.4) and (6.10.5) we decompose by virtue of (6.10.7)

$$U = U(-1) \oplus U(0), \qquad U(\varepsilon) \subset V_w(\varepsilon).$$

Then

$$\dim U(0) = \dim V_w(0) - 1 = n_1(c)$$

and

$$\dim U(-1) = \dim V_w(1) - \dim V_{c,w}(\tfrac{1}{2}) = n_1(c).$$

Next, since $V_c(-1) \subset U$, we see from (6.10.8) that $L_w : V_c(-1) \to U(-1)$ and $L_w + 1 : V_c(-1) \to U(0)$. We claim that L_w and $L_w + 1$ are bijections of $V_c(-1)$ onto $U(-1)$ and $U(0)$, respectively. By virtue of $\dim U(0) = \dim U(-1) = \dim V_c(-1)$ it suffices to show that these maps are injections. To this end, observe that $L_w - a$

acts as an injection on $V_c(-1)$ for any $a \neq \frac{2}{3}$. Indeed, setting $u = \frac{3}{2}w - c$ we have $L_c u = -\frac{1}{2}u$; hence $u \in V_c(-\frac{1}{2})$. Thus for any nonzero $x \in V_c(-1) \cap \ker(L_w - a)$ one has $ax = L_w x = \frac{2}{3}L_u x + \frac{2}{3}L_c x = \frac{2}{3}L_u x - \frac{2}{3}x$, and by Table 6.7.1, $L_u x \in V_c(\frac{1}{2})$; therefore $L_u x = 0$ and $a = -\frac{2}{3}$, which proves our observation and, thus, the claim. Since $V_w(-1) = V_{c,w}(-\frac{1}{2}) \oplus U(-1)$ and $V_w(0) = \mathbb{R}w \oplus U(0)$, we obtain (6.10.4) and (6.10.5), respectively. The lemma follows. $\square$

6.10.2. Proof of Theorem 6.10.1. The "only if" part of the theorem is by Proposition 6.10.1. Let us show that any reducible algebra is Clifford. To this end, one can assume without loss of generality that V is normalized. Let us first consider the case when $n_1(c) = 0$ for some $c \in I(V)$. Then by (6.7.12) and (6.7.13), $n_3(c) = n_2(c) - 2$ and $\dim V = 2n_2(c) - 1$; in particular $n_2(c) \geq 2$. Let w satisfy condition (b) in Theorem 6.9.2 and let $V = V_w(0) \oplus V_w(-1) \oplus V_w(1)$ be the eigendecomposition associated to L_w according to Lemma 6.7.1. By the choice of w, $V_c(1, -\frac{1}{2}) \ominus \mathrm{span}(c, w) \subset V_w(-1)$, and therefore $\dim V_w(-1) \geq n_2(c) - 1$. Since $2n_2(c) - 1 = \dim V$ and since by Lemma 6.7.1, $\dim V_w(-1) = \dim V_w(1)$, we have by virtue of $\dim V_w(0) \geq 1$

$$2n_2(c) - 1 = \dim V = 2\dim V_w(1) + \dim V_w(0) \geq 2n_2(c) - 1,$$

which implies $\dim V_w(1) = \dim V_w(-1) = n_2(c) - 1$ and $\dim V_w(0) = 1$; hence $V_c(1, -\frac{1}{2}) \ominus \mathrm{span}(c, w) = V_w(-1)$. By virtue of $V = \mathbb{R}c \oplus V_c(-\frac{1}{2}) \oplus V_c(\frac{1}{2})$ this yields $V_c(\frac{1}{2}) = V_w(1)$. In particular, in the notation of Lemma 6.7.1, $V'(w) = \{0\}$. Now, let us define $V_0 = V_w(0) = \mathbb{R}w$ and $V_1 = V_w(-1) \oplus V_w(1)$. We claim that $V = V_0 \oplus V_1$ is a polar algebra. Notice that (6.5.1) is trivially satisfied, $\mathrm{tr}\, L_w = 0$ by the harmonicity of a REC algebra, and also $L_w^2|_{V_1} = L_w^2|_{V_w(-1) \oplus V_w(1)} = 1$, which yields (6.5.3) by virtue of $\dim V_0 = 1$. Finally, since $V'(w) = \{0\}$ in our setting, (6.7.15) and (6.7.16) in Lemma 6.7.1 yield $V_1 V_1 \subset \mathbb{R}w = V_0$; hence condition (6.5.2) follows. This yields the desired conclusion in the case $n_1(c) = 0$.

Now let V be a reducible normalized REC algebra such that $n_1(c) \geq 1$ for some $c \in I(V)$. Let $w \in V$ satisfy condition (b) of Theorem 6.9.2. Then we shall show that $V = V_0 \oplus V_1$ with $V_0 = V_w(0)$ and $V_1 = V_w(-1) \oplus V_w(1)$ carries the structure of a polar algebra. Notice that condition (6.5.2) is by virtue of (6.7.15) and (6.7.16) in Lemma 6.7.1. Furthermore, by Lemma 6.10.1, $\dim V_0 = 1 + n_1(c) \geq 2$; hence we only need to verify (6.5.1) and (6.5.3).

In order to prove (6.5.1) we observe that any $x \in V_0$ can be written by (6.10.5) as $x = tw + z$, where $t \in \mathbb{R}$ and $z = w\xi + \xi$ for some $\xi \in V_c(-1)$. By (6.10.8), $L_w z = 0$; hence $x^2 = (tw + z)^2 = z^2$. Using (6.8.1) we have

$$z^2 = (\xi w)^2 + 2L_\xi^2 w + \xi^2 = (\xi w)^2 + 2L_\xi^2 w - \tfrac{4}{3}|\xi|^2 c.$$

We have already seen in the proof of Lemma 6.10.1 that $u = \frac{3}{2}w - c \in V_c(-\frac{1}{2})$; hence using (6.8.15) and (6.8.1) we obtain

$$2L_\xi^2 w = \tfrac{4}{3}L_\xi^2 u + \tfrac{4}{3}L_\xi^2 c = \tfrac{4}{9}|\xi|^2(u + 4c).$$

Furthermore, by virtue of (6.10.4), $w\xi \in V_w(-1)$; hence (6.7.15) yields $(w\xi)^2 = -|w\xi|^2 w$. Using (6.10.8) and (6.8.1)

$$|w\xi|^2 = \langle L_w^2 \xi, \xi \rangle = -\langle L_w \xi, \xi \rangle = -\langle w, \xi^2 \rangle = \tfrac{4}{3}|\xi|^2 \langle w, c \rangle = \tfrac{2}{3}|\xi|^2,$$

where the last equality is by virtue of $\langle w, c\rangle = \frac{2}{3}\langle u+c, c\rangle = \frac{1}{2}$. Combining the relations obtained yields
$$z^2 = \tfrac{2}{3}|\xi|^2(-w + \tfrac{2}{3}u + \tfrac{2}{3}c) = 0,$$
therefore implying (6.5.1). Further, in order to establish (6.5.3) we observe that it suffices to prove that for any $z \in V_0$ such that $|z| = 1$, $L_z^2 = 1$ on V_1. Let such a z be chosen arbitrarily. Then $z^2 = 0$ and Lemma 6.7.1 yields $\dim \ker L_z = \dim \ker L_w$. Moreover, by property (6.5.1) proven above we have $\ker L_z \supset V_0 = \ker L_w$; thus $\ker L_z = \ker L_w$, which implies $V_z(0) = V_w(0) = V_0$. Applying Lemma 6.7.1 again, we see that $L_z^2 = 1$ on $V_z(0)^\perp = V_1$, as desired; thus $V = V_0 \oplus V_1$ is polar.

Finally, we observe that by Corollary 6.5.1, $\dim V_0$ is a similarity invariant of a polar algebra, and it was established in the course of the proof of the theorem that $\dim V_0 = 1 + \dim V_c(-1) = 1 + n_1(c)$ (see also (6.10.5)), which finishes the proof of the theorem.

6.11. Exceptional REC algebras

6.11.1. The REC algebras satisfying $n_2(c) = 0$. In some cases one can characterize REC algebras completely. Below we consider one of them.

THEOREM 6.11.1. *Given a REC algebra, the following conditions are equivalent:*

(A) *V is a Cartan REC algebra.*
(B) *For any $c \in I(V)$, $n_2(c) = 0$.*
(C) *There exists $c \in I(V)$ such that $n_2(c) = 0$.*
(D) *$x^2 \neq 0$ for any $0 \neq x \in V$.*

In particular, any REC algebra having the property $n_2(c) = 0$ is exceptional.

PROOF. The implication (A)$\Rightarrow$(B) is by Corolary 6.7.2 and (B)$\Rightarrow$(C) is trivial. In order to establish (C)$\Rightarrow$(D) we argue by contradiction and assume that for some $c \in I(V)$ with the property $n_2 = 0$ there exists $x \in V$, $x \neq 0$, such that $x^2 = 0$. We can without loss of generality assume that x is normalized, i.e., $x \in \mathcal{N}(V)$. By Lemma 6.7.1, the operator L_x has the three eigenvalues 0 and ± 1, and, in particular, $V_x(1) \neq \{0\}$; hence there exists $z \in V_x(1)$, $|z|^2 = 2$. Then setting $c = \frac{1}{2}(x+z)$ we obtain by virtue of $L_x z = z$ and (6.7.15) that $c^2 = \frac{1}{4}(2xz + z^2) = \frac{1}{4}(2z + 2x) = c$; hence $c \in I(V)$ (notice that x and z are linearly independent so that $c \neq 0$). It is easy to see that $\mathrm{span}(x, z)$ is an invariant subspace of L_c, and moreover by virtue of $L_c x = z$ and $L_c z = \frac{1}{2}z + x$, L_x has the two eigenvalues 1 and $-\frac{1}{2}$ on $\mathrm{span}(x, z)$; i.e., $n_2(c) \geq 1$, a contradiction.

It remains to show that (D)$\Rightarrow$(A). Let V be a normalized REC algebra satisfying condition (D). In order to prove (6.4.13) we assume by contradiction that $k_1 := \min_{x \in S} f(x) < k_2 := \max_{x \in S} f(x)$, where $f(x) = \langle x^2, x^2 \rangle$ and $S = \{x \in V : |x|^2 = 1\}$. Note that by (D), $k_1 > 0$. Let x_1 and x_2 denote the respective points where the values k_1 and k_2 are attained. By the stationary equation $0 = \partial_y f|_{x_i} = 4\langle y, x_i^3\rangle$ for any $y \in V$ satisfying $\langle x_i, y\rangle = 0$, which yields $x_i^3 = m_i x_i$, $m_i \in \mathbb{R}$. We have that $m_i = m_i \langle x_i, x_i\rangle = \langle x_i^3, x_i\rangle = \langle x_i^2, x_i^2\rangle = k_i$ is the corresponding extremum value of f; hence $x_i^3 = k_i x_i$.

Substituting this into (6.7.5) yields $(k_i - \frac{4}{3})\langle x_i, x_i^2\rangle = 0$. By our assumption $k_1 \neq k_2$; hence $k_j \neq \frac{4}{3}$ for some index $j \in \{1, 2\}$, implying that $\langle x_j, x_j^2\rangle = 0$ for the corresponding point x_j. Then (6.7.6) yields $x_j^2 x_j^2 = 4(1 - k_j)x_j^2$. Applying condition (D) we have $x_j^2 \neq 0$ and subsequently $x_j^2 x_j^2 \neq 0$; hence $k_j \neq 1$. Then

$c = \frac{1}{4(1-k_j)} x_j^2 \in I(V)$. We have $L_c x_j = \frac{1}{4(1-k_j)} x_j^3 = \frac{k_j}{4(1-k_j)} x_j$, which shows that $t = \frac{k_j}{4(1-k_j)}$ is an eigenvalue of L_c. Since $\langle x_j, c \rangle = \langle x_j, x_j^2 \rangle = 0$, we have $x_j \in c^\perp$. Since V is normalized, we have

$$\frac{3}{4} = \langle c, c \rangle = \frac{\langle x_j^2, x_j^2 \rangle}{16(1-k_j)^2} = \frac{k_j}{16(1-k_j)^2},$$

showing that $k_j \in \{\frac{4}{3}, \frac{3}{4}\}$, therefore implying by our choice of k_j that $k_j = \frac{3}{4}$. But this yields $t = \frac{k_j}{4(1-k_j)} = \frac{3}{4}$, which contradicts Proposition 6.7.1, thus proving that $k_1 = k_2$, and (A) follows. □

6.11.2. Exceptional algebras.

LEMMA 6.11.1. *If $c \in I(V)$ satisfies $n_2(c) \geq 1$, then there exists an element $\theta \in V_c(-\frac{1}{2})$ such that $|\theta|^2 = \langle \theta, \theta^2 \rangle = \frac{3}{2}$ and $\theta^2 = \theta - c$. In particular, $\frac{2}{3}(c + \theta) \in \mathcal{N}(V)$; hence $\mathcal{N}(V) \neq \emptyset$.*

PROOF. We claim first that under the assumptions made, $\langle x, x^2 \rangle \not\equiv 0$ on $V_c(-\frac{1}{2})$. Indeed, arguing by contradiction, assume $\langle x, x^2 \rangle \equiv 0$; then polarizing the latter identity yields $V_c(-\frac{1}{2}) V_c(-\frac{1}{2}) \subset V_c(-\frac{1}{2})^\perp$. But $V_c(-\frac{1}{2}) V_c(-\frac{1}{2}) \subset V_c(1, -\frac{1}{2})$; hence $V_c(-\frac{1}{2}) V_c(-\frac{1}{2}) \subset V_c(1)$. The latter, in particular, means that there exists a quadrature form q such that $x^2 = q(x)c$ for all $x \in V_2 v$. Since $\langle x^2, c \rangle = \langle x, cx \rangle = -\frac{1}{2} \langle x, x \rangle$, one gets $q(x) = -\frac{2}{3}|x|^2$; hence $x^3 = \frac{1}{3}|x|^2 x$, $x^3 x = -\frac{2}{9}|x|^4$, and $x^2 x^2 = \frac{4}{9}|x|^4 c$. Applying the latter identities to (6.7.5) we obtain $|x|^4 c = 0$; thus $x = 0$, and a contradiction follows.

Now, let us assume $n_2(c) \geq 1$ and consider $W = V_c(-\frac{1}{2})$ as a metrized algebra of V; i.e., define the new (Freudenthal-Springer) multiplication $x \diamond y = \pi_W(xy)$ on W induced by the cubic form $\langle x, x^2 \rangle$ on W (see Subsection 6.3.5). Then by Lemma 6.6.1, there exists $\theta \in I(V)$ on $(W, \diamond)$; i.e., $\theta^{\diamond 2} = \theta$. This yields by the definition of $\diamond$ that $\theta^2 = \theta + kc$ for some $k \in \mathbb{R}$. Then by Lemma 6.9.2, $k = -1$ and $|\theta|^2 = \frac{3}{2}$, and we also have $\langle \theta^2, \theta \rangle = |\theta|^2$. Finally, observe that setting $x = \theta + c$ yields $x^2 = \theta^2 - \theta + c = 0$, where $|x|^2 = |\theta|^2 + |c|^2 = \frac{9}{4}$, thus proving the last statement of the lemma. □

PROPOSITION 6.11.1. *If V is an exceptional REC algebra, then $\operatorname{tr} L_x^2 = \sigma |x|^2$ for any $x \in V$, where $\nu > 0$. In particular, if V is a normalized exceptional REC algebra, then $\sigma = \frac{2}{3}(3n_1(c) + n_2(c) + 1)$.*

PROOF. Note first that the desired identity holds in any Cartan REC algebra because (6.4.14) yields $\operatorname{tr} L_x^2 = 18(2 + \dim V)|x|^2$. Therefore we suppose that V is exceptional but not Cartan; hence Theorem 6.11.1 yields $n_2(c) \geq 1$ for any idempotent $c \in V$ and also $\mathcal{N}(V) \neq \emptyset$. Let us consider the linear operator

$$A = \sum_{i=1}^n L_{e_i}^2,$$

where $\{e_i\}_{1 \leq i \leq n}$, $n = \dim V$, is any orthonormal basis of V. It is easy to see that the sum in the definition of A does not depend of the choice of the orthonormal basis and, furthermore, A is obviously a selfadjoint operator. Observe also that for any $x \in V$,

$$\langle Ax, x \rangle = \sum_{i=1}^n |e_i x|^2 = \operatorname{tr} L_x^2.$$

Thus, the proposition will be proved if we establish that A has a single eigenvalue. First let us consider $x \in \mathcal{N}(V)$ and let the orthonormal bais $\{e_i\}_{1\le i\le n}$ be an eigen-basis of L_x. Then by (6.7.15), $L_{e_i}^2 x = \varepsilon^2 x$, where $e_i \in V_x(\varepsilon)$ and $\varepsilon \in \{0, \pm 1\}$; hence

(6.11.1) $$Ax = \sigma x, \quad x \in \mathcal{N}(V),$$

where $\sigma = (\dim V_w(1) + \dim V_w(-1)) = n - \dim V_w(0)$ and by Lemma 6.7.1 the constant σ does not depend on a choice of $w \in \mathcal{N}(V)$. This shows by the linearity of A that $\mathrm{span}(\mathcal{N}(V)) \subset V_\sigma$, where

$$V_\sigma = \{x \in V : Ax = \sigma x\}.$$

We claim that $\mathrm{span}(I(V)) \subset V_\sigma$. Indeed, let $c \in I(V)$ and let $\{e_i\}_{1\le i \le n}$ be an eigen-basis of L_c. Then, collecting the basis vectors according to the corresponding invariant subspaces, we find

$$Ac = \sum_{i=1}^n e_i(e_i c) = \tfrac{4}{3}c - \mathcal{E}(-1) - \tfrac{1}{2}\mathcal{E}(-\tfrac{1}{2}) + \tfrac{1}{2}\mathcal{E}(\tfrac{1}{2}), \qquad \text{where } \mathcal{E}(t) := \sum_{e_i \in V_c(t)} e_i^2.$$

By (6.7.4), $\mathcal{E}(\tfrac{1}{2}) = -\tfrac{4}{3}c - \mathcal{E}(-1) - \mathcal{E}(-\tfrac{1}{2})$. Furthermore, since V is exceptional, we have by (c) in Theorem 6.9.2 and (6.9.13) that $\mathcal{E}(-\tfrac{1}{2}) = -\tfrac{4}{3}n_2(c)c$. Combining this with (6.8.1) yields

$$Ac = \tfrac{2}{3}c - \tfrac{3}{2}\mathcal{E}(-1) - \mathcal{E}(-\tfrac{1}{2}) = \tfrac{2}{3}c + 2n_1(c)c + \tfrac{2}{3}n_2(c)c = \tfrac{2}{3}(3n_1(c) + n_2(c) + 1)c,$$

which proves that any idempotent is an eigenvector of A. On the other hand, since $n_2(c) \ne 0$, by Lemma 6.11.1 there exists $\theta \in V_c(-\tfrac{1}{2})$ such that $w = \tfrac{2}{3}(c+\theta) \in \mathcal{N}(V)$. By the above, both w and c are eigenvectors of A, and also $\langle w, c\rangle = \tfrac{1}{2} \ne 0$; therefore they belong to the same eigen-subspace, thus proving the claim. Observe that by virtue of (6.8.5) one also concludes that $V_c(-1) \subset V_\sigma$.

Now, given $w \in \mathcal{N}(V)$ and any $w' \in V_w(\varepsilon)$, $\varepsilon^2 = 1$, satisfying $|w'| = \tfrac{1}{\sqrt{2}}$, Corollary 6.7.3 yields $\tfrac{1}{2}(\varepsilon w \pm w') \in I(V)$; hence $w' \in \mathrm{span}(I(V))$, implying

(6.11.2) $$V_w(-1) \oplus V_w(1) \subset V_\sigma, \quad \forall w \in \mathcal{N}(V).$$

Next, let us show that $V_c(-\tfrac{1}{2}) \subset V_\sigma$. To this end observe that the set $M = \{x \in V_c(-\tfrac{1}{2}) : \langle x, x^2\rangle = 0, \; |x|^2 = 1\}$ is nonempty because $\langle x, x^2\rangle$ is an odd continuous function. Then Corollary 6.7.2 yields $|x^2| = |x|^2 = 1$ for any $x \in M$, and moreover (6.9.4) and (6.9.3) yield $x^2 x^2 = 0$ and $x^3 = |x|^2 x = x$, respectively. In particular, for any $x \in M$, $x^2 \in \mathcal{N}(V)$ and $L_{x^2}x = x$, implying by Corollary 6.7.3 that $c_\pm = \tfrac{1}{2}(x^2 \pm x\sqrt{2}) \in I(V)$; therefore $x = \tfrac{1}{\sqrt{2}}(c_+ - c_-) \in \mathrm{span}(I(V))$ and it follows that $\mathrm{span}(M) \subset V_\sigma$. On the other hand, notice that M is an embedded isoparametric minimal hypersurface; hence by the maximum principle [**129**] it cannot be contained in any semisphere of M. Hence $\mathrm{span}(M) = V_c(-\tfrac{1}{2})$, implying our claim.

To finish the proof we suppose by contradiction that A has an eigenvalue $k \ne \sigma$, i.e., $Az = kz$ for some nonzero z. Then by the above $z \in V_c(\tfrac{1}{2})$ for any $c \in I(V)$. In particular, $z \in V_{c_\pm}(\tfrac{1}{2})$, where $x \in M$ and $c_\pm = \tfrac{1}{2}(x^2 \pm x\sqrt{2}) \in I(V)$, which yields $zx^2 = z$; hence $z \in V_{x^2}(1)$, which by (6.11.2) yields $z \in V_\sigma$, a contradiction. The proposition follows. $\square$

6.11.3. The cubic trace identity.

PROPOSITION 6.11.2. *Any normalized REC algebra satisfies the cubic trace identity*

(6.11.3) $$\operatorname{tr} L_x^3 = (1 - n_1(c))\langle x, x^2 \rangle, \qquad \forall c \in I(V), x \in V.$$

PROOF. First assume that V is reducible and normalized. Then Theorem 6.10.1 yields that V is polar with $V_0 = V_w(0)$ for any choice of $w \in \mathcal{N}(V)$; hence (6.5.12) yields $\operatorname{tr} L_x^3 = (2 - \dim V_0)\langle x, x^2 \rangle$. Furthermore, by (6.10.1), $n_1(c) + 1 = \dim V_0$ for any $c \in I(V)$, which proves (6.11.3) in the reducible case.

Now suppose V is a normalized exceptional REC algebra. Applying the trace to (7.7.7) and using (6.7.4) yields

(6.11.4) $$\operatorname{tr} L_x^3 = -\operatorname{tr} L_x L_{x^2} + \tfrac{1}{3}(n+6)\langle x, x^2 \rangle.$$

Further, on polarizing the trace identity in Proposition 6.11.1 we find $\operatorname{tr} L_x L_y = \tfrac{2}{3}(3n_1(c) + n_2(c) + 1)\langle x, y \rangle$ for any $x, y \in V$; thus,

$$\operatorname{tr} L_x L_{x^2} = \tfrac{2}{3}(3n_1(c) + n_2(c) + 1)\langle x, x^2 \rangle.$$

Hence combining the latter with (6.11.4) and (6.7.13) readily yields the desired relation. □

We have an important corollary of (6.11.3).

COROLLARY 6.11.1. *The characteristic dimensions $(n_1(c), n_2(c))$ are similarity invariants of a REC algebra and do not depend on a particular choice of an idempotent c.*

PROOF. Indeed, using (6.7.5) and Proposition 6.11.2 we have for any (not necessarily normalized) REC algebra

$$\tau_3(x) := \frac{4|x|^2 \operatorname{tr} L_x^3}{3\langle x^2, x^3 \rangle} = 1 - n_1(c).$$

On the other hand, the latter identity holds true for any REC algebra by Proposition 6.3.1; hence it is a similarity invariant, thus proving, in particular, that $n_1(c)$ does not depend on a particular choice of an idempotent $c \in I(V)$. □

6.11.4. The finiteness of exceptional REC algebras.

THEOREM 6.11.2. *If V is an exceptional REC algebra, then $n_2 \in \{0, 5, 8, 14, 26\}$ and the possible characteristic dimensions (n_1, n_2) are displayed in Table 6.1.1.*

PROOF. Combining Corollary 6.9.2 and Theorem 6.10.1 $V_c^\bullet$ we have either $V_c^\bullet = \mathbb{R}$ or $V_c^\bullet = \operatorname{Herm}_3(\mathbb{F}_d)$ for some $d \in \{1, 2, 4, 8\}$. This yields the only possible values of $n_2 = \dim V_c(-\tfrac{1}{2})$ claimed in the proposition. Further, by Corollary 6.8.2 we have

(6.11.5) $$n_1 - 1 \leq \rho(n_1 + n_2 - 1).$$

On the other hand, we have for the Hurwitz-Radon function

(6.11.6) $$\rho(m) \leq \tfrac{1}{2}(m+8), \qquad m \geq 1.$$

Indeed, by (6.5.18), $\rho(m) = 8a + 2^b$, where $m = 2^s n_1$ with n_1 an odd number and $s = 4a + b$, $b = 0, 1, 2, 3$. Observe that $2^b \leq 2b + 2$ for the admissible values of b; hence $\rho(m) \leq 8a + 2b + 2$. On the other hand, $8a + 2b + 2 = 2s + 2 \leq 2^{s-1} + 4 \leq \tfrac{1}{2}(2^s n_1 + 8)$, which proves (6.11.6).

Thus, applying (6.11.6) to (6.11.5) we obtain $n_1 \leq 2n_2 + 17$. Thus, given $n_2 \in \{0, 5, 8, 14, 26\}$ and examining (6.11.6) for only finitely many numbers n_1 satisfying $n_1 \leq 2n_2 + 17$, one readily finds the numbers presented in Table 6.1.1. □

COROLLARY 6.11.2. *A REC algebra is exceptional if and only if it permits the quadratic trace identity* $\operatorname{tr} L_x^2 = \sigma |x|^2$ *and* $n_2 \neq 2$.

PROOF. Let V be exceptional. Then Theorem 6.11.2 yields $n_2 \neq 0$ and Proposition 6.11.1 yields the quadratic trace identity. In the converse direction, let a REC algebra V satisfy $\operatorname{tr} L_x^2 = \sigma |x|^2$ and $n_2 \neq 0$. Arguing by contradiction assume V is Clifford; hence it is similar to a special algebra $V_0 \oplus V_1$, and (6.5.11) yields $\dim V_1 = 2 \dim V_0$. Then Corollary 6.7.1 gives $n_2 = \frac{1}{2} \dim V_1 - \dim V_0 + 2 = 2$, a contradiction. □

6.11.5. Mutants: REC algebras with $n_2 = 2$. Corollary 6.11.2 characterize exceptional REC algebras as those possessing the quadratic trace identity with the only exception for Clifford REC algebras having the property $n_2 = 2$, i.e., *mutants*. These algebras occupy an intermediate place in the classification by sharing certain characteristic properties of both polar and exceptional REC algebras. Another property mutants share with exceptional algebras is the vanishing property $\beta(c) = 0$ (cf. condition (c) in Theorem 6.9.2). This can be readily justified by using the classification result established in [**250**].

It is easy to see that there are exactly four (similarity classes of) mutants, each one in dimension $3d$, $d \in \{1, 2, 4, 8\}$. Indeed, $n_2 = 2$ yields by Corollary 6.8.2 that $\rho(n_1 + 1) \geq n_1 - 1$, and using (6.11.6) one immediately identifies that the only possible values are $n_1 = d - 1$, where $d \in \{1, 2, 4, 8\}$. In fact, it is readily verified that the mutant REC algebra in dimension $3d$ is similar to $V = \operatorname{Herm}_3(\mathbb{F}_d) \ominus \mathcal{D}_3$, $d \in \{1, 2, 4, 8\}$, in Table 6.4.1 with a generic element $x \in V$ given as

$$x = \begin{pmatrix} 0 & y_3 & \bar{y}_2 \\ \bar{y}_3 & 0 & y_1 \\ y_2 & \bar{y}_1 & 0 \end{pmatrix}$$

where $y_i \in \mathbb{F}_d$.

The cubic forms $u(x) = \operatorname{Re}(y_1 y_2 y_3)$ associated to mutant REC algebras appears in various mathematical contexts: calibrated geometries (as an associative calibration on $\operatorname{Im} \mathbb{O}$) in [**103**] and Sections 2.3–2.5 above; constructing nonclassical and singular solutions of fully nonlinear elliptic equations in Chapter 4; special Riemannian geometries satisfying the nearly integrabilty condition [**191**], [**95**]; explicit solutions to the Ginzburg-Landau system [**86**], [**93**], and Section 5.5; the harmonic analysis of cubic isoparametric minimal hypersurfaces [**230**], [**231**].

6.11.6. Concluding remarks. Some of the characteristic dimensions in Table 6.1.1 are realizable as exceptional REC algebras. Below we provide some comments about the known examples.

(i) Example #1 with $(n_1, n_2) = (1, 0)$ corresponds to the trivial REC algebra associated with the cubic form $x_1^2 x_2$ in $\mathbb{R}^2$.

(ii) Examples #2, 3, 4, 5 with $(n_1, n_2) = (d + 2, 0)$, $d = 1, 2, 4, 8$, provide the Cartan REC algebras $\operatorname{Herm}_3(\mathbb{F}_d) \ominus \mathcal{D}_1$ in $\mathbb{R}^{3d+2}$.

(iii) The series #6, 10, 16, 20 with $(n_1, n_2) = (0, 3d + 2)$ are realizable for $d = 1, 2, 4$ only; one can show that there is no exceptional REC algebras with $(n_1, n_2) = (0, 26)$ in $\mathbb{R}^{51}$ corresponding to $d = 8$. The three realizable REC algebras in $\mathbb{R}^9$, $\mathbb{R}^{15}$, $\mathbb{R}^{27}$ can be obtained by the "extracting Jordan algebra" procedure considered by Allison and Faulkner (see Theorem 5.4 in [**10**]). More explicitly, the exceptional REC algebra with $(n_1, n_2) = (0, 3d+2)$ is exactly the Freudenthal algebra of the cubic form $\operatorname{tr} x^3$, where x is a generic trace free element in the Jordan rank 4 algebra $\operatorname{Herm}_4(\mathbb{F}_d)$, $d = 1, 2, 4$. The cases $d = 1$ and $d = 2$ correspond exactly to the two nonhomogeneous cubic minimal cones in $\mathbb{R}^9$ and $\mathbb{R}^{15}$, respectively, constructed by Hsiang in [**110**].

(iv) The series #7, 11, 17, 21 with $(n_1, n_2) = (1, 3d+2)$ are realizable for all $d = 1, 2, 4, 8$. The four REC algebras in $\mathbb{R}^{12}$, $\mathbb{R}^{18}$, $\mathbb{R}^{30}$, and $\mathbb{R}^{54}$ can be thought of as the "complexifications" of the simple Jordan algebras $\operatorname{Herm}_3(\mathbb{F}_d)$, $d = 1, 2, 4, 8$. More precisely, if $V = \operatorname{Herm}_3(\mathbb{F}_d)$ and $N(x)$ is the generic norm of V, then the cubic form $N(x+y\sqrt{-1}) + N(x-y\sqrt{-1})$ gives rise to an exceptional REC algebra on $V \times V$ having $(n_1, n_2) = (1, 3d+2)$, and, vice versa, one can prove that any such exceptional algebra with $n_1 = 1$ is obtained in this way.

Finally, we also wish to mention interesting applications of Jordan algebras and Jordan systems to integrable systems, e.g., nonlinear Schrödinger equations, recently developed by S. Svinolupov [**238**], [**239**] and Svinolupov and Sokolov [**240**], [**241**]. For instance, a one-to-one correspondence between Jordan algebras and certain systems that possess at least one nondegenerate higher symmetry or conservation law is shown to exist in [**239**].

CHAPTER 7

Singular Solutions in Calibrated Geometries

In this final chapter we briefly describe singular solutions of elliptic equations arising in calibrated geometry. These equations are not uniformly, but only strictly, elliptic. However, their constructions are rather close in spirit to those of previous chapters, and moreover, one can hope that a more profound understanding of their interrelations would give interesting consequences.

In Section 7.1 we recall the basic definitions by Harvey and Lawson as well as some fundamental examples, and we discuss briefly the corresponding partial differential equations; note that the corresponding polylinear algeba is surveyed in Sections 2.4 and 2.5. Section 7.2 is devoted to a construction of singular solutions for the coassociator equation. Section 7.3 contains a construction of singular solutions to the special Lagrangian equation. In Section 7.4 we describe constructions of more general singular solutions to the special Lagrangian equations, leading to examples of the failure of the maximum principle for the Hessian of a solution to a uniformly elliptic equation in three and more dimensions as well as to examples of special interesting solutions to the minimal surface system.

7.1. Calibrated geometries

In the present section we recall some basic definitions and examples concerning calibrations and calibrated geometries. We assume some basic facts of the differential geometry as background. Let (M, g) be a Riemannian manifold, and let N be a compact submanifold of M. Let $\iota : N \longrightarrow M$ be the corresponding immersion. Then $(N, \iota^*(g))$ is a Riemannian manifold (we only need the case of $M = \mathbb{R}^n$). We recall that N is a *minimal submanifold* if its volume is stationary under small variations of the immersion $\iota : N \longrightarrow M$, which is equivalent to the condition that N has zero mean curvature.

Harvey and Lawson [**103**] defined calibrated submanifolds as follows.

Let (M, g) be a Riemannian manifold. An oriented tangent k-plane V on M is a vector subspace $V \subset T_x M$ of some tangent space to M with $\dim V = k$, equipped with an orientation. If V is an oriented tangent k-plane on M, then $g|V$ is a Euclidean metric on V, which gives a natural volume form vol_V on V, which is a k-form on V. Now let φ be a closed k-form on M. We say that φ is a *calibration* on M if for every oriented k-plane V on M we have $\varphi_{|V} \leq vol_V$, i.e., $\varphi_{|V} = \alpha \cdot vol_V$ with $\alpha \leq 1$. Let N then be an oriented submanifold of M of dimension k. Each tangent space $T_x N, x \in N$, is an oriented tangent k-plane; N is called a *calibrated submanifold* if $\varphi_{|T_x N} = vol_{V|T_x N}, \forall x \in N$. Calibrated submanifolds are automatically minimal; see Th. II.4.2 in [**103**].

Notice that the condition that N is a calibrated submanifold depends upon the tangent spaces of N, i.e., depends on the immersion and its first derivative. Therefore, this condition can be expressed as a first-order p.d.e., which we are going to write down in some particular cases. This is for comparison with the condition of minimality, which is a second-order p.d.e., and thus in principle is more complicated than first-order ones. Hence calibrated geometry gives many examples of minimal submanifolds.

Let us give some examples. The most classical one is of course that of the complex Kähler manifold M and its complex subvarieties, which are calibrated with respect to (suitably normalized) powers of the Kähler form of M, the corresponding p.d.e. being, naturally, the Cauchy-Riemann equations. In all of the examples which follow, the ambient manifold M is simply $\mathbb{R}^n$.

7.1.1. Lagrangian geometry. Let $n = 2m$ be even, and let $M = \mathbb{R}^n = \mathbb{C}^m$. Denote by $(z_1, \ldots, z_m)$ the complex coordinates on $\mathbb{C}^m$ and let

$$g = |dz_1|^2 + \cdots + |dz_m|^2, \quad \omega = \frac{i}{2}(dz_1 \wedge d\bar{z}_1 + \cdots + dz_m \wedge d\bar{z}_m),$$

and

(7.1.1) $$\psi = dz_1 \wedge \cdots \wedge dz_m$$

be the corresponding metric, Kähler form, and volume form, respectively.

Then $\text{Re}(\psi)$ and $\text{Im}(\psi)$ are calibrations on $\mathbb{C}^m$. If L is an oriented real m-submanifold of $\mathbb{C}^m$, then L is a *special Lagrangian submanifold* if L is calibrated with respect to $\text{Re}(\psi)$. More generally, for $\theta \in \mathbb{R}$ the manifold L is called the special Lagrangian with phase $e^{i\theta}$ if it is calibrated with respect to $\cos\theta \,\text{Re}\,\psi + \sin\theta \,\text{Im}\,\psi = \text{Re}(e^{i\theta}\psi)$.

Another characterization of special Lagrangian submanifolds is given by Corollary III.1.11 in [**103**]:

PROPOSITION 7.1.1. *Let L be a real m-dimensional submanifold of $\mathbb{C}^m$. Then L admits an orientation making it into a special Lagrangian submanifold if and only if $\omega|_L = 0$ and $\text{Im}(\psi)|_L = 0$. More generally, it admits an orientation making it into a special Lagrangian submanifold with phase $e^{i\theta}$ if and only if $\omega|_L = 0$ and $\text{Im}(e^{-i\theta}\psi)|_L = 0$.*

Locally any Lagrangian submanifold is a graph:

LEMMA 7.1.1. *Suppose $\Omega \subset \mathbb{R}^m$ is open and $f : \Omega \longrightarrow \mathbb{R}^m$ is a C^1 mapping. Let M denote the graph of f in $\mathbb{C}^m = \mathbb{R}^m + i\mathbb{R}^m$. Then the graph M is Lagrangian if and only if the Jacobian matrix $(\partial f^k / \partial x_j)$ is symmetric. In particular, if Ω is simply connected, it is Lagrangian if and only if $f = \nabla F$ is the gradient of some potential function $F \in C^2(\Omega)$.*

Applying this lemma one easily deduces the following result ([**103**], Th.II.2.3):

THEOREM 7.1.1. *Let $F \in C^2(\Omega)$, let $f = \nabla F$ be the gradient, and let M denote the graph of f in $\mathbb{C}^m = \mathbb{R}^m + i\mathbb{R}^m$. Then M (with the correct orientation) is special Lagrangian if and only if*

(7.1.2) $$\sum_{0 \leq 2k \leq m-1} (-1)^k \sigma_{2k+1}(F) = 0,$$

which is equivalent to

(7.1.3) $$\text{Im}\{\det(I + iD^2 u)\} = 0$$

with $\sigma_l(F)$ *being the elementary symmetric functions of the eigenvalues of the Hessian* $D^2 F$.

For the more general case of the calibration with phase $e^{i\theta}$ equation (7.1.3) becomes
$$\mathrm{Im}\{e^{-i\theta}\det(I+iD^2 u)\}=0$$
and can be rewritten in the equivalent form

(7.1.4) $$\sum_{j=1}^{m}\arctan(\lambda_j)=\theta,$$

where the $\{\lambda_1,\ldots,\lambda_m\}$ are the eigenvalues of $D^2 F$.

Equation (7.1.4) is strictly, but not uniformly, elliptic and due to the minimality of the graph M and a regularity result by Morrey [**168**], any of its C^2 solutions are analytic.

7.1.2. Associative calibration. Here and in Subsection 7.1.3 we suppose then that $n = 7$, and in Subsection 7.1.4 we set $n = 8$. We identify $\mathbb{R}^7$ with the space $\mathrm{Im}\,\mathbb{O}$ of purely imaginary octonions and denote by $(x_1,\ldots,x_7)$ the canonical coordinates corresponding to the basis $(e_1,\ldots,e_7)$ introduced in Section 2.2. Recall that the form $\varphi(x,y,z) = \langle x, yz \rangle$ is alternating on $\mathbb{O}^3$ and hence on $\mathrm{Im}\,\mathbb{O}^3$ and thus defines a form $\varphi \in \Lambda^3(\mathrm{Im}\,\mathbb{O})^*$ which is called the *associative calibration* on $\mathrm{Im}\,\mathbb{O}$. Consulting the multiplication table for $\mathbb{O}$ in Section 2.2 one easily deduces that

(7.1.5) $$\varphi = dx_{124} + dx_{137} + dx_{156} + dx_{235} + dx_{267} + dx_{346} + dx_{457},$$

where $dx_{ijl} := dx_i \wedge dx_j \wedge dx_l$.

Note that the difference in indices and signs between this formula (and the formulas below) and those in [**103**] as well as in Section 2.4 is due to the different choice of a basis in $\mathrm{Im}\,\mathbb{O}^3$; namely in [**103**] the basis $(e'_2,\ldots,e'_8)$ is given by

$$e'_2 = e_1,\ e'_3 = e_2,\ e'_4 = e_4,\ e'_5 = e_3,\ e'_6 = e_7,\ e'_7 = e_5,\ e'_8 = -e_6.$$

One has the following fundamental identity which easily implies that φ gives a calibration on $\mathrm{Im}\,\mathbb{O}$ ($=$ Proposition 2.4.3):

(7.1.6) $$\varphi(x,y,z)^2 + |[x,y,z]|^2/4 = |x \times y \times z|^2$$

for any triple $(x,y,z) \in \mathrm{Im}\,\mathbb{O}^3$, $[x,y,z]$ being the *associator* $[x,y,z] := (xy)z - x(yz)$.

An oriented 3-plane P in $\mathrm{Im}\,\mathbb{O}^3$ with $\varphi_{|P} = vol_{|P}$ is called an associative 3-plane. An oriented 3-dimensional submanifold of $\mathbb{R}^7$ calibrated with respect to φ is called an associative 3-fold. All tangent planes of such a 3-fold are associative. One can show that these planes are exactly 3-planes of the form $\mathrm{Im}\,\mathbb{H}$ for a quaternion subalgebra in $\mathbb{O}$ (with a suitable orientation).

One can write down a partial differential equation for a function f in order that its graph be an associative 3-fold. More precisely let $\Omega \subset \mathrm{Im}\,\mathbb{H}$ be an open domain, and let $f : \Omega \longrightarrow \mathbb{H}$ be a C^1 function. We write $x = x_1 i + x_2 j + x_3 k \in \mathrm{Im}\,\mathbb{H}$.

PROPOSITION 7.1.2. *The graph of* f *is an associative 3-fold in* $\mathrm{Im}\,\mathbb{H} \oplus \mathbb{H} = \mathrm{Im}\,\mathbb{O}$ *if and only if*

$$\frac{\partial f}{\partial x_1}i + \frac{\partial f}{\partial x_2}j + \frac{\partial f}{\partial x_3}k + \frac{\partial f}{\partial x_1} \times \frac{\partial f}{\partial x_2} \times \frac{\partial f}{\partial x_3} = 0.$$

Since associative 3-folds are automatically minimal, we get that any C^1 solution of that equation is analytic.

7.1.3. Coassociative calibration.
One defines the 4-form $\psi \in \Lambda^4(\operatorname{Im}\mathbb{O})^*$ by
$$\psi(x,y,z,w) := \langle x, [y,z,w]\rangle/2, \ \forall x,y,z,w \in \operatorname{Im}\mathbb{O};$$
ψ is alternating in y,z,w since the associator is alternating, and
$$2\psi(x,x,z,w) = \langle x,(xz)w\rangle - \langle x,x(zw)\rangle = |x|^2(\langle \bar{w},z\rangle - \langle \bar{w},z\rangle) = 0.$$
Also, one can prove that (see Proposition 2.4.5)
$$\psi(x,y,z,w)^2 + |[x,y,z,w]|^2/4 = |x \times y \times z \times w|^2$$
for the *coassociator*
$$[x,y,z,w] := -2(\langle y,zw\rangle x + \langle z,xw\rangle y + \langle x,yw\rangle z + \langle y,xz\rangle w).$$
It is called the *coassociative calibration*. One easily verifies that

(7.1.7) $\quad \psi = *\varphi = dx_{1236} - dx_{1257} - dx_{1345} - dx_{1467} + dx_{2347} - dx_{2456} - dx_{3567}.$

It is indeed a calibration, which follows from Proposition 2.4.5.

An oriented 4-plane P in $\operatorname{Im}\mathbb{O}^3$ with $\psi_{|P} = vol_{|P}$ is called a coassociative 4-plane. An oriented 4-dimensional submanifold of $\mathbb{R}^7$ calibrated with respect to ψ is called a coassociative 4-fold. All tangent planes of such a 4-fold are coassociative. One can show that these planes are exactly 4-planes with $[x,y,z,w] = 0$ for any basis (x,y,z,w); see Lemma 2.4.7.

Let us then write down the corresponding coassociative partial differential equation. Let $\Omega \subset \mathbb{H}$ be an open domain, and let $f : \Omega \longrightarrow \operatorname{Im}\mathbb{H}$ be a C^1 function. We write $x = x_0 + x_1 i + x_2 j + x_3 k \in \mathbb{H}$ and $f = f^1 i + f^2 j + f^3 k \in \operatorname{Im}\mathbb{H}$.

PROPOSITION 7.1.3. *The graph of f is a coassociative 4-fold in $\operatorname{Im}\mathbb{H} \oplus \mathbb{H} = \operatorname{Im}\mathbb{O}$ if and only if*
$$D(f) := -\nabla f^1 i - \nabla f^2 j - \nabla f^3 k = \nabla f^1 \times \nabla f^2 \times \nabla f^3 =: \sigma(f).$$

Since coassociative 4-folds are automatically minimal, we get that any C^1 solution of that equation is analytic.

7.1.4. Cayley calibration on $\mathbb{O}$.
Define a 4-form $\Phi \in \Lambda^4 \mathbb{O}^*$ by $\Phi(x,y,z,w) = \langle x, y \times z \times w\rangle$; it is clearly alternating in (y,z,w), and for orthogonal x,z,w one has
$$\langle x, x \times z \times w\rangle = \langle x, x(\bar{z}w)\rangle = |x|^2 \langle z,w\rangle = 0,$$
which implies that Φ is alternating.

Let the coordinates $(x_0, \ldots, x_7)$ in $\mathbb{R}^8 = \mathbb{O}$ correspond to the basis $(e_0, \ldots, e_7)$; by a brute force calculation one gets

$$\Phi = dx_{0124} + dx_{0137} + dx_{0156} + dx_{0235} + dx_{0267} + dx_{0346} + dx_{0457}$$

(7.1.8) $\qquad + dx_{1236} - dx_{1257} - dx_{1345} - dx_{1467} + dx_{2347} - dx_{2456} - dx_{3567}$

$$= dx_0 \wedge \varphi + \psi.$$

Schwarz inequality, more precisely, the formula (2.5.3), implies that Φ is a calibration on $\mathbb{R}^8$, and an oriented 4-dimensional submanifold of Ω calibrated with respect to Ω is called a *Cayley* 4-fold in $\mathbb{R}^8$. Its tangent planes are Cayley 4-planes characterized by the equation $\operatorname{Im}(x \times y \times z \times w) = 0$ for any basis (x,y,z,w); see

Lemma 2.5.2. One can also give an equation characterizing functions with Cayley graphs, but we do not need it here.

In a sense, Cayley manifolds generalize special Lagrangian, associative, and coassociative manifolds since one has

PROPOSITION 7.1.4. (i) *Let $M \subset \mathrm{Im}\,\mathbb{O}$ be a 4-fold. Then M is Cayley if and only if M is coassociative.*

(ii) *Let $M = \mathbb{R} \times N$ with $N \subset \mathrm{Im}\,\mathbb{O}$. Then M is Cayley if and only if N is associative.*

(iii) *Any special Lagrangian submanifold of $\mathbb{C}^4 = \mathbb{O}$ is a Cayley manifold.*

7.2. Singular coassociative 4-folds

One of the most interesting examples of a coassociative submanifold is found by looking for symmetric solutions which lead to singular solutions. The group $Sp(1) = \mathbb{S}^3 \subset \mathbb{H}$ acts on $\mathbb{O}$ in several ways. Consider the action given by

$$(q, a + be) \mapsto qa\bar{q} + b\bar{q}e, \; \forall q \in Sp(1) \subset \mathbb{H}.$$

This action embeds $Sp(1)$ into G_2, and therefore it preserves the coassociative calibration. The orbit of any point $a + be \in \mathbb{O}$, with $b \neq 0$, is diffeomorphic to $\mathbb{S}^3$.

To obtain a coassociative 4-fold invariant under $Sp(1)$ we choose a fixed unit vector $f \in \mathrm{Im}\,\mathbb{H}$ and seek a curve in the half-plane $\mathbb{R} \times \mathbb{R}^+ \simeq \mathbb{R}f \times \mathbb{R}^+e \in \mathrm{Im}\,\mathbb{O}$ which is swept out into a coassociative submanifold (of dimension 4) under the action of $Sp(1)$. Let (s, r) denote coordinates in $\mathbb{R} \times \mathbb{R}^+$.

THEOREM 7.2.1. *Suppose $f \in \mathrm{Im}\,\mathbb{H}$ is a fixed unit vector and $c \in \mathbb{R}$. Then*

$$M_c = \left\{ sqf\bar{q} + r\bar{q}, q \in Sp(1),\; s(4s^2 - 5r^2)^2 = c \right\}$$

is a coassociative submanifold invariant under $Sp(1)$. In particular, if $c = 0$, $s = \sqrt{5}r/2$, then

$$M_0 = \left\{ r(\sqrt{5}qf\bar{q}/2 + r\bar{q}) : q \in Sp(1) = \mathbb{S}^3,\; r \in \mathbb{R}^+ \right\},$$

which is the cone on the graph of the Hopf map $\eta : \mathbb{S}^3 \longrightarrow \mathbb{S}^2$ defined by $\eta(q) = \sqrt{5}qf\bar{q}/2$, is a coassociative 4-fold.

The cone M_0 above is the graph of the function $\eta : \mathbb{H} \longrightarrow \mathrm{Im}\,\mathbb{H}$ defined by

$$(7.2.1) \qquad \eta(q) = \frac{\sqrt{5}qf\bar{q}}{2|q|}.$$

This function represents a Lipschitz solution to the nonparametric minimal surface system which is not C^1, and its graph is absolutely area minimizing. Furthermore, the function given by (7.2.1) is a Lipschitz solution to the coassociator equation which is not C^1, making sharp the regularity result that C^1 solutions to that equation are real analytic.

One notes also that $\eta(q)$ is in some sense a specialization (up to a rescaling) to four dimensions of the function $w_{12}(x) = P_{12}(x)/|x|$ in Chapter 4, which is a nonclassical solution to a uniformly elliptic equation. Indeed $P_{12} = \mathrm{Re}(q_1q_2q_3)$, and the product $q_1q_2q_3$ becomes $qf\bar{q}$ if $q_1 = q$, $q_2 = f$, $q_3 = \bar{q}$. It would be very interesting to understand this coincidence.

To prove Theorem 7.2.1 one needs the following two lemmas.

LEMMA 7.2.1. *Suppose that $M = \{g(x) + xe : x \in \mathbb{H}\} \subset \operatorname{Im} \mathbb{O}$ is the graph of a function $g : \mathbb{H} \longrightarrow \operatorname{Im} \mathbb{H}$. Then M is invariant under the above action of $Sp(1)$ if and only if*

$$(7.2.2) \qquad g(x) = \frac{\bar{x} g(|x|) x}{|x|^2}$$

for all $x \in \mathbb{H}$.

PROOF. If M is $Sp(1)$-invariant, then for each $q \in Sp(1) \subset \mathbb{H}$ and each $x \in \mathbb{H}$, the vector $qg(x)\bar{q} + x\bar{q}e$ also belongs to M. Consequently,

$$g(x\bar{q}) = qg(x)\bar{q}$$

for all $q \in Sp(1)$ and all $x \in \mathbb{H}$. Replace x by $|x|$ and q by $x/|x|$ to obtain (7.2.2). □

LEMMA 7.2.2. *Suppose $f \in \operatorname{Im} \mathbb{H}$ is a fixed unit vector and $\phi : \mathbb{R}^+ \longrightarrow \mathbb{R}$ is a given function, $s = \phi(r)$. Define $g : \mathbb{H} \longrightarrow \operatorname{Im} \mathbb{H}$ by*

$$g(x) = \frac{\bar{x} f x\, \phi(|x|)}{|x|^2}$$

so that $M = graph(g)$ is $Sp(1)$-invariant. Then M is coassociative if and only if

$$(7.2.3) \qquad \phi'(r) = \frac{4r\phi(r)}{4\phi(r)^2 - r^2}.$$

PROOF. We must compute $D(g) - \sigma(g)$. Since $g(x\bar{q}) = qg(x)\bar{q}$ for $q \in Sp(1)$, one gets

$$dg_{x\bar{q}}(u) = q\, dg_x(uq)\, \bar{q},$$

where dg_x denotes the differential of g at x. A brute force calculation then gives that $(Dg)x\bar{q} = q(Dg)(x)$ and $(\sigma g)x\bar{q} = q(\sigma g)(x)$. Consequently, it suffices to compute $Dg - \sigma g$ at $x = |x| = r \in \mathbb{R}^+$. For convenience, and without loss of generality, we set $f = i$ so that

$$g(x) = \frac{\bar{x} i x\, \phi(r)}{r^2}.$$

Straightforward calculation then shows that

$$\frac{\partial g}{\partial x_0}(r) = \phi'(r)i, \quad \frac{\partial g}{\partial x_1}(r) = 0, \quad \frac{\partial g}{\partial x_2}(r) = \frac{2\phi(r)}{r}k, \quad \frac{\partial g}{\partial x_3}(r) = -\frac{2\phi(r)}{r}j.$$

Computing directly from the definitions of D and σ, we find that

$$(Dg)(r) = \left(\frac{4\phi(r)}{r} + \phi'(r)\right) i \quad \text{and} \quad (\sigma g)(r) = \frac{4\phi(r)^2 \phi'(r)}{r^2} i.$$

Therefore, we have that

$$(Dg - \sigma g)(r\bar{q}) = \frac{4r\phi + (r^2 - 4\phi^2)\phi'(r)}{r^2} \bar{q} i$$

and the proof is complete. □

One can then finish the proof of Theorem 7.2.1. Since the right-hand side of (7.2.3) is homogeneous of degree zero in r and $s = \phi(r)$, the standard substitution $z = r/s$ yields the integral curves

$$s(4s^2 - 5r^2)^2 = c,$$

thereby completing the proof of the theorem.

7.3. Singular solutions of special Lagrangian equations

In this section we construct singular solutions to special Lagrangian equations in $\mathbb{R}^3$.

7.3.1. Special Lagrangian equations in $\mathbb{R}^3$.
We consider a fully nonlinear second-order elliptic equation of the form (where $h \in \mathbb{R}$)

$$(7.3.1) \qquad \mathbf{F}_h(D^2u) = \det(D^2u) - \operatorname{tr}(D^2u) + h\sigma_2(D^2u) - h = 0$$

defined in a smooth-bordered domain of $\Omega \subset \mathbb{R}^3$, $\sigma_2(D^2u) = \lambda_1\lambda_2 + \lambda_2\lambda_3 + \lambda_1\lambda_3$ being the second symmetric function of the eigenvalues $\lambda_1, \lambda_2, \lambda_3$ of D^2u. This equation is equivalent to the special Lagrangian potential equation with phase θ:

$$SLE_\theta: \qquad \operatorname{Im}\{e^{-i\theta}\det(I + iD^2u)\} = 0$$

for $h := -\tan(\theta)$, which can be rewritten as

$$\mathbf{F}_\theta = \arctan\lambda_1 + \arctan\lambda_2 + \arctan\lambda_3 - \theta = 0.$$

The set

$$\{A \in S^2(\mathbb{R}^3): \mathbf{F}_h(A) = 0\} \subset S^2(\mathbb{R}^3)$$

has three connected components, C_i, $i = 1, 2, 3$, which correspond to the values

$$\theta_1 = -\arctan(h) - \pi, \quad \theta_2 = -\arctan(h), \quad \theta_3 = -\arctan(h) + \pi.$$

We study the Dirichlet problem in a bounded domain with smooth boundary $\partial\Omega$.

For $\theta_1 = -\arctan(h) - \pi$ and $\theta_3 = -\arctan(h) + \pi$ the operator $\mathbf{F}_\theta$ is concave or convex, and the Dirichlet problem in these cases was treated in [**48**]; smooth solutions are established there for smooth boundary data on appropriately convex domains.

The middle branch C_2, $\theta_2 = -\arctan(h)$, is never convex (neither concave), and in this case the results of [**48**] cannot be applied. We will show that the classical solvability for special Lagrangian equations *does not* hold (cf. [**177**]).

More precisely, we show the existence for any $\theta \in\,]-\pi/2, \pi/2[$ of a small ball $B \subset \mathbb{R}^3$ and of an analytic function ϕ on ∂B for which the unique Harvey-Lawson solution u_θ of the Dirichlet problem satisfies the following:

(i) $u_\theta \in C^{1,1/3}$.

(ii) $u_\theta \notin C^{1,\delta}$ for $\delta > 1/3$.

Our construction uses the Legendre transform for solutions of $\mathbf{F}_{\frac{1}{h}}(D^2u) = 0$, which gives solutions of $\mathbf{F}_h(D^2u) = 0$; in particular, for $h = 0$ it transforms solutions of $\sigma_2(D^2u) = 1$ (with $h = \infty$, $\theta = \frac{\pi}{2}$) into solutions of $\det(D^2u) = \operatorname{tr}(D^2u)$ (with $h = 0 = \theta$).

7.3.2. Some properties of special Lagrangian equations.
Let us give some properties of the special Lagrangian equation

$$(7.3.2) \qquad \mathbf{F}_{-c}(D^2u) = \det(D^2u) - \operatorname{tr}(D^2u) - c\sigma_2(D^2u) + c = 0.$$

Note first that the set $\{\mathbf{F}_{-c,\Lambda} = 0\}$ is a real cubic surface $\mathbf{S}_{c,\Lambda}$ with three components ("branches") which can be represented as a graph:

$$(7.3.3) \qquad \lambda_3 = \frac{c(1 - \lambda_1\lambda_2) + \lambda_1 + \lambda_2}{\lambda_1\lambda_2 - 1 + c(\lambda_1 + \lambda_2)}.$$

One easily proves the following by brute force computations:

LEMMA 7.3.1. 1) *The components of* $\mathbf{S}_{c,\Lambda}$ *are given by*
$$C_1 = \{(\lambda_1, \lambda_2, \lambda_3) : \lambda_1\lambda_2 - 1 + c(\lambda_1 + \lambda_2) > 0, \ \lambda_1 > -c, \lambda_2 > -c\},$$
$$C_2 = \{(\lambda_1, \lambda_2, \lambda_3) : \lambda_1\lambda_2 - 1 + c(\lambda_1 + \lambda_2) < 0\},$$
$$C_3 = \{(\lambda_1, \lambda_2, \lambda_3) : \lambda_1\lambda_2 - 1 + c(\lambda_1 + \lambda_2) > 0, \ \lambda_1 < -c, \lambda_2 < -c\};$$

equivalently,
$$C_1 = \{(\lambda_1, \lambda_2, \lambda_3) : \arctan\lambda_1 + \arctan\lambda_2 + \arctan\lambda_3 = \pi + \arctan c\},$$
$$C_2 = \{(\lambda_1, \lambda_2, \lambda_3) : \arctan\lambda_1 + \arctan\lambda_2 + \arctan\lambda_3 = \arctan c\},$$
$$C_3 = \{(\lambda_1, \lambda_2, \lambda_3) : \arctan\lambda_1 + \arctan\lambda_2 + \arctan\lambda_3 = -\pi + \arctan c\}.$$

2) *For any* $c \in \mathbb{R}$, C_1 *is convex,* C_3 *is concave,* C_2 *is neither.*

PROOF. 1) is straightforward; 2) follows from the Hessian of λ_3 in (7.3.3):
$$\frac{\partial^2 \lambda_3}{\partial \lambda_1^2} = \frac{2(\lambda_2 + c)(\lambda_2^2 + 1)(1 + c^2)}{(\lambda_1\lambda_2 - 1 + c\lambda_1 + c\lambda_2)^3}, \quad \frac{\partial^2 \lambda_3}{\partial \lambda_2^2} = \frac{2(\lambda_1 + c)(\lambda_1^2 + 1)(1 + c^2)}{(\lambda_1\lambda_2 - 1 + c\lambda_1 + c\lambda_2)^3},$$
$$\det(D^2\lambda_3) = \frac{4(c^2 + 1)^2(c\lambda_1 + c\lambda_2 + \lambda_1^2\lambda_2^2 + \lambda_1\lambda_2 + \lambda_2^2 + \lambda_1^2)}{(\lambda_1\lambda_2 - 1 + c\lambda_1 + c\lambda_2)^5},$$

which implies, e.g., that the point with $\lambda_1 = \lambda_2 = -c - 1/10$ for $c \geq 0$, $\lambda_1 = \lambda_2 = -c + 1/10$ for $c \leq 0$ is a saddle point on C_2. □

The corresponding Dirichlet sets F_c^i, $i = 1, 2, 3$, are given (via $F_{c,\Lambda}^i$) by
$$F_{c,\Lambda}^1 = \{(\lambda_1, \lambda_2, \lambda_3) : \arctan\lambda_1 + \arctan\lambda_2 + \arctan\lambda_3 \geq \pi + \arctan c\},$$
$$F_{c,\Lambda}^2 = \{(\lambda_1, \lambda_2, \lambda_3) : \arctan\lambda_1 + \arctan\lambda_2 + \arctan\lambda_3 \geq \arctan c\},$$
$$F_{c,\Lambda}^3 = \{(\lambda_1, \lambda_2, \lambda_3) : \arctan\lambda_1 + \arctan\lambda_2 + \arctan\lambda_3 \geq -\pi + \arctan c\}.$$

A simple calculation yields that the derivatives
$$\frac{\partial F_c^j}{\partial x_i} = \frac{1}{\lambda_i^2 + 1} > 0$$

tend to 0 at infinity; hence we have

LEMMA 7.3.2. F_c^j *for* $j = 1, 2, 3$ *is strictly, but not uniformly, elliptic.*

REMARK 7.3.1. If we (artificially) impose the uniform ellipticity condition, we get a smooth solution. Indeed, if $\mathbf{F}_{-c}$ verifies this condition on u, the derivatives
$$\frac{1}{\lambda_1^2 + 1}, \ \frac{1}{\lambda_2^2 + 1}, \ \frac{1}{\lambda_3^2 + 1} = \frac{(\lambda_1\lambda_2 - 1 + c\lambda_1 + c\lambda_2)^2}{(1 + \lambda_2^2)(\lambda_1^2 + 1)(c^2 + 1)} \in \left[\frac{1}{M}, M\right]$$
for some ellipticity constant M, which implies that $u \in C^{1,1}$ and thus is smooth by [**279**].

We now give the principal technical result which permits us to construct a singular solution of SLE.

7.3. SINGULAR SOLUTIONS OF SPECIAL LAGRANGIAN EQUATIONS

PROPOSITION 7.3.1. *There exists a ball $B = B(0, \varepsilon)$ centered at the origin such that the following are true.*

1) *The equation*
$$\lambda_1 \lambda_2 + \lambda_2 \lambda_3 + \lambda_1 \lambda_3 = \sigma_2(D^2 u) = 1$$
has an analytic solution u_0 in B satisfying

(i) $u_0 = -\frac{y^4}{3} + 5y^2 z^2 - x^4 + 7x^2 z^2 - \frac{z^4}{3} + 2y^2 z - 2zx^2 + \frac{y^2}{2} + \frac{x^2}{2} + O(r^5)$,

(ii) $\lambda_1 = 1 + O(r)$, $\lambda_2 = 1 + O(r)$, $\lambda_3 = -\frac{x^2}{2} - \frac{3y^2}{2} - z^2 + O(r^3)$.

2) *The equation*
$$\lambda_1\lambda_2+\lambda_2\lambda_3+\lambda_1\lambda_3+c(\lambda_1\lambda_2\lambda_3-\lambda_1-\lambda_2-\lambda_3) = \sigma_2(D^2u)+c(\det(D^2u)-\mathrm{tr}(D^2u)) = 1$$
for $c \neq 0, -1$ has an analytic solution u_c in B satisfying

(i)
$$\begin{aligned}
u_c = &\frac{-z^4}{(c+1)(c^2+2c+2)(c^2+c+1)(c^2+1)} \\
&+ \frac{2(4c^5+4c^4+8c^3+5c^2+4c+4)}{(c+1)(c^2+c+1)}z^2y^2 \\
&+ \frac{2x^2z^2(4c^2+4c+3)}{(c+1)(c^2+c+1)} + \frac{y^4(c^2+1)(3c^4+2c^3+2c^2-4c-4)}{(c+1)(c^2+c+1)} \\
&- \frac{x^4(3c^2+2c+2)}{(c+1)(c^2+c+1)} - 2(c^2+1)zy^2 + 2zx^2 \\
&+ \frac{(c^2+c+1)y^2}{2} + \frac{(c+1)x^2}{2} + O(r^5),
\end{aligned}$$

(ii)
$$\lambda_1 = c^2 + c + 1 + O(r), \quad \lambda_2 = c + 1 + O(r),$$
$$\lambda_3 = -\frac{x^2(c^2+1)(c^2+2c+2) + 3y^2c^2(c^2+1)(c^2+2c+2) + 3z^2}{2(c+1)(c^2+c+1)(c^2+1)(c^2+2c+2)} + O(r^3).$$

3) *The equation ($c = -1$)*
$$\lambda_1\lambda_2 + \lambda_2\lambda_3 + \lambda_1\lambda_3 - \lambda_1\lambda_2\lambda_3 + \lambda_1 + \lambda_2 + \lambda_3 = \sigma_2(D^2u) - \det(D^2u) + \mathrm{tr}(D^2u) = 1$$
has an analytic solution u_{-1} in B satisfying

(i) $u_{-1} = 48y^2x^2 - 12y^2z^2 - \frac{119x^4}{2} + 93x^2z^2 + \frac{z^4}{2} + 2y^2z - 9x^2z - \frac{y^2}{6} + x^2 + O(r^5)$,

(ii) $\lambda_1 = 2 + O(r)$, $\lambda_2 = 6y^2 + 6x^2 + \frac{3z^2}{2} + O(r^3)$, $\lambda_3 = -\frac{1}{3} + O(r)$,

for $r = |(x, y, z)|$.

PROOF. Let us note that
$$v_0 := -\frac{y^4}{3} + 5y^2 z^2 - x^4 + 7x^2 z^2 - \frac{z^4}{3} + 2y^2 z - 2zx^2 + \frac{y^2}{2} + \frac{x^2}{2},$$
$$v_{-1} := 48y^2x^2 - 12y^2z^2 - \frac{119x^4}{2} + 93x^2z^2 + \frac{z^4}{2} + 2y^2z - 9x^2z - \frac{y^2}{6} + x^2,$$

and
$$v_c = \frac{-z^4}{(c+1)(c^2+2c+2)(c^2+c+1)(c^2+1)}$$
$$+ \frac{2z^2y^2(4c^5+4c^4+8c^3+5c^2+4c+4)}{(c+1)(c^2+c+1)}$$
$$+ \frac{2x^2z^2(4c^2+4c+3)}{(c+1)(c^2+c+1)} + \frac{y^4(c^2+1)(3c^4+2c^3+2c^2-4c-4)}{(c+1)(c^2+c+1)}$$
$$- \frac{x^4(3c^2+2c+2)}{(c+1)(c^2+c+1)} - 2(c^2+1)zy^2$$
$$+ 2zx^2 + \frac{(c^2+c+1)y^2}{2} + \frac{(c+1)x^2}{2},$$

verify their respective equations up to second order; i.e.,

$$\sigma_2(D^2v_0) - 1 = O(r^3),$$

$$\sigma_2(D^2v_{-1}) - \det(D^2v_{-1}) + \text{tr}(D^2v_{-1}) - 1 = O(r^3),$$

$$\sigma_2(D^2v_c) + c(\det(D^2u_c) - \text{tr}(D^2u_c)) - 1 = O(r^3),$$

which can be proven by a brute force (e.g., MAPLE) calculation, e.g.,

$$\sigma_2(D^2v_0) - 1$$
$$= 4(-10y^4 - 32y^2x^2 - 50y^2z^2 - 42x^4 - 130x^2z^2 + 11z^4 - 36y^2z + 4x^2z + 4z^3).$$

To prove 1) one considers the following Cauchy problem for the equation $F_0 = \sigma_2(D^2u) - 1 = 0$:

$$u|_{z=0} = v_0|_{z=0} = -\frac{y^4}{3} - x^4 + \frac{y^2}{2} + \frac{x^2}{2},$$

$$\left(\frac{\partial u}{\partial z}\right)_{z=0} = \left(\frac{\partial v_0}{\partial z}\right)_{z=0} = 2y^2 - 2x^2.$$

Since the equation is elliptic, we get by the Cauchy-Kowalevskaya theorem a unique local analytic solution u_0 which should coincide with v_0 within to 4-th order.

The same argument is valid for

$$F_c = \sigma_2(D^2u) + c(\det(D^2u) - \text{tr}(D^2u)) - 1 = 0$$

and the Cauchy problem

$$u|_{z=0} = v_c|_{z=0}$$
$$= \frac{y^4(c^2+1)(3c^4+2c^3+2c^2-4c-4)}{3(c+1)(c^2+c+1)}$$
$$- \frac{x^4(3c^2+2c+2)}{3(c+1)(c^2+c+1)} + \frac{y^2(c^2+c+1)}{2} + \frac{x^2(c+1)}{2},$$

$$\left(\frac{\partial u}{\partial z}\right)_{z=0} = \left(\frac{\partial v_c}{\partial z}\right)_{z=0} = -2(c^2+1)y^2 + 2x^2.$$

The claim on the eigenvalues follows directly from the formulas

$$\det(D^2 v_0) = -\frac{x^2}{2} - \frac{3y^2}{2} - z^2 + O(r^3),$$

$$\det(D^2 v_c) = -\frac{x^2}{2} - \frac{3y^2 c^2}{2} - \frac{3z^2}{(c^2+1)(c^2+2c+2)} + O(r^3),$$

which are straightforward, e.g.,

$$\begin{aligned}\det(D^2 v_0) = {} & 120y^4 x^2 - 140y^4 z^2 + 168y^2 x^4 + 1440y^2 x^2 z^2 - 994y^2 z^4 - 420x^4 z^2 \\ & - 1350 x^2 z^4 + 40 y^4 z - 140 z^6 + 192 y^2 x^2 z - 136 y^2 z^3 - 168 x^4 z \\ & - 120 x^2 z^3 - 16 z^5 - 10 y^4 + 20 y^2 x^2 + 28 y^2 z^2 - 42 x^4 + 28 x^2 z^2 \\ & - 8 z^4 - 24 y^2 z + 40 x^2 z - \frac{3y^2}{2} - \frac{x^2}{2} - z^2.\end{aligned}$$

The argument works for 3) as well. □

7.3.3. Legendre transform. Let us recall some essential properties (see, e.g., Section 1.6 in [**69**]) of the Legendre transform (for simplicity of notation we consider here only the case of three dimensions used below). Let f be a C^2 function defined in a domain $D \subset \mathbb{R}^3$ such that its gradient map $\nabla f : D \longrightarrow \mathbb{R}^3$ maps D bijectively onto a domain G. Let $g = (\nabla f)^{-1} = (P, Q, R) : G \longrightarrow D$ be the map inverse to the gradient. Then the Legendre transform $\tilde{f} : G \longrightarrow \mathbb{R}$ is given by

(7.3.4) $\quad \tilde{f}(u,v,w) := uP(u,v,w) + vQ(u,v,w) + wR(u,v,w) - f(g(u,v,w)).$

Suppose also that $\det(D^2 f) \neq 0$ except for a point $a \in D$ with $b = (\nabla f)(a)$. Then

$$D^2 \tilde{f} = (D^2 f)^{-1} \text{ on } G \setminus \{b\}.$$

We want then to apply the Legendre transform to the solutions u_c on a small ball centered at the origin. We need thus to verify that ∇u_c is injective. One finds $\nabla u_c = [U(x,y,z)x+(c+1)x, V(x,y,z)y+(c^2+c+1)y, -4z^3 m_c + x^2 W_1(z) + y^2 W_2(z)]$, where

$$U(x,y,z), V(x,y,z) \in \mathbb{R}\{\{x,y,z\}\},$$
$$W_1(z), W_2(z) \in \mathbb{R}\{\{z\}\},$$
$$U(0,0,0) = V(0,0,0) = 0,$$

and

$$m_c := \begin{cases} 1/((c+1)(c^2+2c+2)(c^2+c+1)(c^2+1)) > 0 & \text{for } c \neq 0, -1, \\ 1/2 & \text{for } c = -1, \\ 1/3 & \text{for } c = 0. \end{cases}$$

To prove the injectivity of the gradient map we use Theorem 1.1 of [**79**], which says in our situation that the degree of ∇u_c equals $\dim_{\mathbb{R}} Q(\nabla u_c) - 2 \dim_{\mathbb{R}} I$ where I is an ideal of $Q(\nabla u_c)$ which is maximal with respect to the property $I^2 = 0$, the ring $Q(\nabla u_c)$ being defined as

$$Q(\nabla u_c) := \mathbb{R}\{\{x,y,z\}\}/((u_c)_x, (u_c)_y, (u_c)_z).$$

Therefore, to prove the injectivity it is sufficient to prove

LEMMA 7.3.3. *The ring $Q(\nabla u_c)$ is isomorphic to $\mathbb{R}[h]/(h^3)$.*

PROOF. For simplicity of notation we consider only the case $c = 0$, the general case being completely similar. Then

$$\nabla u_0 = \left[U(x,y,z)x + x, V(x,y,z)y + y, -\frac{4z^3}{3} + x^2 W_1(z) + y^2 W_2(z) \right]$$

and

$Q(\nabla u_c)$

$$= \mathbb{R}\{\{x,y,z\}\}/\left(U(x,y,z)x + x, V(x,y,z)y + y, -\frac{4z^3}{3} + x^2 W_1(z) + y^2 W_2(z) \right).$$

If one sets $p := x + U(x,y,z)x$ and $q := y + V(x,y,z)y$, one sees that the ring $Q(\nabla u_0)$ is isomorphic to

$$\mathbb{R}\{\{p,q,z\}\}/\left(p, q, -\frac{4z^3}{3} + p^2 W_1'(p,q,z) + q^2 W_2'(p,q,z) \right)$$

and thus to $\mathbb{R}\{\{z\}\}/\left(-\frac{4z^3}{3}\right)$, which implies the result. □

We can now prove the main result of the section.

THEOREM 7.3.1. *Let* $\theta \in]-\frac{\pi}{2}, \frac{\pi}{2}[$, *and let*

$$\mathbf{F}_\theta(u) = \arctan \lambda_1 + \arctan \lambda_2 + \arctan \lambda_3 - \theta = 0.$$

Then for some ball B_ε *centered at the origin there exists an analytic function* f_θ *on* ∂B_ε *such that the unique (Harvey-Lawson) solution* u_θ *of the Dirichlet problem satisfies the following:*

(i) $u_\theta \in C^{1,1/3}$.
(ii) $u_\theta \notin C^{1,\delta}$ *for any* $\delta > 1/3$.

PROOF. We begin with (i). One can apply the Legendre transform to u_c with $c = \cot(\theta)$ for $\theta \neq 0$ and to u_0 for $\theta = 0$ due to the injectivity of ∇u_c. Set $u_\theta := \widetilde{u}_c$ in this situation. Since u_c with $c \neq 0$ satisfies the equation

$$\sigma_2(D^2 u) + c(\det(D^2 u) - \text{tr}(D^2 u)) - 1 = 0,$$

its Legendre transform $\widetilde{u}_c$ satisfies

$$c(\sigma_2(D^2 u) - 1) + \det(D^2 u) - \text{tr}(D^2 u) = 0,$$

the signature of $(\lambda_1(\widetilde{u}_c), \lambda_2(\widetilde{u}_c), \lambda_3(\widetilde{u}_c))$ being $(+,+,-)$ for $c \geq -1$ and $(-,-,+)$ for $c < -1$, which implies that $\widetilde{u}_c$ lies on the middle branch of this equation. The same is true for $\widetilde{u}_0$ and the equations

$$\sigma_2(D^2 u) - 1 = 0, \quad \det(D^2 u) - \text{tr}(D^2 u) = 0.$$

The function $\widetilde{u}_c$ is analytic outside zero and belongs to $C^{1,1/3}(B_\varepsilon)$, which proves (i). Indeed, it is sufficient to prove the boundedness of the $C^{1,1/3}$-norm of $\widetilde{u}_c$ on a small ball. Since one has $\forall f \in C^1$, $3|f|^2|\nabla f| = |\nabla(f^3)|$, it is sufficient to prove the boundedness of the product $|D\widetilde{u}_c|^2 |D^2 \widetilde{u}_c|$. To prove this last assertion we note that

$$\frac{\partial \widetilde{u}_c}{\partial u} = \frac{\partial}{\partial u}(uP + vQ + wR - u_c(P,Q,R)) = P,$$

$$\frac{\partial \widetilde{u}_c}{\partial v} = \frac{\partial}{\partial v}(uP + vQ + wR - u_c(P,Q,R)) = Q,$$

$$\frac{\partial \widetilde{u}_c}{\partial w} = \frac{\partial}{\partial w}(uP + vQ + wR - u_c(P,Q,R)) = R$$

since $\nabla u_c = (u,v,w)$. Thus, $|D\widetilde{u}_c|^2 = P^2 + Q^2 + R^2$. On the other hand, since the Hessian of $\widetilde{u}_c$ satisfies $D^2(\widetilde{u}_c) = D^2(u_c)^{-1}$, the matrix $\det(D^2(u_c))D^2(\widetilde{u}_c)$ has bounded entries and thus

$$\|\widetilde{u}_c\|_{C^2} \leq \frac{C}{|\det(D^2(u_c))|}$$

or an absolute constant C (e.g., for $C = 10$ in the case $c = 0$). Since by Proposition 7.3.1, $|\det(D^2(u_c))| \geq C'(c)(P^2+Q^2+R^2)$ for an absolute constant $C'(c)$ (e.g., $C'(0) = 1/2 - \varepsilon$), we get the boundedness of the product. To finish the proof of (i) one needs then to prove that $\widetilde{u}_c$ is a Harvey-Lawson solution of the corresponding Dirichlet problem.

This is implied by the following form of the Alexandrov maximum principle [7].

PROPOSITION 7.3.2. *Let F be a Dirichlet domain, and let $u = v + w$ where v is of $\widetilde{F}$ type. If $u \in C^2(B \setminus \{0\}) \cap C^1(B)$ and D^2u is nonnegatively defined on $B \setminus \{0\}$, then*

$$\sup_B u \leq \sup_{\partial B} u.$$

For (ii) one sets $u = 0$, $v = 0$, $w \neq 0$. Then

$$\lambda_3(u_\theta) = -2m_c^{-1} w^{-2/3}/3 + o(w^{-2/3}),$$

which contradicts the condition $u_\theta \in C^{1,\delta}$ for $\delta > 1/3$ and finishes the proof of (ii). $\square$

REMARK 7.3.2. Let us consider the special Lagrangian submanifold $L_{u,c} \subset \mathbb{C}^3$ corresponding to our singular solution u_c, i.e., the graph of the map

$$i\nabla u_c : B \to i\mathbb{R}^3.$$

It is easy to show that it is smooth and that the singularity of u_c implies only that the projection

$$L_{u,c} \to B$$

is singular (maps between smooth manifolds).

The interest to singularities of special Lagrangian manifolds is motivated in particular by the SYZ conjecture of Strominger, Yau, and Zaslow [236] on the structure of mirror-dual manifolds. The conic singularities of special Lagrangian manifolds were intensively studied; see, e.g., [127], [105], [35].

Let $u \in C^2(\Omega)$, $\Omega \subset \mathbb{R}^n$ be a solution of a special Lagrangian equation. Then the gradient graph $M = (x, \nabla u)$ is a special Lagrangian submanifold of $\mathbb{R}^{2n}$; i.e., M is a minimal Lagrangian submanifold in $\mathbb{R}^{2n}$. For solutions of the special Lagrangian equations which are not in C^2 the problem of minimality of the gradient graph is not entirely solved. The gradient graph of the singular solutions of the special Lagrangian equation given above is a smooth manifold by Remark 7.3.2. That observation raises questions: Are gradient graphs of solutions to special Lagrangian equations always smooth manifolds; are they minimal manifolds in terms of minimal currents; etc.? We suspect that the following question can be answered in the affirmative.

QUESTION. Let $\Omega = B_1 = \{x : |x| < 1\} \subset \mathbb{R}^3$. Does there exists a smooth function g on $\partial B = \mathbb{S}^2$ such that (say, for for $h = 0$) the Dirichlet problem with g as the boundary data has no solution $u \in W^{2,1}(B) \cap C^1(B)$?

REMARK 7.3.3. If so, for the purported unique viscosity solution u to the Dirichlet problem with the boundary date g we have either $u \notin C^1$ or the gradient graph $(u, \nabla u)$ is not a smooth minimal surface in $\mathbb{R}^6$. However, this does not say whether the gradient graph of a viscosity solution of SLE is a rectifiable minimal set, which is also a very interesting question.

7.4. More singular solutions and a failure of the maximum principle

7.4.1. The main result and its consequences.
We give now a construction of singular solutions for special Lagrangian equation in $\mathbb{R}^3$ due to D. Wang and Y. Yuan [**270**]. This construction gives more singular solutions, which are also "more singular" than those we have just constructed above in Section 7.3. It is based on the idea of a generalization of the *Lewy rotation*, i.e., of the contact transformation of a smooth surface $z = z(x,y)$ in $\mathbb{R}^3$ given by

$$\xi := x + z_x(x,y), \quad \eta := y + z_y(x,y).$$

Hans Lewy introduced this transformation to obtain a priori bounds for the Monge-Ampère equation in two dimensions [**148**].

THEOREM 7.4.1. *Let $\theta \in]-\frac{\pi}{2}, \frac{\pi}{2}[$, and let $m \geq 2$ be an integer. Then there exists a solution u_θ^m of the Dirichlet problem for the equation*

$$\mathbf{F}_\theta(u) = \arctan \lambda_1 + \arctan \lambda_2 + \arctan \lambda_3 - \theta = 0$$

in the unit ball $B_1 \subset \mathbb{R}^3$ satisfying
  (i) $u_\theta^m \in C^{1,1/(2m+1)}(B_1) \cap C^\infty(B_1 \setminus \{0\})$,
  (ii) $u_\theta^m \notin C^{1,\delta}(B_1)$ *for any* $\delta > \frac{1}{2m+1}$.

Using the above-mentioned rotations forth and back one obtains the following corollary for smooth solutions of the equation:

THEOREM 7.4.2. *Let $\theta \in]-\frac{\pi}{2}, \frac{\pi}{2}[$. Then there exists a family $u_\theta^\varepsilon, \varepsilon > 0$ of smooth solutions to the equation $\mathbf{F}_\theta(u) = 0$ in $B_1 \subset \mathbb{R}^3$ such that*

$$\|Du_\theta^\varepsilon\|_{L^\infty(B_1)} \leq C, \quad \lim_{\varepsilon \to 0} |D^2 u_\theta^\varepsilon(0)| = \infty,$$

for a constant C.

Therefore we see that the maximum principle for the Hessian of a solution to $\mathbf{F}_\theta(u) = 0$ does not hold; moreover one can show that the equation is uniformly elliptic on the solutions u_θ^ε, thus showing that the maximum principle for the Hessian fails for uniformly elliptic equations as well (recall that in two dimensions it holds). Note also that, adding appropriate quadratic terms, one immediately transposes these results to all dimensions $n \geq 3$ for special Lagrangian equations with subcritical phases θ, i.e., for $|\theta| < \frac{(n-2)\pi}{2}$.

Another application of the main result concerns the *minimal surface system*

(7.4.1) $$\Delta_g U := \sum_{i,j=1}^n \frac{1}{\sqrt{g}} (\sqrt{g} g^{ij} U_{x_i})_{x_i} = 0$$

on a vector function $U : \mathbb{R}^n \longrightarrow \mathbb{R}^n$, g being the induced metric $g := I + D^t U \cdot DU$.

THEOREM 7.4.3. *Let $m \geq 2$ be an integer, and let $n = 3$. Then there exists a weak solution U^m of (7.4.1) in $B_1 \subset \mathbb{R}^3$ satisfying*
  (i) $\forall p < \frac{2m+2}{2m-1}$, $U^m \in W^{1,p}(B_1)$,
  (ii) $U^m \notin W^{1, \frac{2m+2}{2m-1}}(B_1)$.
Furthermore, there exists a family U^ε, $\varepsilon > 0$, of smooth solutions to (7.4.1) in $B_1 \subset \mathbb{R}^3$ such that
$$\|U^\varepsilon\|_{L^\infty(B_1)} \leq C, \quad \lim_{\varepsilon \to 0} |DU^\varepsilon(0)| = \infty,$$
for a constant C.

Therefore, the $W^{2,1}$ regularity for solutions of special Lagrangian equations and the $W^{1,1}$ regularity for solutions of the minimal surface system do not admit substantial improvements.

7.4.2. Solution of the σ_2-equation and rotating to $|\theta| < \frac{\pi}{2}$. We begin with the critical phase $\theta = \frac{\pi}{2}$ case where the special Lagrangian equation becomes especially simple, $\sigma_2(D^2 u) = 1$, which permits its elementary treatment by Cauchy-Kowalevskaya.

One approximates a solution to

(7.4.2) $\quad \begin{cases} \sigma_2(D^2 u) = 1, \\ u_3(x_1, x_2, 0) = P_3(x_1, x_2, 0), \\ u(x_1, x_2, 0) = P(x_1, x_2, 0) \end{cases}$

by a $2m$-degree polynomial

$$P = \frac{x_1^2 + x_2^2}{2} + b_m x_3 + \frac{m^2}{4} \rho^{2m-2} x_3^2 + \varepsilon \sum_{j=0}^{m} a_j x_3^{2m-2j} \rho^{2j},$$

with $Z := x_1 + x_2 i$, $\rho := |Z|$, $b_m := \operatorname{Re}(Z^m)$ and appropriate coefficients ε, a_j.

PROPOSITION 7.4.1. *One can choose $\varepsilon > 0$ and $a_j, j = 1, \ldots, m$, in such a way that the following properties hold:*
  (i) $\sigma_2(D^2 P) = 1 + O(r^{3m-3})$.
  (ii) $\lambda_1 = 1 + O(r^{m-1}), \lambda_2 = 1 + O(r^{m-1}), -\delta_2(m) r^{2m-2} \leq \lambda_3 \leq -\delta_1(m) r^{2m-2}$.
  (iii)
$$DP = \begin{bmatrix} x_1 + O(\rho r^{m-1} + r^{2m}) \\ x_2 + O(\rho r^{m-1} + r^{2m}) \\ b_m + \frac{m^2}{2} \rho^{2m-2} x_3 - 2m\varepsilon x_3^{2m-1} + \varepsilon \rho^2 O(r^{2m-3}) + O(r^{2m}) \end{bmatrix}.$$

  (iv) $\delta_1(m) r^{2m-1} \leq |DP| \leq \delta_2(m) r$.

Condition (i) is satisfied for
$$a_0 = -1, \quad j^2 a_{j+1} = a_j (j - m)(2m - 2j - 1), \quad j = 0, \ldots, m-1.$$
An appropriate choice of ε permits us to satisfy the other conditions as well. Applying Cauchy-Kowalevskaya one gets a solution u of (7.4.2) on a ball of some radius r_m. Replacing it by $u(r_m x) r_m^{-2}$ one obtains a solution on B_1.

Let us then apply rotations from the group $U(3)$ to obtain solutions for $|\theta| < \frac{\pi}{2}$. One notes that the $\frac{\pi}{2}$ rotation is just the Lewy rotation, while the Legendre transform is essentially the composition of the Lewy rotation with complex conjugation.

These $U(3)$ rotations are very natural in calibrated geometry: A solution of the special Lagrangian equation gives a minimal graph in $\mathbb{C}^3$ and we simply rotate it unitarily, conserving all of its geometric properties. What is highly nontrivial is to show that under appropriate rotations the graph of the solution remains the graph of a function (perhaps defined in a smaller neighborhood).

REMARK 7.4.1. One notes the geometric "secret" of Legendre, Lewy, and the generalization of Lewy rotation which various arguments for special Lagrangian equations are based upon: Legendre transformation is just a half-pi rotation followed by conjugation in $\mathbb{C}^n$; Lewy rotation is nothing more than a Legendre transformation on functions added by a quadratic; the generalization of Lewy rotation is just $U(n)$ rotation. The early development of that for special Lagrangian equations can be found in [**280**], [**279**].

For $\alpha \in [0, \frac{\pi}{4}[$ we consider the $U(3)$ rotation, which is the rotation by α in the z_1 and z_2 directions in $\mathbb{C}^3$ and which does not change z_3:
$$z'_1 := e^{i\alpha} z_1, \quad z'_2 := e^{i\alpha} z_2, \quad z'_3 := z_3.$$

For the special Lagrangian 3-fold $\mathcal{M}$ parametrized over B_1 by the gradient of the solution u to SLE we get a new parametrization:
$$(7.4.3) \quad \begin{cases} \widetilde{x} = (x_1 \cos\alpha + u_1(x)\sin\alpha, x_2 \cos\alpha + u_2(x)\sin\alpha, x_3) \\ \widetilde{y} = (-x_1 \sin\alpha + u_1(x)\cos\alpha, -x_2 \sin\alpha + u_2(x)\cos\alpha, u_3(x)). \end{cases}$$

One can show that Proposition 7.4.1(ii) implies

PROPOSITION 7.4.2. $\mathcal{M}$ is a special Lagrangian graph $(\widetilde{x}, D\widetilde{u}(\widetilde{x}))$ over the ball $B_{\frac{1}{\sqrt{2}}}$ in the $\widetilde{x}$-space.

The Hessian is calculated as
$$D^2 \widetilde{u} = \frac{\partial \widetilde{y}}{\partial \widetilde{x}} = \frac{\partial \widetilde{y}}{\partial x}\left(\frac{\partial \widetilde{x}}{\partial x}\right)^{-1} = diag\left(\tan\left(\frac{\pi}{4} - \alpha\right), \tan\left(\frac{\pi}{4} - \alpha\right), 0\right) + O\left(r^{m-1}\right)$$
and its determinant is calculated as
$$\det D^2 \widetilde{u} = \tan\left(\frac{\pi}{4} - \alpha\right) \left(\frac{-m^2 \rho^{2m-2}}{2} + \tan\left(\frac{\pi}{4} - \alpha\right) H_{33}\right) + O\left(r^{2m-1}\right).$$

These formulas easily imply

PROPOSITION 7.4.3. (i) $\det D^2 \widetilde{u}(\widetilde{x}) > 0$ on a small punctured ball.
(ii) For the upper 2×2 minor D' of $\det D^2 \widetilde{u}(\widetilde{x})$ one has
$$2\tan\left(\frac{\pi}{4} - \alpha\right) I_2 \geq D' \geq \frac{\tan\left(\frac{\pi}{4} - \alpha\right) I_2}{2}$$
on a small punctured ball.
(iii) Let $\widetilde{\lambda}_i, i = 1, 2, 3$, be the eigenvalues of $D^2 \widetilde{u}(\widetilde{x})$ and let $\widetilde{\theta}_i := \arctan \widetilde{\lambda}_i, i = 1, 2, 3$. Then
$$\widetilde{\theta}_1 = \left(\frac{\pi}{4} - \alpha\right)\left(1 + O\left(r^{m-1}\right)\right),$$
$$\widetilde{\theta}_2 = \left(\frac{\pi}{4} - \alpha\right)\left(1 + O\left(r^{m-1}\right)\right),$$
$$\widetilde{\theta}_3 \approx -\cotan\left(\frac{\pi}{4} - \alpha\right) r^{2m-2}.$$

This proposition permits us to apply the following main technical property of Legendre's transform (up to complex conjugation).

PROPOSITION 7.4.4. *Let $\mathcal{L}$ be a special Lagrangian graph (x, Df) with the potential funcion f smooth on a ball B_ρ satisfying the following:*

(i) $Df(0) = 0$, $\det D^2 f(x) < 0, \forall x \in B_\rho \setminus \{0\}$.

(ii) *For the upper 2×2 minor D' of $D^2 f(x)$ one has on B_ρ*

$$k^{-1} I_2 \geq D' \geq k I_2 \quad \text{and also} \quad |(f_{13}(x), f_{23}(x))| \leq \frac{1}{2}.$$

Then $\mathcal{L}$ can be represented again as a graph $(\widetilde{x}, \widetilde{y}) = (\widetilde{x}, D\widetilde{f}(\widetilde{x}))$ on the open set $\Omega := Df(B_{\rho k^2/2})$, with

$$\widetilde{x} + i\widetilde{y} = -i(x + iy), \quad \widetilde{f} \in C^1(\Omega) \cap C^\infty(\Omega \setminus \{0\}).$$

Indeed, applying Proposition 7.4.4 to $f = \widetilde{u}$, $\rho = \widetilde{r}_{m,\alpha}$, $k = 2\tan\left(\frac{\pi}{4} - \alpha\right)$ for $\alpha = -\theta/2$, $0 < \theta < \frac{\pi}{2}$ being the phase of the initial SLE, we get a function $\widetilde{\widetilde{u}}(\widetilde{\widetilde{x}})$ defined in a neighborhood of 0, and a simple calculation gives for its Hessian eigenvalues $\widetilde{\widetilde{\lambda}}_i, i = 1, 2, 3$:

$$\arctan \widetilde{\widetilde{\lambda}}_1 + \arctan \widetilde{\widetilde{\lambda}}_2 + \arctan \widetilde{\widetilde{\lambda}}_3 = -2\alpha = \theta,$$

as necessary. One can also verify that $\widetilde{\widetilde{u}}$ is a viscosity solution around 0. To finish the proof of Theorem 7.4.1 it is then sufficient to show that $\widetilde{\widetilde{u}}$ is of corresponding regularity. This can be done in a manner completely similar to the regularity statements in Theorem 7.3.1.

7.4.3. Smooth solutions and the minimal surface system.

To get smooth solutions with the desired properties one applies yet another family of $U(3)$ rotations to the solutions obtained in the previous subsection.

For a fixed phase $\theta \in \left[0, \frac{\pi}{2}\right)$ one sets $\gamma = \frac{\pi}{8} - \frac{\theta}{4}$ and chooses a small $\varepsilon \in (0, \gamma)$. Let $\alpha = \frac{\theta}{2} - \frac{3\varepsilon}{2}$. Applying the α rotation (as before, only to the first two coordinates and leaving the third unchanged) one obtains a solution on certain balls of radius, say, r_θ in such a way (shrinking r_θ further if necessary) that for the corresponding Hessian eigenvalues $\widetilde{\lambda}_i^\varepsilon, i = 1, 2, 3$, the following properties hold: If $\widetilde{\theta}_i^\varepsilon = \arctan \widetilde{\lambda}_i^\varepsilon, i = 1, 2, 3$, then

$$\widetilde{\theta}_1^\varepsilon \text{ and } \widetilde{\theta}_2^\varepsilon = \left(\frac{\pi}{4} - \frac{\theta}{2} + \frac{3\varepsilon}{2}\right)\left(1 + O\left(r^{m-1}\right)\right) \geq \gamma,$$

$$\widetilde{\theta}_3^\varepsilon = -\delta_{m,\alpha}(\widetilde{x}) \cot\left(\frac{\pi}{4} - \frac{\theta}{2} + \frac{3\varepsilon}{2}\right) r^{2m-2}\left(1 + O\left(r^{m-1}\right)\right).$$

Then one applies Proposition 7.4.4 for $\rho = r_\theta$, $k = \tan\left(\frac{\pi}{4} - \alpha\right)/2$ to $(\widetilde{x}, D\widetilde{u}(\widetilde{x}))$, which leads to the potential $\widetilde{\widetilde{u}}^\varepsilon$ with the graph $\widetilde{\widetilde{u}}^\varepsilon(\widetilde{\widetilde{x}})$, $D\widetilde{\widetilde{u}}^\varepsilon(\widetilde{\widetilde{x}})$ on $D\widetilde{u}(B_{r_0 k^2/2})$, with the Hessian eigenvalues $\widetilde{\widetilde{\lambda}}_i^\varepsilon$, $i = 1, 2, 3$, satisfying

$$\arctan \widetilde{\widetilde{\lambda}}_1^\varepsilon + \arctan \widetilde{\widetilde{\lambda}}_2^\varepsilon + \arctan \widetilde{\widetilde{\lambda}}_3^\varepsilon = -\theta - 3\varepsilon.$$

One verifies also that

$$c_1 r \geq |D\widetilde{u}(\widetilde{x})| \geq c_2 r^{m-1},$$

which guarantees the noncollapsing of neighborhoods used in the proof and the uniformity of O terms for ε going to 0.

To finish the construction of the solutions in Theorem 7.4.2 one makes the final rotation $z' = e^{i\varepsilon}\tilde{\tilde{z}}$ and applies Proposition 7.4.4 once more; this gives the necessary solutions for the phase $-\theta$ and by symmetry for the phase θ as well.

Let us then explain how to get a proof of Theorem 7.4.3. To prove the theorem one only needs to take $\theta = 0$ and let $U^m = Du_0^m$, $U^\varepsilon = Du^\varepsilon$. Indeed, from Proposition 7.4.1(ii) and Proposition 7.4.2 one finds that
$$|DU^m| = |D^2 u_0^m| \approx |Du_0^m|^{2-2m}.$$
Therefore, one obtains
$$\int_{B_1} |DU^m|^p dy \approx \int_{B_1} |Du_0^m|^{(2-2m)p} dy$$
$$\approx \int_{Du^m(B_1)} \frac{|x|^{(2-2m)p}}{|\det(D^2 u_0^m)|} dx \approx \int_{Du^m(B_1)} |x|^{(2-2m)(p-1)} dx,$$
thus proving points (i) and (ii) of Theorem 7.4.3. To prove that U^m is a distributional solution one notes that the bounds for the eigenvalues guarantee that the expression
$$\sum_{i,j=1}^n \sqrt{g} g^{ij} \langle U_{x_i}, \phi_{x_j} \rangle$$
is integrable for any smooth compactly supported ϕ; since it is zero pointwise outside the origin, we get the conclusion. The statement about U^ε is straightforward.

Bibliography

[1] U. Abresch, *Isoparametric hypersurfaces with four or six distinct principal curvatures. Necessary conditions on the multiplicities*, Math. Ann. **264** (1983), 283–302. MR714104 (85g:53052b).

[2] J. F. Adams, *Lectures on exceptional Lie groups*, Univ. Chicago Press, Chicago, 1996. MR1428422 (98b:22001).

[3] G. M. Adel'son-Vel'skii, *Generalization of a geometrical Bernstein theorem*, Dokl. Akad. Nauk SSSR **49**(1945), 309–401.

[4] M. A. Akivis and V. V. Goldberg *Differential geometry of varieties with degenerate Gauss maps*, Springer, 2004. MR2014407.

[5] A. A. Albert, *On a certain algebra of quantum mechanics*, Ann. of Math. **35** (1934), 65–73. MR1503142.

[6] A. D. Alexandroff, *Sur les théorèmes d'unicité pour les surfaces fermées*, Dokl. Acad. Nauk **22** (1939), 99–102.

[7] A. D. Alexandrov, *Some theorems on partial differential equations of the second order*, Vest. Len. Univ. **9** (1954), 3–17.

[8] A. D. Alexandrov, *A characteristic property of sphere*, Ann. Mat. Pura Appl. **58** (1962), 303–315. MR0143162 (26 ♯ 722).

[9] A. D. Alexandrov, *Uniqueness conditions and estimates for the solutions of the Dirichlet problem*, Amer. Math. Soc. Trans. **68** (1968), 89–119.

[10] B. N. Allison and J. R. Faulkner, *A Cayley-Dickson process for a class of structurable algebras*, Trans. Amer. Math. Soc. **283** (1984), 185–210. MR0735416.

[11] F. J. Almgren, Jr., *Some interior regularity theorems for minimal surfaces and an extension of Bernstein's theorem*, Ann. of Math. (2) **84** (1966), 277–292. MR0200816 (34 ♯ 702).

[12] S. N. Armstrong, L. Silvestre, and C. K. Smart, *Partial regularity of solutions of fully nonlinear uniformly elliptic equations*, Comm. Pure Appl. Math. **65** (2012), 1169–1184. MR2928094.

[13] S. N. Armstrong, B. Sirakov, and C. K. Smart, *Fundamental solutions of homogeneous fully nonlinear elliptic equations*, Comm. Pure Appl. Math. **64** (2011), 737–777. MR2663711.

[14] S. N. Armstrong, B. Sirakov, and C. K. Smart, *Singular solutions of fully nonlinear elliptic equations and applications*, Arch. Ration. Mech. Anal. **205** (2012), 345–394. MR2947535.

[15] M. F. Atiyah and J. Berndt, *Projective planes, Severi varieties and spheres*, Surv. Diff. Geom., VIII, Int. Press, Somerville, MA, 2003, pp. 1–27. MR2039984 (2005b:53080).

[16] M. F. Atiyah, R. Bott, and A. Shapiro, *Clifford modules*, Topology **3** (1964), 3–38. MR0167985 (29:5250).

[17] M. Atiyah, V. G. Drinfel'd, N. J. Hitchin, and Yu. I. Manin, *Construction of Instantons*, Phys. Lett. A **65** (1978), 185–187. MR0598562 (82g:81049).

[18] T. Aubin, *Équations du type Monge-Ampère sur les variétés kähleriennes compactes*. C. R. Acad. Sci. Paris Sér. A-B **283** (1976), A119–A121. MR0433520 (55 ♯ 6496).

[19] J. C. Baez, *The octonions*, Bull. Amer. Math. Soc. (N.S.) **39** (2002), 145–205. MR1886087(2003f:17003).

[20] P. Baird and J. C. Wood, *Harmonic morphisms between Riemannian manifolds*, London Math. Soc. Mon., 29, Oxford Univ. Press, Oxford, 2003. MR2044031 (2005b:53101).

[21] J. Ball, *Convexity conditions and existence theorems in nonlinear elasticity*, Arch. Rat. Mech. Anal. **63** (1977), 337–403. MR0475169 (57 ♯ 14788).

[22] J. Bao, J. Chen, B. Guan, and M. Ji, *Liouville property and regularity of a Hessian quotient equation*, Amer. J. Math. **125** (2003), 301-316. MR1963687 (2004b:35079).

[23] P. Bauman, *Positive solutions of elliptic equations in non-divergence form and their adjoints*, Ark. Mat. **22** (1984), 153–173. MR0765409 (86m:35008).

[24] A. Belousov, L. Doskolovich, and S. Kharitonov, *A gradient method of designing optical elements for forming a specified irradiance on a curved surface*, J. Optic. Tech. **75** (2008), 161–165.

[25] S. Bernstein, *Über ein geometrisches Theorem und seine Anwendung auf die partiellen Differentialgleichungen vom elliptischen Typus*, Math. Z. **26** (1927), 551–558. MR1544873.

[26] J. Bolton, *Transnormal hypersurfaces*, Proc. Cambridge Phil. Soc. **74** (1973), 43–48. MR0328827 (48:7169).

[27] E. Bombieri and E. Giusti, *Harnack's inequality for elliptic differential equations on minimal surfaces*, Invent. Math. **15** (1972), 24–46. MR0308945 (46:8057).

[28] E. Bombieri, E. De Giorgi, and E. Giusti, *Minimal cones and the Bernstein problem*, Invent. Math. **7** (1969), 243–268. MR0250205 (40:3445).

[29] M. Bordemann, *Nondegenerate invariant bilinear forms on nonassociative algebras*, Acta Math. Univ. Comenian. (N.S.) **66** (1997), 151–201. MR1620480 (99k:17005).

[30] A. A. Borisenko, *A Liouville-type theorem for the equation of special Lagrangian manifolds*, Math. Notes **52** (1992), 1094–1096. MR1201944 (93k:53055).

[31] S. Brendle, *Embedded minimal tori in S^3 and the Lawson conjecture*, Acta Math. **211** (2013), 177–190. MR3143888.

[32] Y. Brenier, *Polar factorization and monotone rearrangement of vector-valued functions*, Comm. Pure Appl. Math. **44** (1991), 375–417. MR1100809 (92d:46088).

[33] H. Brézis, *Symmetry in nonlinear PDE's*, Proc. Symp. Pure Math. **65** (1999), 1–12. MR1662746 (99k:35044).

[34] H. Brézis, F. Merle, and T. Rivière, *Quantization effects for $-\Delta u = u(1-|u|^2)$ in $\mathbb{R}^2$*, Arch. Rat. Mech. Anal. **126** (1994), 123–148. MR1268048 (95d:35042).

[35] R. L. Bryant, *Second order families of special Lagrangian 3-folds*, Perspectives in Riemannian Geometry, CRM Proc. Lecture Notes **40** (2006), 63–98. MR2237106 (2007e:53063).

[36] X. Cabre, *Elliptic PDEs in Probability and Geometry. Symmetry and regularity of solutions*. Discr. Cont. Dyn. Syst. **20** (2008), 425–457. MR2373200 (2008m:35094).

[37] L. Caffarelli, *Interior regularity for fully nonlinear equations*, Ann. Math. **130** (1989), 189–213. MR1005611 (90i:35046).

[38] L. Caffarelli, *A localization property of viscosity solutions to the Monge-Ampère equation and their strict convexity*, Ann. of Math. **131** (1990), 129–134. MR1038359 (91f:35058).

[39] L. Caffarelli, *Interior $W^{2,p}$ estimates for solutions of the Monge-Ampère equation*, Ann. of Math. **131** (1990), 135–150. MR1038360 (91f:35059).

[40] L. Caffarelli, *Some regularity properties of solutions of Monge Ampère equation*, Comm. Pure Appl. Math. **44** (1991), 965–969. MR1127042 (92h:35088).

[41] L. Caffarelli, *A note on the degeneracy of convex solutions to Monge Ampère equation*, Comm. Part. Diff. Eq. **18** (1993), 1213-1217. MR1233191 (94f:35045).

[42] L. Caffarelli and X. Cabre, *Fully Nonlinear Elliptic Equations*, Amer. Math. Soc., Providence, R.I., 1995. MR1351007 (96h:35046).

[43] L. Caffarelli and X. Cabre, *Interior $C^{2,\alpha}$ regularity theory for a class of nonconvex fully nonlinear elliptic equations*, J. Math. Pures Appl. **82** (2003), 573–612. MR1995493 (2004f:35049).

[44] L. Caffarelli, Y. Y. Li, and L. Nirenberg, *Some remarks on singular solutions of nonlinear elliptic equations I.* J. Fixed P. Th. Appl. **5** (2009), 353–395. MR2529505 (2010i:35089).

[45] L. Caffarelli, Y. Y. Li, and L. Nirenberg, *Some remarks on singular solutions of nonlinear elliptic equations III: viscosity solutions including parabolic operators*, Comm. Pure Appl. Math. **66** (2013), 109–143. MR2994551.

[46] L. Caffarelli, J. J. Kohn, L. Nirenberg, and J. Spruck, *The Dirichlet problem for nonlinear second-order elliptic equations II. Complex Monge-Ampère, and uniformly elliptic, equations*, Comm. Pure Appl. Math. **38** (1985), 209–252. MR0739925 (87f:35096).

[47] L. Caffarelli, L. Nirenberg, and J. Spruck, *The Dirichlet problem for nonlinear second-order elliptic equations I. Monge-Ampère equation*, Comm. Pure Appl. Math. **37** (1984), 369–402. MR0780073 (87f:35097).

[48] L. Caffarelli, L. Nirenberg, and J. Spruck; *The Dirichlet problem for nonlinear second order elliptic equations III. Functions of the eigenvalues of the Hessian*, Acta Math. **155** (1985), 261–301. MR0806416 (87f:35098).

BIBLIOGRAPHY

[49] L. Caffarelli and L. Silvestre, *On the Evans-Krylov theorem*, Proc. Amer. Math. Soc. **138** (2010), 263-265. MR2550191 (2011g:35106).

[50] L. Caffarelli and Y. Yuan, *A priori estimates for solutions of fully nonlinear equations with convex level set*, Indiana Univ. Math. J. **49** (2000), no. 2, 681–695. MR1793687 (2002b:35049).

[51] E. Calabi, *Improper affine hyperspheres of convex type and a generalization of a theorem by K. Jörgens*, Michigan Math. J. **5** (1958), 105–126. MR0106487 (21:5219).

[52] E. Calabi, *Examples of Bernstein problems for some nonlinear equations*, Global Analysis (Proc. Sympos. Pure Math., Vol. XV, Berkeley, Calif., 1968), Amer. Math. Soc., Providence, R.I., 1970, pp. 223–230. MR0264210 (41:8806).

[53] M. do Carmo and H. B. Lawson Jr., *On Alexandrov-Bernstein theorems in hyperbolic space*, Duke Math. J. **50** (1983), 995–1003. MR0726314 (85f:53009).

[54] E. Cartan, *Familles de surfaces isoparamétriques dans les espaces à courbure constante*, Ann. Mat. Pura Appl. **17** (1938), 177–191. MR1553310.

[55] E. Cartan, *Sur des familles remarquables d'hypersurfaces isoparamétriques dans les espaces sphériques*, Math. Z. **45** (1939), 335–367. MR0000169.

[56] E. Cartan, *Sur des familles d'hypersurfaces isoparamétriques des espaces sphériques à 5 et à 9 dimensions*, Univ. Nac. Tucumán. Revista A. **1** (1940), 5–22. MR0004519 (3,18g).

[57] Th. E. Cecil, *Isoparametric and Dupin hypersurfaces*, SIGMA Symmetry Integrability Geom. Methods Appl. **4** (2008), Paper 062, 28. MR2434936 (2009i:53044).

[58] T. E. Cecil, *Lie sphere geometry*, Springer, New York, 2008. MR2361414 (2008h:53091).

[59] T. E. Cecil, Q.-S. Chi, and G. R. Jensen, *Isoparametric hypersurfaces with four principal curvatures*, Ann. Math. **166** (2007), 1–76. MR2342690 (2008m:53150).

[60] S.-Y. A. Chang and Y. Yuan, *A Liouville problem for the σ_2 equation*, Discr. Cont. Dyn. Syst 28 (2010), 659-664. MR2644763 (2012d:35089).

[61] B.-Y. Chen, *Riemannian submanifolds*, Handbook of differential geometry, Vol. I, North-Holland, Amsterdam, 2000, 187–418. MR1736854 (2001b:53064).

[62] J. Chen, M. Warren, and Y. Yuan, *A priori estimate for convex solutions to special Lagrangian equations and its application*, Comm. Pure Appl. Math. **62** (2009), 583–595. MR2492708 (2010b:35116).

[63] S. Y. Cheng and S.-T. Yau, *Complete affine hypersurfaces. I. The completeness of affine metrics*, Comm. Pure Appl. Math. **39** (1986), 839–866. MR0859275 (87k:53127).

[64] S.-S. Chern, *On the curvature of a piece of hypersurface in Euclidean space*, Abh. Math. Sem. Univ. Hamburg **29** (1965), 77-99. MR0188949 (32 ♯ 6376).

[65] S.-S. Chern and R. Lashof, *On the total curvature of immersed manifolds*, Amer. J. Math. **79** (1957), 306–318. MR0084811 (18,927a).

[66] Q.-S. Chi, *Isoparametric hypersurfaces with four principal curvatures, II*, Nagoya Math. J. **204** (2011), 1–18. MR2863363.

[67] Q.-S. Chi, *Isoparametric hypersurfaces with four principal curvatures, III*, J. Diff. Geom. **94** (2013), 469–504. MR3080489.

[68] T. H. Colding and W. P. Minicozzi, II, *A course in minimal surfaces*, Amer. Math. Soc., Providence, RI, 2011. MR2780140.

[69] R. Courant and D. Hilbert, *Methods of Mathematical Physics. Vol. 2, Partial Differential Equations*, Wiley, N.Y., 1989. MR1013360 (90k:35001).

[70] M. G. Crandall, L. C. Evans, and P-L. Lions, *Some properties of viscosity solutions of Hamilton-Jacobi equations*, Trans. Amer. Math. Soc. **282** (1984), 487–502. MR0732102 (86a:35031).

[71] M. G. Crandall, H. Ishii, and P-L. Lions, *User's guide to viscosity solutions of second order partial differential equations*, Bull. Amer. Math. Soc. (N.S.) **27** (1992), 1-67. MR1118699 (92j:35050).

[72] M. G. Crandall and P-L. Lions, *Viscosity solutions of Hamilton-Jacobi equations*, Trans. Amer. Math. Soc. **277** (1983), 1–42. MR0690039 (85g:35029).

[73] E. De Giorgi, *Una estensione del teorema di Bernstein*, Ann. Sc. Norm. Sup. Pisa (3) **19** (1965), 79–85. MR0178385 (31:2643).

[74] R. De Sapio, *On Spin(8) and Triality: A Topological Approach*, Expo. Math. **19** (2001), 143–161. MR1835965 (2002g:22025a).

[75] S. K. Donaldson, *Nahm's equations and free-boundary problems*, The many facets of geometry, Oxford Univ. Press, Oxford, 2010, 71–91. MR2681687 (2011h:58023).

[76] J. Dorfmeister, *Theta functions for special, formally real Jordan algebras, a remark on a paper of H. L. Resnikoff: "Theta functions for Jordan algebras" (Invent. Math.* **31** *(1975), 87–104)*, Invent. Math. **44** (1978), 103–108. MR0486675 (58:6378).

[77] J. Dorfmeister and E. Neher, *An algebraic approach to isoparametric hypersurfaces in spheres, I,* Tohoku Math. J. (2) **35** (1983), 187–224. MR0699927 (84k:53049).

[78] J. Douglas,*The higher topological form of Plateau's problem*, Ann. Sc. Norm. Super. Pisa (2) **8** (1939), 195–218. MR0002477 (2,60e).

[79] D. Eisenbud and H. Levin, *An algebraic formula for the degree of a C^∞ map germ*, Ann. Math. **106** (1977), 19–44. MR0467800 (57:7651).

[80] A. Elduque, *Vector cross products*, 7 pp., available at http://www.unizar.es/matematicas/algebra/elduque/Talks/crossproducts.pdf.

[81] A. Elduque and S. Okubo, *On algebras satisfying $x^2 x^2 = N(x)x$*, Math. Z. **235** (2000), 275–314. MR1795509 (2001h:17003).

[82] P. Etingof, D. Kazhdan, and A. Polishchuk, *When is the Fourier transform of an elementary function elementary?* Selecta Math. (N.S.) **8** (2002), 27–66. MR1890194 (2003g:42008).

[83] L. C. Evans, *Classical solutions of fully nonlinear, convex, second-order elliptic equations*, Comm. Pure Appl. Math. **35** (1982), 333–363. MR0649348 (83g:35038).

[84] L. C. Evans, *Partial Differential Equations, 2nd edition*, Amer. Math. Soc., Providence, RI, 2010. MR2597943 (2011c:35002).

[85] J. Faraut and A. Korányi, *Analysis on symmetric cones*, Oxford Mathematical Monographs, The Clarendon Press, Oxford University Press, New York, 1994. MR1446489 (98g:17031).

[86] A. Farina, *Two results on entire solutions of Ginzburg-Landau system in higher dimensions*, J. Funct. Anal. **214** (2004), 386–395. MR2083306 (2005g:35093).

[87] D. Ferus, *Notes on isoparametric hypersurfaces*, preprint, Escola de Geometria Diferencial, Universidade Estadual de Campinas, 1980.

[88] D. Ferus, H. Karcher, and H. F. Münzner, *Cliffordalgebren und neue isoparametrische Hyperflächen*, Math. Z. **177** (1981), 479–502. MR624227 (83k:53075).

[89] W. H. Fleming, *Flat chains over a finite coefficient group*, Trans. Amer. Math. Soc. **121** (1966), 160–186. MR0063358 (16,108b).

[90] A. T. Fomenko, *Variational principles of topology*, Mathematics and its Applications (Soviet Series) 42, Kluwer Academic Publishers Group, Dordrecht, 1990. MR1057340 (91k:58020).

[91] H. Freudenthal, *Beziehungen der $\mathfrak{E}_7$ und $\mathfrak{E}_8$ zur Oktavenebene. I*, Nederl. Akad. Wetensch. Proc. Ser. A. **57** (1954), 218–230 = Indag. Math. **16** (1954), 218–230. MR0063358 (16,108b).

[92] H. Freudenthal, *Beziehungen der $\mathfrak{E}_7$ und $\mathfrak{E}_8$ zur Oktavenebene. II*, Nederl. Akad. Wetensch. Proc. Ser. A. **57** (1954), 363–368 = Indag. Math. **16** (1954), 363–368. MR0068549 (16,900d).

[93] J. Ge and Y. Xie, *Gradient map of isoparametric polynomial and its application to Ginzburg-Landau system*, J. Funct. Anal. **258** (2010), 1682–1691. MR2566315 (2011d:53140).

[94] D. Gilbarg and N. Trudinger, *Elliptic Partial Differential Equations of Second Order, 2nd ed.*, Springer-Verlag, Berlin-Heidelberg-New York-Tokyo, 1983. MR0737190 (86c:35035).

[95] M. Godlinski and P. Nurowski, *$GL(2,\mathbb{R})$ geometry of ODE's*, J. Geom. Phys. **60** (2010), 991–1027. MR2647299 (2011e:53027).

[96] R. Goodman and N. R. Wallach, *Representations and invariants of the classical groups*, Encyclopedia of Mathematics and its Applications, 68, Cambridge Univ. Press, Cambridge, 1998. MR1606831 (99b:20073).

[97] P. Guan, N. S. Trudinger, and X.-J. Wang, *On the Dirichlet problem for degenerate Monge-Ampère equations*, Acta Math. **182** (1999), 87–104. MR1687172 (2000h:35051).

[98] M. Günaydin and O. Pavlyk, *Generalized spacetimes defined by cubic forms and the minimal unitary realizations of their quasiconformal groups*, J. High Energy Phys. **101** (2005), 31 pp. (electronic). MR2166016 (2006g:81075).

[99] M. Günaydin, G. Sierra, and P. K. Townsend, *The geometry of $N = 2$ Maxwell-Einstein supergravity and Jordan algebras*, Nuclear Phys. B **242** (1984), 244–268. MR756588 (86j:83059a).

[100] J. Hahn, *Isoparametric hypersurfaces in the pseudo-Riemannian space forms*, Math. Z. **187** (1984), 195–208. MR0753432 (87a:53094).

[101] P. Hartman and L. Nirenberg, *On spherical image whose Jacobians do not change sign*, Amer. J. Math. **81** (1959), 901–920. MR0126812 (23:A4106).

[102] Q. Han, N. Nadirashvili, and Y. Yuan, *Linearity of homogeneous order-one solutions to elliptic equations in dimension three*, Comm. Pure Appl. Math. **56** (2003), 425–432. MR1949137 (2003k:35043).

[103] R. Harvey and H. B. Lawson, Jr., *Calibrated geometries*, Acta Math. **148** (1982), 47–157. MR666108 (85i:53058).

[104] F. R. Harvey and H. B. Lawson, Jr., *Dirichlet duality and the nonlinear Dirichlet problem*, Comm. Pure Appl. Math. **62** (2009), 396-443. MR2487853 (2010d:35097).

[105] M. Haskins *Special Lagrangian cones*, Amer. J. Math. **126** (2004), 845–871. MR2075484 (2005e:53074).

[106] H. He, H. Ma, and F. Xu, *On eigenmaps between spheres*, Bull. London Math. Soc. **35** (2003), 344–354. MR1960944 (2003m:53109).

[107] H. Hironaka, *Normal cones in analytic Whitney stratification*, Publ. Inst. Hautes Études Sci. **36** (1969), 127–139. MR0277759 (43:3492).

[108] R. A. Horn and C. R. Johnson, *Matrix analysis*, Camb. Univ. Press, Cambridge, 1985. MR0832183 (87e:15001).

[109] W. Hsiang, *On the compact homogeneous minimal submanifolds*, Proc. Nat. Acad. Sci. U.S.A. **56** (1966), 5–6. MR0205203 (34:5037).

[110] W. Hsiang, *Remarks on closed minimal submanifolds in the standard Riemannian m-sphere*, J. Differential Geometry **1** (1967), 257–267. MR0225244 (37:838).

[111] W. Hsiang and H. B. Lawson, Jr., *Minimal submanifolds of low cohomogeneity*, J. Diff. Geom. **5** (1971), 1–38. MR0298593 (45:7645).

[112] A. Hurwitz, *Über die Komposition der quadratischen Formen*, Math. Ann. **88** (1922), 1–25. MR1512117.

[113] D. Husemoller, *Fibre bundles*, Graduate Texts in Mathematics, 20, Springer-Verlag, N.Y., 1994. MR1249482 (94k:55001).

[114] R. Immervoll, *On the classification of isoparametric hypersurfaces with four distinct principal curvatures in spheres*, Ann. of Math. **168** (2008), 1011–1024. MR2456889 (2010b:53110).

[115] R. Iordanescu, *Jordan structures in analysis, geometry and physics*, Editura Academiei Române, Bucharest, 2009. MR2515186 (2010e:17026). Also arXiv:1106.4415, 2011.

[116] H. Ishii, *On uniqueness and existence of viscosity solutions of fully nonlinear second-order elliptic PDEs*, Comm. Pure Appl. Math. **42** (1989), 15–45. MR0973743 (89m:35070).

[117] A. O. Ivanov, *Calibration forms and new examples of globally minimal surfaces*, in: *Minimal surfaces*, Amer. Math. Soc., Providence, RI, 1993, pp. 235–267. MR1233576 (94g:58045).

[118] N. M. Ivochkina, *A priori estimate of $|u|_{C_2(\overline{\Omega})}$ of convex solutions of the Dirichlet problem for the Monge-Ampère equation* (Russian), Zap. Nauchn. Sem. Len. Otdel. Mat. Inst. Stekl. (LOMI) **96** (1980), 69–79. MR0579472 (82b:35056).

[119] N. Jacobson, *Some groups of transformations defined by Jordan algebras. I*, J. Reine Angew. Math. **201** (1959), 178–195. MR0106936.

[120] N. Jacobson, *Generic norm of an algebra*, Osaka Math. J. **15** (1963), 25–50. MR0153719.

[121] N. Jacobson, *Structure and representations of Jordan algebras*, Amer. Math. Soc. Colloquium Publ., Vol. XXXIX, Amer. Math. Soc., Providence, R.I., 1968. MR0251099 (40:4330).

[122] R. Jensen, *The maximum principle for viscosity solutions of fully nonlinear second order partial differential equations*, Arch. Rat. Mech. An. **101** (1988), 1–27. MR0920674 (89a:35038).

[123] P. Jordan, *Über Verallgemeinerungsmöglichkeiten des Formalismus der Quantenmechanik*, Nachr. Akad. Wiss. Göttingen. Math. Phys. **41** (1933), 209-217.

[124] P. Jordan, J. von Neumann, and E. Wigner, *On an algebraic generalization of the quantum mechanical formalism*, Ann. Math. **35** (1934), 29–64. MR1503141.

[125] K. Jörgens, *Über die Lösungen der Differentialgleichung $rt-s^2 = 1$*, Math. Ann. **127** (1954), 130–134. MR0062326 (15,961e).

[126] J. Jost and Y.-L. Xin, *A Bernstein theorem for special Lagrangian graphs*, Calc. Var. Part. Diff. Eq. **15** (2002), 299–312. MR1938816 (2003i:53075).

[127] D. Joyce, *Special Lagrangian Submanifolds with Isolated Conical Singularities. V. Survey and Applications*, J. Diff. Geom. **63** (2003), 279–347. MR2015549 (2005e:53076).

[128] M.-A. Knus, A. Merkurjev, M. Rost, and J.-P. Tignol, *The book of involutions*, Amer. Math. Soc., Providence, RI, 1998. MR1632779 (2000a:16031).

[129] S. Kobayashi and K. Nomizu, *Foundations of differential geometry. Vol. II*, Wiley, N. Y., 1996. MR1393941 (97c:53001b).

[130] M. Koecher, *The Minnesota notes on Jordan algebras and their applications*, Lect. Notes Mat. **1710**, Springer-Verlag, Berlin, 1999. MR1718170 (2001e:17040).

[131] D. Koutroufiotis, *On a conjectured characterization of the sphere*, Math. Ann. **205** (1973), 211–217. MR0348684 (50:1181).

[132] S. Krutelevich, *Jordan algebras, exceptional groups, and Bhargava composition*, J. Algebra **314** (2007), 924–977. MRS2344592 (2008k:20103).

[133] N. V. Krylov, *Nonlinear elliptic and parabolic equations of the second order*, Reidel, Dordrecht, 1987. MR0901759 (88d:35005).

[134] N. V. Krylov, *On the General Notion of Fully Nonlinear Second-Order Elliptic Equations*, Trans. Amer. Math. Soc. **347** (1995), 857–895. MR1284912 (95f:35075).

[135] N. V. Krylov, *On weak uniqueness for some diffusions with discontinuous coefficients,* Stoch. Proc. Appl. **113** (2004), 37–64. MR2078536 (2005e:60119).

[136] N. V. Krylov, *Smoothness of the payoff function for a controllable diffusion process in a domain* (Russian), Izv. Akad. Nauk SSSR Ser. Mat. **53** (1989), 66–96; translation in Math. USSR-Izv. **34** (1990), 65–95. MR0992979 (90f:93040).

[137] N. V. Krylov and M. V. Safonov, *A property of solutions of parabolic equations with measurable coefficients,* Izv. Akad. Nauk SSSR Ser. Mat. **44** (1980), 161–175. MR0563790 (83c:35059).

[138] D. A. Labutin, *Isolated singularities for fully nonlinear elliptic equations*, J. Diff. Eq. **177** (2001), 49–76. MR1867613 (2003a:35065).

[139] D. A. Labutin, *Potential estimates for a class of fully nonlinear elliptic equations*, Duke Math. J. **111** (2002), 1–49. MR1876440 (2002m:35053).

[140] E. M. Landis, *Second order equations of elliptic and parabolic type*, Amer. Math. Soc., 1998. MR1487894 (98k:35034).

[141] R. Langevin, G. Levitt, and H. Rosenberg, *Hérissons et multihérissons, (enveloppes paramétrées par leur application de Gauss)*, Singularities, Warsaw, 1985, Banach Center Publ. 20, PWN, Warsaw, 1988, 245–253. MR1101843 (92a:58015).

[142] H. B. Lawson, Jr., *Complete minimal surfaces in S^3*, Ann. Math. **92** (1970), 335–374. MR0270280 (42:5170).

[143] H. B. Lawson, Jr., *Lectures on minimal submanifolds. Vol. I*, Publish or Perish, Wilmington, Del., 1980. MR0576752 (82d:53035b).

[144] H. B. Lawson and R. Osserman, *Non-existence, non-uniqueness and irregularity of solutions to the minimal surface system*, Acta Math. **139** (1977), 1–17. MR0452745 (80b:35059).

[145] P. Lévay and P. Vrana, *Special entangled quantum systems and the Freudenthal construction*, J. Phys. A **42** (2009), 285–303. MR2519736 (2010i:81097).

[146] T. Levi-Civita, *Famiglie di superficie isoparametrische nel'ordinario spacio euclideo*, Atti. Accad. naz. Lincei. Rend. Cl. Sci. Sci. Fis. Mat. Natur **26** (1937), 350–362.

[147] J. L. Lewis, *Smoothness of certain degenerate elliptic equations*, Proc. Amer. Math. Soc. **80** (1980), 259–265. MR577755 (82a:35047).

[148] H. Lewy, *A priori limitations for solutions of Monge-Ampère equations. II.* Trans. Amer. Math. Soc. **41** (1937), 365-374. MR1501906.

[149] B. Q. Li, *Entire solutions of eiconal [eikonal] type equations*, Arch. Math. **89** (2007), 350–357. MR2355154 (2008h:32058).

[150] T. Liu, *Algebraic minimal submanifolds in the spheres*, J. of Beijing Teacher College **10** (1989), 10–17.

[151] C. Loewner and L. Nirenberg, *Partial differential equations invariant under conformal or projective transformations. Contributions to analysis (a collection of papers dedicated to Lipman Bers)*, Ac. Pr., N.Y., 1974, 245–272. MR0358078 (50:10543).

[152] M. A. Magid, *Lorentzian isoparametric hypersurfaces*, Pacific J. Math. **118** (1985), 165–197. MR0783023 (87b:53097).

[153] M. Marcus and H. Minc *A Survey of Matrix Theory and Matrix Inequalities*, Dover, 1992. MR1215484.

[154] Y. Martinez-Maure, *Contre-exemple à une caractérisation conjecturée de la sphère*, C. R. Math. Acad. Sci. Paris **332** (2001), 41–44. MR1805625 (2002a:53084).

[155] Y. Martinez-Maure, *Théorie des hérissons et polytopes,* C. R. Math. Acad. Sci. Paris **336** (2003), 241–244. MR1968266 (2004c:52003).

[156] U. Massari, M. Miranda, and M. Miranda, *The Bernstein theorem in higher dimensions*, Boll. Unione Mat. Ital. (9) **1** (2008), 349–359. MR2424298 (2009e:49052).

[157] K. McCrimmon, *The Freudenthal-Springer-Tits constructions of exceptional Jordan algebras*, Trans. Amer. Math. Soc. **139** (1969), 495–510. MR0238916 (39:276).

[158] K. McCrimmon, *A taste of Jordan algebras*, Springer, N.Y., 2004. MR2014924 (2004i:17001).

[159] K. Meyberg and J. M. Osborn, *Pseudo-composition algebras*, Math. Z. **214** (1993), 67–77. MR1234598 (94i:17003).

[160] K. Meyberg, *Trace formulas in vector product algebras*, Commun. Alg. **30**(2002), 2933–2940. MR1908248 (2003d:17002.

[161] J. S. Milne, *Elliptic curves*, BookSurge Publishing, 2006. MR2267743 (2007h:14044).

[162] J. W. Milnor, *Singular points of complex hypersurfaces*, Ann. Math. Stud., 61, Princeton Univ. Press, Princeton, N.J., 1968. MR0239612 (39:969).

[163] J. W. Milnor, *On manifolds homeomorphic to the 7-sphere*, Ann. Math. **64** (1956), 399–405. MR0082103 (18,498d).

[164] M. Miranda, *Recollections on a conjecture in mathematics*, Mat. Contemp. **35** (2008), 143–150. MR2584181.

[165] P. Mironescu, *On the stability of radial solutions of the Ginzburg-Landau equation*, J. Funct. Anal. **130** (1995), 334–344. MR1335384 (96i:35124).

[166] R. Miyaoka, *Transnormal functions on a Riemannian manifold*, Diff. Geom. Appl. **31** (2013), 130–139. MR3010083.

[167] R. Miyaoka, *Isoparametric hypersurfaces with $(g,m)=(6,2)$*, Ann. Math. **177** (2013), 53-110. MR2999038.

[168] C. B. Morrey, *Second order elliptic systems of partial differential equations*, Contributions to the Theory of Partial Differential Equations, Ann. of Math. Studies, 33, Princeton Univ. Press, Princeton, 1954, 101–160. MR0068091 (16,827e).

[169] H. F. Münzner, *Über Flächen mit einer Weingartenschen Ungleichung*. Math. Z. **97** (1967), 123–139. MR0208541 (34:8351).

[170] H. F. Münzner, *Isoparametrische Hyperflächen in Sphären*, Math. Ann. **251** (1980), 57–71. MR583825 (82a:53058).

[171] H. F. Münzner, *Isoparametrische Hyperflächen in Sphären. II. Über die Zerlegung der Sphäre in Ballbündel*, Math. Ann. **256** (1981), 215–232. MR620709 (82m:53053).

[172] N. Nadirashvili, *Nonuniqueness in the martingale problem and the Dirichlet problem for uniformly elliptic operators*. An. Sc. Norm. Sup. Pisa, Cl. Sc. **24** (1997), 537–550. MR1612401 (99b:35042).

[173] N. Nadirashvili, *On derivatives of viscosity solutions to fully nonlinear elliptic equations*, Mosc. Math. J. **11** (2011), 149–155. MR2808216 (2012e:35063).

[174] N. Nadirashvili, V. G. Tkachev, and S. Vlăduţ, *Non-classical solution to Hessian equation from Cartan isoparametric cubic*, Adv. Math. **231** (2012), 1589–1597. MR2964616.

[175] N. Nadirashvili and S. Vlăduţ, *Nonclassical solutions of fully nonlinear elliptic equations*, Geom. Func. An. **17** (2007), 1283–1296. MR2373018 (2008m:35121).

[176] N. Nadirashvili and S. Vlăduţ, *Singular solutions to fully nonlinear elliptic equations*, J. Math. Pures Appl. **89** (2008), 107–113. MR2391642 (2009a:35080).

[177] N. Nadirashvili and S. Vlăduţ, *Singular solution to special Lagrangian equations*. Ann. Inst. H. Poincaré Anal. Non Linéaire **27** (2010), 1179–1188. MR2683755 (2011k:35051).

[178] N. Nadirashvili and S. Vlăduţ, *Weak solutions of nonvariational elliptic equations*, Proceedings of the International Congress of Mathematicians, Volume III, 2001–2018, Hindustan Book Agency, New Delhi, 2010. MR2840858 (2012h:35076).

[179] N. Nadirashvili and S. Vlăduţ, *Singular solutions of Hessian fully nonlinear elliptic equations*, Adv. Math. **230** (2011), 1589–1597. MR2824567 (2012h:35077).

[180] N. Nadirashvili and S. Vlăduţ, *Octonions and nonclassical solutions of fully nonlinear elliptic equations*, Geom Funct. An. **21** (2011), 483-498. MR2795515 (2012e:35062).

[181] N. Nadirashvili and S. Vlăduţ, *On axially symmetric solutions of Hessian fully nonlinear elliptic equations*, Math. Zeit. **270** (2012), 331–336. MR2875836.

[182] N. Nadirashvili and S. Vlăduţ, *On homogeneous solutions of nonlinear elliptic equations in four dimensions*, Comm. Pure Appl. Math. **66** (2013), 1653–1662. MR3084701.

[183] N. Nadirashvili and S. Vlăduţ, *Singular solutions of Hessian elliptic equations in five dimensions*, J. Math. Pures Appl. **100** (2013), 769–784. MR3125267.

[184] N. Nadirashvili and Y. Yuan, *Homogeneous solutions to fully nonlinear elliptic equation*, Proc. Amer. Math. Soc. **134** (2006), 1647–1649. MR2204275 (2007b:35129).

[185] J. von Neumann, *On an algebraic generalization of the quantum mechanical formalism (Part I)*, Rec. Math. [Mat. Sbornik] **1(43)** (1936), 415–484.
[186] L. Nirenberg, *On nonlinear elliptic partial differential equations and Hölder continuity*, Comm. Pure Appl. Math. **6** (1953), 103–156. MR0064986 (16,367c).
[187] L. Nirenberg, *On Singular Solutions of Nonlinear Elliptic and Parabolic Equations*, Milan Math. J. **79** (2011), 3–12. MR2831435 (2012i:35032).
[188] J. C. C. Nitsche, *Lectures on minimal surfaces. Vol. 1*, Cambridge University Press, Cambridge, 1989. MR1015936 (90m:49031).
[189] K. Nomizu, *Some results in E. Cartan's theory of isoparametric families of hypersurfaces*, Bull. Amer. Math. Soc. **79** (1973), 1184–1188. MR0326625 (48:4968).
[190] K. Nomizu, *On isoparametric hypersurfaces in the Lorentzian space forms*, Japan. J. Math. (N.S.) **7** (1981), 217–226. MR0728336 (84k:53050).
[191] P. Nurowski, *Distinguished dimensions for special Riemannian geometries*, J. Geom. Phys. **58** (2008), 1148–1170. MR2451275 (2009k:53108).
[192] R. Osserman, *The minimal surface equation*, Seminar on nonlinear partial differential equations (Berkeley, Calif., 1983), Math. Sci. Res. Inst. Publ., 2, Springer, N. Y., 1984, 237–259. MR765237 (86b:58128).
[193] H. Ozeki and M. Takeuchi, *On some types of isoparametric hypersurfaces in spheres. I*, Tôhoku Math. J. (2) **27** (1975), 515–559. MR0454888 (56:13132a).
[194] H. Ozeki and M. Takeuchi, *On some types of isoparametric hypersurfaces in spheres. II*, Tôhoku Math. J. (2) **28** (1976), 7–55. MR0454889 (56:13132b).
[195] G. Panina, *New counterexamples to A. D. Alexandrov's hypothesis*, Adv. Geometry **5** (2005), 301–317. MR2131822 (2006d:52002).
[196] C.-K. Peng and L. Xiao, *On the classification of cubic minimal cones*, J. of Grad. School **10** (1993), 1–19.
[197] J. Peng and Z. Tang, *Brouwer degrees of gradient maps of isoparametric functions*, Sci. China Ser. A **39** (1996), 1131–1139. MR1442672 (98c:58039).
[198] O. M. Perdomo, *Characterization of order 3 algebraic immersed minimal surfaces of S^3*, Geom. Dedicata **129** (2007), 23–34. MR2353979 (2008g:53011).
[199] O. M. Perdomo, *Non-existence of regular algebraic minimal surfaces of spheres of degree 3*, J. Geom. **84** (2005), 100–105. MR2215368 (2007a:53016)
[200] O. Perron, *Eine neue Behandlung der ersten Randwertaufgabe $\Delta u = 0$*, Math. Zeit. **18** (1923), 42–54. MR1544619.
[201] H. P. Petersson and M. L. Racine, *Jordan algebras of degree 3 and the Tits process*, J. Algebra **98** (1986), 211–243. MR0825144 (87h:17038a).
[202] H. P. Petersson, *Structure theorems for Jordan algebras of degree three over fields of arbitrary characteristic*, Comm. Algebra **32** (2004), 1019–1049. MR2063796 (2005h:17061).
[203] A. V. Pogorelov, *Extension of a general uniqueness theorem of A. D. Aleksandrov to the case of nonanalytic surfaces* (Russian), Doklady Akad. Nauk SSSR (N.S.) **62** (1948), 297–299. MR0027569 (10,325a).
[204] A. V. Pogorelov, *On the improper convex affine hyperspheres*, Geom. Dedic. **1** (1972), 33–46. MR0319126 (47:7672).
[205] A. V. Pogorelov, *Mnogomernaya problema Minkovskogo*, "Nauka", Moscow, 1975, translation: *The Minkowski multidimensional problem*, Wiley, N.Y., 1978. MR0500748 (58:18296).
[206] T. Radó, *The problem of least area and the problem of Plateau*, Math. Z. **32** (1930), 763–796. MR1545197.
[207] J. Radon, *Lineare scharen orthogonaler matrizen*, Abh. Math. Sem. Univ. Hamburg **1** (1922), 1–14. MR3069384.
[208] R. C. Reilly, *On the Hessian of a function and the curvatures of its graph*, Mich. Math. J. **20** (1974), 373–383. MR0334045 (48:12364).
[209] H. L. Resnikoff, *Theta functions for Jordan algebras*, Invent. Math. **31** (1975), 87–104. MR0412495 (54:618).
[210] S. A. Robertson, *On transnormal manifolds*, Topology **6** (1967), 117–123. MR0206977(34:6793).
[211] R. T. Rockafellar, *Convex analysis*, Princeton Univ. Press, 1996. MR1451876 (97m:49001).
[212] H. Röhrl and S. Walcher, *Algebras of complexity one*, Alg. Gr. Geom. **5** (1988), 61–107. MR0959584 (89i:17004).

[213] H. H. Rose, *Geometrical Charged-Particle Optics*, Springer Ser. Opt. Sc., Springer, Amsterdam, 2009.

[214] M. V. Safonov, *On the classical solution of nonlinear elliptic equations of second order*, Mathematics of the USSR-Izvestiya **33** (1989), 597–612. MR0984219 (90d:35104).

[215] M. V. Safonov, *Unimprovability of estimates of Hölder constants for solutions of linear elliptic equations with measurable coeficients*, Math. USSR-Sbornik **60** (1988), 269–281. MR0882838 (88e:35049).

[216] M. V. Safonov, *Nonuniqueness for Second-Order Elliptic Equations with Measurable Coefficients*, SIAM J. Math. Anal. **30** (1999), 879–895. MR1684729 (2000c:35035).

[217] D. A. Salamon and Th. Walpuski, *Notes on the octonians*, arXiv:1005.2820 (2010).

[218] O. Savin, *Small perturbation solutions for elliptic equations*, Comm. Part. Diff. Eq. **32** (2007), 557–578. MR2334822 (2008k:35175).

[219] R. D. Schafer, *An introduction to nonassociative algebras*, Dover, N. Y., 1966. MR1375235 (96j:17001)

[220] B. Segre, *Famiglie di ipersuperficie isoparametrische negli spazi euclidei ad un qualunque numero di demesioni*, Atti. Accad. Linc. Rend. Cl. Sci. Fis. Mat. Natur. **21** (1938), 203–207.

[221] V. V. Sergienko and V. G. Tkachev, *New examples of higher-dimensional minimal hypersurfaces*, in prep., 2013.

[222] J. Serrin, *Isolated singularities of solutions of quasi-linear equations*, Acta Math. **113** (1965), 219– 240. MR0176219 (31:494).

[223] D. B. Shapiro, *Compositions of quadratic forms*, de Gruyter Expositions in Mathematics, 33, Walter de Gruyter, Berlin, 2000. MR1786291 (2002f:11046).

[224] L. Silvestre and B. Sirakov, *Boundary regularity for viscosity solutions of fully nonlinear elliptic equations*, arXiv: 1306.6672v1, 2013.

[225] L. Simon, *The minimal surface equation*, Geometry, V, Enc. Math. Sci. 90, Springer, Berlin, 1997, 239–272. MR1490041 (99b:53014).

[226] L. Simon, *Entire solutions of the minimal surface equation*, J. Diff. Geom. **30** (1989), 643–688. MR1021370 (91a:35068).

[227] L. Simon and B. Solomon, *Minimal hypersurfaces asymptotic to quadratic cones in $\mathbb{R}^{n+1}$*, Invent. Math. **86** (1986), 535–551. MR860681 (87k:49047).

[228] J. Simons, *Minimal varieties in riemannian manifolds*, Ann. Math. **88** (1968), 62–105. MR0233295 (38:1617).

[229] K. Smoczyk, *On algebraic selfsimilar solutions of the mean curvature flow*, Analysis (Munich) **31** (2011), 91–102. MR2752787.

[230] B. Solomon, *The harmonic analysis of cubic isoparametric minimal hypersurfaces. I. Dimensions 3 and 6*, Amer. J. Math. **112** (1990), 157–203. MR1047297(91e:58203).

[231] B. Solomon, *The harmonic analysis of cubic isoparametric minimal hypersurfaces. II. Dimensions 12 and 24*, Amer. J. Math. **112** (1990), 205–241. MR1047298 (91e:58204).

[232] T. A. Springer, *On a class of Jordan algebras*, Nederl. Akad. Wetensch. Proc. Ser. A 62 = Indag. Math. **21** (1959), 254–264. MR0110739 (22:1607).

[233] T. A. Springer and F. D. Veldkamp, *Octonions, Jordan algebras and exceptional groups*, Springer, Berlin, 2000. MR1763974 (2001f:17006).

[234] S. Sternberg, *Lectures on differential geometry*, Prentice-Hall, 1964. MR0193578 (33:1797).

[235] S. Stolz, *Multiplicities of Dupin hypersurfaces*, Invent. Math. **138** (1999), 253–279. MR1720184(2001d:53065).

[236] A. Strominger, S.-T. Yau, and E. Zaslow, *Mirror symmetry is T-duality*, Nuclear Physics B **479** (1996), 243–259. MR1429831 (97j:32022).

[237] D. Stroock and S. R. S. Varadhan, *Theory of diffusion processes*, Stochastic differential equations, C.I.M.E. Summer Sch. **77** (2010), Springer, Heidelberg, 149–191. MR2830392.

[238] S. I. Svinolupov, *Jordan algebras and generalized Korteweg-de Vries equations*, Teoret. Mat. Fiz. **87** (1991), 391–403. MR1129673 (92k:58129).

[239] S. I. Svinolupov, *Jordan algebras and integrable systems*, Funk. An. Appl. **27** (1993), 40–53, 96. MR1264317 (95b:35203).

[240] S. I. Svinolupov and V. V. Sokolov, *Vector-matrix generalizations of classical integrable equations*, Teoret. Mat. Fiz. **100**(1994), 214–218. MR1311194 (95k:35154).

[241] S. I. Svinolupov and V. V. Sokolov, *Deformations of nonassociative algebras and integrable differential equations*, Acta Appl. Math. **41** (1995), 323–339. MR1362135 (97e:17002).

[242] T. Takahashi, *Minimal immersions of Riemannian manifolds*, J. Math. Soc. Japan **18** (1966), 380–385. MR0198393(33:6551).

[243] Z. Tang, *Harmonic Hopf constructions and isoparametric gradient maps*, Diff. Geom. Appl. **25** (2007), 461–465. MR2351423 (2008j:53113).

[244] Z. Tang, *New constructions of eigenmaps between spheres*, Internat. J. Math. **12** (2001), 277–288. MR1841516 (2002j:58027).

[245] G. Thorbergsson, *A survey on isoparametric hypersurfaces and their generalizations*, Handbook of differential geometry, Vol. I, North-Holland, Amsterdam, 2000, 963–995. MR1736861 (2001a:53097).

[246] G. Thorbergsson, *Singular Riemannian foliations and isoparametric submanifolds*, Milan J. Math. **78** (2010), 355–370. MR2684784 (2011g:53047).

[247] J. Tits, *Algèbres alternatives, algèbres de Jordan et algèbres de Lie exceptionnelles. I. Construction*, Nederl. Akad. Wetensch. Proc. Ser. A 69 = Indag. Math. **28** (1966), 223–237. MR0219578 (36:2658).

[248] V. G. Tkachev, *A remark on the Jörgens-Calabi-Pogorelov theorem* (Russian), Dokl. Akad. Nauk **340** (1995), 317–318.

[249] V. G. Tkachev, *On a classification of minimal cubic cones in $\mathbb{R}^n$*, submitted, arxiv:1009.5409, 2010, 42 pp.

[250] V. G. Tkachev, *A generalization of Cartan's theorem on isoparametric cubics*, Proc. Amer. Math. Soc. **138** (2010), 2889–2895. MR2644901.

[251] V. G. Tkachev, *Minimal cubic cones via Clifford algebras*, Comp. An. Op. Th. **4** (2010), 685–700. MR2719795.

[252] V. G. Tkachev, *Jordan algebra approach to the cubic eiconal equation*, J. Algebra **419** (2014), 34–51. MR3253278

[253] N. Trudinger, *Hölder gradient estimates for fully nonlinear elliptic equations*, Proc. Roy. Soc. Edinburgh Sect. A **108** (1988), 57–65. MR0931007 (90b:35041).

[254] N. Trudinger, *Weak solutions of Hessian equations*, Comm. Part. Diff. Eq. **22** (1997), 1251–1261. MR1466315 (99a:35077).

[255] N. S. Trudinger, *On regularity and existence of viscosity solutions of nonlinear second order, elliptic equations*, in: Progr. Nonlinear Differential Equations Appl., 2, Birkhäuser, Boston, MA, 1989, 939–957. MR1034037 (90m:35041).

[256] N. S. Trudinger, *The Dirichlet problem for the prescribed curvature equations*, Arch. Rat. Mech. Anal. **111** (1990), 153–170. MR1057653 (91g:35118).

[257] N. S. Trudinger, *On the twice differentiability of viscosity solutions of nonlinear elliptic equations*, Austral. Math. Soc. **39** (1989), 443–447. MR0995142 (90f:35038).

[258] N. S. Trudinger and X.-J. Wang, *Boundary regularity for the Monge-Ampère and affine maximal surface equations*, Ann. Math. **167** (2008), 993–1028. MR2415390 (2010h:35168).

[259] J. I. E. Urbas, *On the existence of nonclassical solutions for two classes of fully nonlinear elliptic equations*, Indiana Univ. Math. J. **39** (1990), 355–382. MR1089043 (92h:35074).

[260] J. I. E. Urbas, *An approximation result for solutions of Hessian equations*, Calc. Var. Part. Diff. Eq. **29** (2007), 219–230. MR2307773 (2008j:35070).

[261] J. I. E. Urbas, *Regularity of generalized solutions of Monge-Ampère equations*, Math. Z. **197** (1988), 365–393. MR0926846 (89b:35046).

[262] E. B. Vinberg, *Homogeneous cones*, Soviet Math. Dokl. **1** (1960), 787–790. MR0158414 (28:1637).

[263] E. B. Vinberg, *The theory of homogeneous convex cones*, Trudy Mosk. Mat. Ob. **12** (1963), 303–358. MR0158414 (28:1637).

[264] E. B. Vinberg, *Construction of the exceptional simple Lie algebras*, Lie groups and invariant theory, Amer. Math. Soc. Transl. Ser. 2, 213, Amer. Math. Soc., Providence, RI, 2005, pp. 241–242. MR2140725.

[265] B. L. van der Waerden, *Algebra. Vol. I*, Springer, N.Y., 1991. MR1080172 (91h:00009a).

[266] S. Walcher *On algebras of rank three*, Comm. Algebra **27** (1999), 3401–3438. MR1695570 (2000d:17001).

[267] Q. M. Wang, *Isoparametric functions on Riemannian manifolds. I*, Math. Ann. **277** (1987), 639–646. MR0901710 (88h:53056).

[268] J. S. Wang, Degree 3 algebraic minimal surfaces in the 3-sphere, Acta Math. Sc. **6** (2012), 2065–2084. MR2989397.

[269] X.-J. Wang, *Some counterexamples to the regularity of Monge-Ampère equations*, Proc. Amer. Math. Soc. **123** (1995), 841–845. MR1223269 (95d:35025).

[270] D. Wang and Yu. Yuan, *Singular solutions to special Lagrangian equations with subcritical phases and minimal surface systems*, Amer. J. Math. **135** (2013), 1157–1177. MR3117304.

[271] M. Warren and Yu. Yuan, *A Liouville type theorem for special Lagrangian Equations with constraints*, Comm. Part. Diff. Eq. **33** (2008), 922–932. MR2424382 (2009f:35108).

[272] G. Weyl, *Das asymptotische Verteilungsgezets des Eigenwerte lineare partieller Differentialgleichungen*, Math. Ann. **71** (1912), 441–479.

[273] H. Whitney, *Local properties of analytic varieties*, Differential and combinatorial topology, Princeton Univ., 1965, 205–244. MR0188486 (32:5924).

[274] R. Wood, *Polynomial maps from spheres to spheres*, Invent. Math. **5** (1968), 163–168. MR0227999 (37:3583).

[275] L. Xiao, *Lorentzian isoparametric hypersurfaces in H_1^{n+1}*, Pacific J. Math. **189** (1999), 377–397. MR1696128 (2000h:53090).

[276] S.-T. Yau, *On the Ricci curvature of a compact Kähler manifold and the complex Monge-Ampère equation. I*, Comm. Pure Appl. Math. **31** (1978), 339–411. MR0480350 (81d:53045).

[277] S.-T. Yau, *Review of geometry and analysis*, Asian J. Math. **4** (2000), 235–278. MR1803723 (2002e:53002).

[278] P. Yiu, *Quadratic forms between Euclidean spheres*, Manuscripta Math. **83** (1994), 171–181. MR1272181 (95b:55013).

[279] Yu Yuan, *A priori estimates for solutions of fully nonlinear special lagrangian equations*, Ann. Inst. H. Poincaré C **18** (2001), 261–270. MR1808031 (2002e:35088).

[280] Yu Yuan, *Global solutions to special Lagrangian equations*, Proc. Amer. Math. Soc. **134** (2006), 1355–1358. MR2199179 (2006k:35111).

[281] Yu Yuan, *A Bernstein problem for special Lagrangian equations*, Invent. Math. **150** (2002), 117–125. MR1930884 (2003k:53060).

[282] F. L. Zak, *Tangents and secants of algebraic varieties*, Translations of Mathematical Monographs, 127, Amer. Math. Soc., Providence, RI, 1993. MR1234494 (94i:14053).

Notation

Below is a list of some notation used in different chapters.

Chapter 1.

$D^2 u$	the Hessian of u
∇u	the gradient of u
M^t	the transpose of a matrix M
M^*	the conjugate transpose of a matrix M, $M^* = \overline{M^t}$
$\text{Sym}_n(\mathbb{R})$	the space of real symmetric $n \times n$ matrices, $M^t = M$
$F(D^2 w) = 0$	a simplest fully nonlinear uniformly elliptic equation
Δ_p	the p-Laplacian operator
Δ_∞	the ∞-Laplacian operator
$\sum a_{ij} \frac{\partial^2}{\partial x_i \partial x_j}$	a linear elliptic operator
$\mathcal{M}^+, \mathcal{M}^+$	Pucci's extremal operators
$\sigma_k(\lambda)$	symmetric functions of eigenvalues of a matrix
$(\varkappa_1, \ldots, \varkappa_n)$	the principal curvatures of a graph
A^u	the conformal Hessian
Ω	a domain in $\mathbb{R}^n$
U^+, U^-	the set of C^2-supersolutions, C^2-subsolutions
$W^{p,q}$	Sobolev spaces
$\text{meas}(E)$	the Lebesgue measure of $E \subset \mathbb{R}^n$
$\mathcal{P}$	the subset of nonnegative matrices
Σ_n	the permutation group on n symbols
B	the unit ball in $\mathbb{R}^n$ centered at the origin
B_r	a ball in $\mathbb{R}^n$ of radius r
$\mathbb{S}^{n-1}$	the unit sphere in $\mathbb{R}^n$
$\mathbb{S}^{n-1}_r$	the sphere in $\mathbb{R}^n$ of radius r
u_v	for a function u and a vector v the derivative of u in the direction v, $u_v = \langle \nabla u, v \rangle$
$\lambda(A)$	the (ordered) spectrum $\text{Spec}(A)$ of the linear operator A

Chapter 2.

$\mathbb{H}$	the algebra of quaternions
$\mathbb{O}$	the algebra of octonions
$(e_0, \ldots, e_7)$	the canonical basis of $\mathbb{O}$
$SO(n)$	the special orthogonal group in $\mathbb{R}^n$
$\text{Spin}(n)$	the spinor group, the universal covering of $SO(n)$

236 NOTATION

C_n	the real Clifford algebra on n generators
$A[n]$	the $n \times n$ matrix algebra $M_n(A)$ over an associative algebra A
$U(n)$	the unitary group in $\mathbb{C}^n$
$SU(n)$	the special unitary group in $\mathbb{C}^n$
$Sp(n)$	the simpletic group in $\mathbb{R}^{2n}$
P_n^+, P_n^-	the space of positive or negative pinors
S_n	the spinor representation of $\mathrm{Spin}(n)$
S_n^+, S_n^-	for $n = 4$ or 8, the positive and negative spinor representations of $\mathrm{Spin}(n)$
G_2	the simple exceptional group, isomorphic to $\mathrm{Aut}_{\mathbb{R}-alg}(\mathbb{O})$
t_d	for $d = 2, 4$, or 8, the normed triality on $\mathbb{C}, \mathbb{H}$, or $\mathbb{O}$
P_s	for $s = 6, 12$, or 24, the harmonic triality polynomials in s variables
$\mathrm{Aut}(t_d)$	for $d = 2, 4$, or 8, the automorphism group of a normed triality
$\langle\,,\,\rangle$	the inner product in $\mathbb{R}^k$
$u \times v$	a cross product on a space of dimension 3 or 7
φ	an associative calibration on a space of dimension 3 or 7
$[u, v, w]$	the associator bracket of the vectors u, v, w
$\mathrm{Gram}(u, v, w)$	the Gram matrix of the vectors u, v, w
ψ	a coassociative calibration on a space of dimension 7
$[u, v, w, x]$	the coassociator bracket of the vectors u, v, w, x
$G(V, \varphi)$	the automorphism group of an associative calibration on V
W	an 8-dimensional normed division algebra
$u \times v \times w$	the triple cross product of W
Φ	a Cayley calibration on W
$G(W, \Phi)$	the automorphism group of Φ on W

Chapter 3.

$\mathbb{F}_d$	the classical division algebra of real dimension $d = 1, 2, 4, 8$, $\mathbb{F}_4 = \mathbb{H}$, $\mathbb{F}_8 = \mathbb{O}$
$\mathbb{F}$	the field $\mathbb{F}_1 = \mathbb{R}$ or $\mathbb{F}_2 = \mathbb{C}$
C_d	for $d = 1, 2, 4, 8$, the Cartan isoparametric cubic in $3d + 2$ variables
(W, Q)	a quadratic subspace
$x \bullet y$	the Jordan product
V	a Jordan algebra
$\mathrm{Herm}_n(\mathbb{F}_d)$	the set of the Hermitian matrices over $\mathbb{F}_d$, $d = 1, 2, 4$, $n \geq 3$, $\mathrm{Herm}_n(\mathbb{R}) = \mathrm{Sym}_n(\mathbb{R})$
$\mathrm{Herm}_3(\mathbb{O})$	the Albert algebra of Hermitian 3×3 matrices over octonions
$e^\perp$	the trace free subspace of V
$m_x(\lambda)$	the minimum polynomial of the regular element x
N	the generic norm in V
T	the bilinear trace form in V
Tr	the generic trace in V
tr	the trace of a linear endomorphism
$x \to x^\#$	the quadratic sharp map
$M_n(A)$	the $n \times n$ matrix algebra over an associative algebra A

NOTATION

$\mathcal{S}(V, Q, c)$	a spin factor
$\mathcal{J}(V, N, c)$	a Springer Jordan algebra
$\operatorname{hess} u(x, y)$	the mixed Hessian

Chapter 4.

$Q(\mathbb{R}^n)$	the vector space of quadratic forms over $\mathbb{R}$
H	the Hessian map $\mathbb{S}^{n-1} \to Q(\mathbb{R}^n)$, $a \mapsto D^2 w(a)$
Q_d	for $d \in \mathbb{R}^{12}$, the quadratic form $(P_{12})_d$
χ_d	the characteristic polynomial of the form $2Q_d$
$\operatorname{Spec}(A)$	the spectrum of a matrix (quadratic form) A
$w_{j,\delta}(x)$	for $j = 3, 6, 12, 24$, $\delta \in [1, 2[$, the basic function $P_j(x)/|x|^\delta$
$w_{12}(x)$	the basic function $P_{12}(x)/|x|$
Λ	the eigenvalue map defined by $D^2 w$
$G(\lambda)$	the characteristic polynomial of the form $D^2 P_{24}$
K_λ	the convex λ-cone
$\operatorname{Res}(F_1, F_2)$	the resultant of the polynomials F_1, F_2
$H(a)$	the Hessian $D^2 w_{12}(a)$ for $a \in \mathbb{R} \setminus \{0\}$
$P_A(\xi)$	the characteristic polynomial of $A := H(a)$
$P_{A,\delta}(\xi)$	the characteristic polynomial of $H_\delta(a) := D^2 w_{12,\delta}(a)$

Chapter 5.

$u(x)$	the normalized Jordan norm on the space of trace free elements
$w_\alpha(x)$	the homogeneous function $u(x)/|x|^\alpha$
$\operatorname{Hess}_x(u)$	the Hessian $D^2 u(x)$
A_1, A_2, A_3	the generators of the automorphism of the Cartan cubic C_1
$w_{3d+2}(x)$	for $d = 1, 2, 4, 8$, the function $C_d(x)/|x|$
$g_{3d+2} = 0$	for $d = 1, 2, 4, 8$, a locally uniformly elliptic polynomial equation for $w_{3d+2}(x)$
$P_6(x, y)$	the OT-FKM type isoparamretric quartic in six dimensions
$w_6(x, y)$	a rational nonclassical solution $P_6(x, y)/|(x, y)|^2$ to a uniformly elliptic equation

Chapter 6.

$Q(x, y)$	a nondegenerate bilinear form
$Q(x) = Q(x, x)$	the corresponding quadratic form
$V(u)$	the Freudenthal-Springer algebra of a cubic form u
c	an idempotent in a REC algebra
L_x	the multiplication operator by $x \in V$
$V_x(a)$	the invariant subspace of L_x corresponding to the egienvalue a
$V_0 \oplus V_1$	a polar decomposition
ρ	the Hurwitz-Radon function (6.5.18)
$\{A, B\}$	the Jordan product of operators A and B, $\{A, B\} = AB + BA$
$\mathcal{N}(V)$	the set of normalized square zero elements

238   NOTATION

$V_c^\bullet$	the Jordan algebra on $V_c(1; -\tfrac{1}{2})$ associated to a REC algebra
$n_i(c), i = 1, 2, 3$	the characteristic dimensions of the idempotent c

Chapter 7.

φ	the associative calibration on $\operatorname{Im}\mathbb{O}$
ψ	the coassociative calibration on $\operatorname{Im}\mathbb{O}$
Φ	the Cayley calibration on $\mathbb{O}$
M_c	the coassociative $Sp(1)$-invariant singular 4-fold in $\operatorname{Im}\mathbb{O}$
$\mathbf{F}_h(u)$	the special Lagrangian operator with the parameter $h = -\tan\theta$
u_c	a special analytic solution to the equation $\mathbf{F}_{-c}(u) = 0$ on a small ball
$f \mapsto \widetilde{f}$	the Legendre transform
$\mathbb{R}\{\{x, y, z\}\}$	the ring of power series in $\mathbb{R}^3$ converging near zero
$Q(\nabla u_c)$	the local ring of the Legendre transform of u_c
u_θ	the singular solution to the equation $\mathbf{F}_\theta(u) = 0$ on a small ball
u_θ^m	even more singular solutions to the equation $\mathbf{F}_\theta(u) = 0$ on a small ball
u_θ^ε	a family of smooth solutions violating the maximum principle for the Hessian
U^m	weak solutions of the minimal surface system which are not $W^{1,1+\varepsilon}$ for $m \to \infty$

Index

4-fold cross product, 59
K-cone condition, 96
Δ_1, 156
∞-Laplacian, 3
λ-cone condition, 87
$\mathcal{N}(V)$, 183
σ_k-equation, 6
p-Laplace equation, 2

a priori bound, 7
ABP maximum principle, 11
Alexandrov's theorem, 28
alternative algebra, 46
anticommutator product, 67
associative bilinear form, 161
associative calibration, 54, 207
associative subspace, 56
associator, 46
associator bracket, 55

Bernstein's theorem, 27, 159
Bott periodicity, 49

calibrated geometries, 205
calibrated submanifold, 205
calibration, 205
Cartan isoparametric cubic, 64
Cartan-Münzner system, 64
Cayley calibration, 59
Cayley form, 61
Cayley submanifold, 208
Cayley subspace, 61
Cayley-Dickson construction, 46
characteristic dimensions, 182
classical solution, 6
Clifford algebra, 48
coassociative calibration, 56, 208
coassociative subspace, 57
coassociator bracket, 57
complex Monge-Ampère equation, 4
conformal Hessian equation, 6
congruent forms, 66
continuity method, 7
coordinate plane, 96
cross product, 54
cubic Jordan algebra, 74
curvature equation, 6

degenerate elliptic linear equation, 2
Dirichlet duality, 22
Dirichlet problem, 6
Dirichlet set, 22
division algebra, 46
Donaldson equation, 5

eiconal equation, 64
eigenfunction, 160
ellipticity criterion, 86
Euclidean Jordan algebra, 72
Evans-Krylov's theorem, 16
exceptional Albert algebra, 71
exceptional isomorphisms, 49

failure of maximum principle, 218
formally real Jordan algebra, 68
Freudenthal multiplication, 162
Freudenthal-Springer algebra, 162

Gauss curvature equation, 6
generic norm, 69
generic trace, 69
geometrically equivalent Clifford systems, 173
Ginzburg-Landau system, 151
group G_2, 51

Hamilton-Jacobi-Bellman equation, 3
harmonic curvature, 6
harmonic REC algebra, 165
Harnack inequality, 13
Hartman-Nirenberg's theorem, 32
hérisson, 28
Hermitian Jordan algebra, 70
Hessian ellipticity criterion, 96
Hessian equation, 5
Hessian of the generic norm, 117
Hessian problem, 86
homogeneous isoparametric hypersurface, 143
homogeneous polynomial map, 118
homogeneous solution, 31
Hsiang problem, 159

idempotent, 73
irreducible representation, 50
irreducible symmetric Clifford system, 173

Isaacs equation, 3
isoparametric function, 63
isoparametric hypersurface, 63

Jordan algebra, 68
Jordan cubic form, 75
Jordan frame, 73
Jordan identity, 67
Jordan matrix algebra, 70
Jordan product, 67
Jörgen-Calabi-Pogorelov theorem, 39

Lagrangian geometry, 206
Laplace's equation, 2
Lawson's cubic minimal cone, 161, 172
Legendre transform, 215
level hypersurface, 65
Lewy rotation, 218
linear in symmetric functions σ_k, equations, 5
linear trace, 74
Liouville's theorem, 38

martingale problem, 13
maximum principle, 7
mean curvature, 6
metrized algebra, 161
metrized subalgebra, 163
minimal hypercones, 159
minimal submanifold, 205
minimal surface system, 218
Minkowski spin factor, 71
Monge-Ampère equation, 4

nicely normed algebra, 47
Nirenberg's theorem, 18
normed division algebra, 46
normed triality, 52

octonion construction, 96
octonions, 47
OT-FKM type isoparametric hypersurface, 144

Peirce decomposition, 73
Peirce decomposition of a REC algebra, 181
polar algebra, 169
polar decomposition, 169
primitive idempotent, 73
Pucci's equation, 3
Pucci's extremal operator, 3

quadratic representation, 72
quadratic sharp map, 75
quadratic spur, 74
quasilinear elliptic equations in divergence form, 2
quasilinear elliptic equations in nondivergence form, 3
quaternions, 47

radial eigencubic, 156
radial eigencubic algebra (REC algebra), 165
rank of a Jordan algebra, 69
reduced cubic factor, 76
reducible REC algebra, 194
regular eigenfunction, 160
regular element in a Jordan algebra, 69
removable singularity, 42
resultant, 101

saddle function, 27
scalar curvature, 6
second approximative differential, 15
sedenions, 47
semisimple Jordan algebra, 72
similar eigenfunctions, 160
similar metrized algebras, 162
similarity invariant, 162
simple Jordan algebra, 72
singular coassociative 4-fold, 209
special Cartan REC algebra, 166
special Jordan algebra, 70
special Lagrangian equations, 211
special Lagrangian submanifold, 206
spin factor of a quadratic form, 71
spin representation, 50
spin type representation, 50
spinor group, 49
Springer construction, 74
strictly elliptic equation, 2
strictly hyperbolic matrix, 115
structural constant, 165
subaffine function, 23
subsolution, 8
supersolution, 8
symmetric Clifford system, 171

third approximative differential, 15
trace form, 75
transnormal function, 65
transport Monge-Ampère equation, 4
triality, 51
triality system, 186
triple cross product, 60

uniformly elliptic equation, 2
uniformly hyperbolic matrix family, 101

vector representation, 50
viscosity solution, 8

Weingarten equation, 6
Whitney's stratification theorem, 35